Teaching Mathematics Using Interactive Mapping

Teaching Mathematics Using Interactive Mapping offers novel ways to learn basic math topics such as simple relational measures or measuring hierarchies through customized interactive mapping activities. These activities focus on interactive web-based Geographic Information System (GIS) and are relevant to today's problems and challenges. Written in a guided, hands-on, understandable manner, all activities are designed to build practical and problem-solving skills that rest on mathematical principles and move students from thinking about maps as references that focus solely on "where is" something, to analytical tools, focusing primarily on the "whys of where." Success with this transition through interaction permits most readers to master mathematical concepts and GIS tools.

FEATURES

- Offers custom-designed geographical activities to fit with specific mathematical topics
- Helps students become comfortable using mathematics in a variety of professions
- Provides an innovative, engaging, and practical set of activities to ease readers through typically difficult, often elementary, mathematical topics: fractions, the distributive law, and much more
- Uses web-based GIS maps, apps, and other tools and data that can be accessed on any device, anywhere, at any time, requiring no prior GIS background
- Written by experienced teachers and researchers with lifelong experience in teaching mathematics, geography, and spatial analysis

This textbook applies to undergraduate and graduate students in universities and community colleges including those in basic mathematics courses, as well as upper-level undergraduate and graduate students taking courses in geographic information systems, remote sensing, photogrammetry, geography, geodesy, information science, engineering, and geology. Professionals interested in learning techniques and technologies for collecting, analyzing, managing, processing, and visualizing geospatial datasets will also benefit from this book as they refresh their knowledge in mathematics.

Teaching Mathematics Using Interactive Mapping

Sandra L. Arlinghaus
Joseph J. Kerski
William C. Arlinghaus

CRC Press
Taylor & Francis Group
Boca Raton London New York

CRC Press is an imprint of the
Taylor & Francis Group, an **informa** business

Designed cover image: © the authors

First edition published 2024
by CRC Press
2385 NW Executive Center Drive, Suite 320, Boca Raton FL 33431

and by CRC Press
4 Park Square, Milton Park, Abingdon, Oxon, OX14 4RN

CRC Press is an imprint of Taylor & Francis Group, LLC

Library of Congress Cataloging-in-Publication Data
Names: Arlinghaus, Sandra L. (Sandra Lach), author. | Kerski, Joseph J., author. | Arlinghaus, William C., 1944- author.
Title: Teaching mathematics using interactive mapping / Sandra L. Arlinghaus, Joseph J. Kerski, and William C. Arlinghaus.
Description: Boca Raton, FL : CRC Press, 2024. | Includes bibliographical references and index.
Identifiers: LCCN 2023026734 (print) | LCCN 2023026735 (ebook) | ISBN 9781032305325 (hardback) | ISBN 9781032305332 (paperback) | ISBN 9781003305613 (ebook) | ISBN 9781032614021 (ebook other)
Subjects: LCSH: Mathematics—Study and teaching (Higher)—Activity programs. | Geography—Mathematics—Study and teaching (Higher) | Geographic information systems—Study and teaching (Higher)
Classification: LCC QA139 .A35 2024 (print) | LCC QA139 (ebook) | DDC 510.71/1—dc23/eng/20231012
LC record available at https://lccn.loc.gov/2023026734
LC ebook record available at https://lccn.loc.gov/2023026735

ISBN: 978-1-032-30532-5 (hbk)
ISBN: 978-1-032-30533-2 (pbk)
ISBN: 978-1-003-30561-3 (ebk)
ISBN: 978-1-032-61402-1 (eBook+)

DOI: 10.1201/9781003305613

Typeset in Times
by codeMantra

Access the Support Material: https://www.routledge.com/9781032305325

Contents

Preface..........xi
Introduction..........xii
Authors..........xxii
Acknowledgments..........xxiv

Chapter 1 Classifying Numbers and the Distributive Law..........1

Chapter Outline..........1
Classification..........2
Conceptual View..........2
Intuitive View: Interactive Mapping Activity..........2
Classifying Numbers..........4
Natural Numbers..........4
Whole Numbers..........4
Integers..........5
Rational Numbers..........5
Irrational Numbers..........5
Real Numbers..........5
Nested Hierarchy..........5
Infinity..........6
The Number Line: Extent and Direction..........6
Different Classes of Numbers Positioned on the Number Line..........6
Absolute Value..........7
Prime Numbers..........7
Using Numbers with Maps to Analyze Demographics..........7
Operations..........9
Law of Excluded Middle..........9
Operations with Numbers..........10
Operations with Maps..........10
Order..........11
Order of Mathematical Operations..........11
Order of Mapping Operations..........11
Legend: Layers and Organization..........13
The Distributive Law..........14
Terms and Factors..........14
Parentheses and the Distributive Law..........15
Geography and Thinking Spatially in Everyday Life: Enlarging Your World!..........17

Chapter 2 Fractions and Decimals..........19

Chapter Outline..........19
Fractions..........20
Adding and Subtracting Fractions with Like Denominators..........20
Multiplying Fractions..........20
The Importance of Multiplying by 1..........21
Adding and Subtracting Fractions with Different Denominators..........21
The Importance of Cancellation..........22

- Divisibility of Numbers ... 23
- Divisibility, Prime Numbers, and Unique Factorization ... 23
- Dividing Fractions ... 24
- Fractions Larger Than One ... 24
- Fractions, Decimals, and Maps ... 24
- Arithmetic with Fractions: Rapid Mental Calculation ... 24

Exponents and Logarithms ... 24
- Exponents ... 24
- Logarithms ... 25

Decimals ... 25
- Convert Fractions to Decimals ... 25
- Convert Decimals to Fractions ... 25
- Adding and Subtracting Decimals ... 26
- Multiplying Decimals ... 26
- Dividing Decimals ... 26
- Rounding and Estimating ... 26

Bases Other Than 10 ... 30
- Binary ... 30
- Octal ... 31
- Hexadecimal ... 31
- Operations ... 32

Time and Circular Measurement ... 32
- Metric ... 32
- English ... 32
- Time ... 32
- Circular Measurement ... 33

Measuring Distances in Latitude Longitude Degrees ... 38

Time and Longitude: Circling Back ... 40

Looking Ahead ... 40

Chapter 3 Simple Relational Measures and Measures of Central Tendency and Variation ... 41

Chapter Outline ... 41

Simple Relational Measures ... 42
- Ratio ... 42
- Proportion ... 42
- Percent ... 42
- Estimation and Rounding ... 43
- Problems—Remember, "Percent" Means "per 100" ... 44
- Associated Interactive Mapping Activity: Mapping Urban Litter Proportions ... 44

Math Word/Story Problems ... 48
- General Strategy ... 48
- Worked Problems ... 49
- Distance, Rate, Time Interactive Mapping Activity ... 49
- Deeper Thought Problem: Hope Village for Children ... 50

Measures of Central Tendency ... 51
- Mean and Related Measures ... 51
- Median ... 51
- Mode ... 52
- Problem ... 52
- Associated Interactive Mapping Activities ... 52

Variation 55
Histogram 56
Scatter Diagram 56
Correlation 57
Simple Linear Regression 58
Associated Interactive Mapping Activity: The Histogram as a Mapping Tool 59
Goodness of Fit and Statistical Significance 61
Coefficient of Determination 61
z-Score 61
Statistical Significance 62
p-Value 62
Associated Interactive Mapping Activities 63
Integrating Elements of This Chapter 67
Fibonacci Numbers and the Golden Ratio 67
Percentages, Estimation, and Rounding in the World of Sculpture 68

Chapter 4 Earth Measurement: Coordinate Systems and Trigonometry 70

Chapter Outline 70
Coordinate Systems in Common Use 71
Cartesian Coordinates 71
Earth Model: Sphere and Plane 72
Earth Model Coordinates: Parallels and Meridians 72
Earth Model Coordinates: Latitude and Longitude 74
Geocoding 75
Ordinal and Cardinal Directions 77
Other Coordinate Systems 79
Universal Transverse Mercator 79
Polar Coordinates 80
Trigonometry: Visual Review of Functions and Applications 81
Visual Review of Trigonometric Functions 82
Applications: Conceptual and Real Worlds 84
Scale 91
Map Scale 91
Activity: Studying Map Scale with USGS Topographic Maps 92
Global and Local Scale 94
Activity: Determining Map Scale 94
Activity: Comparing Sizes of Countries 96
Map Projection 97
Tissot Indicatrix 97
Scale and Measurement on Mars 99
Questions to Think About 102

Chapter 5 Data, Variables, and Thematic Maps 103

Chapter Outline 103
Data 104
Data Scale 104
Data Types 104
Attributes 105
Nominal Attribute 105

Ordinal Attribute 105
Interval Attribute 105
Ratio Attribute 105
Generalization 105
Selection/Filtering 106
Classification 108
Simplification 109
Symbolization 114
Surveys 117
Nominal Scale Survey Questions 117
Ordinal Scale Survey Questions 117
Interval Scale Survey Questions 118
Ratio Scale Survey Questions 118
Thematic Maps 118
Reference Information 119
Standardization of Data 119
Uncertainty 119
Single Variable Mapping and Data Types: Choropleth Maps 120
Nominal Data Type 120
Ordinal Data Types 120
Interval Data Type 120
Ratio Data Type 121
World Earthquakes 121
Exploring Data Using Bivariate Maps, Symbology and Classification 121
Heat Map 124
Relationship Map 124
To Think About 126

Chapter 6 Mathematics' Foundations: Set Theory and Algebra 127

Chapter Outline 127
Existence and Foundational Mathematics 128
Law of Excluded Middle 128
The Axiom of Infinity 128
The Axiom of Choice 128
Set Theory 129
Set Fundamentals 129
Subsets 129
The Empty Set and the Universal Set 129
Basic Set Operations 130
Venn Diagrams 132
Set Theory in Hazards Analysis 134
Algebra 137
Fundamental Proofs 137
Simple Linear Equations and Graphs; Inequalities 138
A Unifying Map 142
Simple Quadratic Equations and Graphs; Inequalities 143
Real-World Shapes and Challenges 148

Chapter 7 Mathematics' Foundations: Dimension and Geometry 150

Chapter Outline 150
Synthetic Geometry 152
Euclid: Selections 152
Axioms for Euclidean Geometry in the Plane 152
Selected Concepts and Theorems 152
Solid Geometry 155
Whole-Earth Measurements 157
Transforming Dimensions: Approximations 158
Area under a Curve: Approximation Techniques 158
Representing Numbers as Isolines 162
Dimension 164
Overview 164
Beyond the Usual—Fractional Dimension 165
Testing the Coastline Paradox: Fractals 166
Non-Euclidean Geometry 167
Error Issues 168
Digitizing 168
Jordan Curve Theorem 168
Geometric Models of the Real World? 169

Chapter 8 Proximity and Adjacency: Measurement 170

Chapter Outline 170
Overview 171
Graph Theory 171
Topology 171
Measuring Proximity 172
Graph Theory: Tracing a Walk through Königsberg 172
Proximity: Tracing Downstream 173
Topology: One-Point Compactification of the Line 175
Proximity Zones and Wildfire 176
Measuring Adjacency 179
Graph Theory: Coloring and Adjacency 179
Topology 182
Terrestrial Surface and Green Infrastructure 187
Map Projection 190

Chapter 9 Measuring Hierarchies and Patterns 193

Chapter Outline 193
Overview 193
The Dot Density Map 196
Interactive View 196
Conceptual Principles 200
Nested and Non-Nested Hierarchies 200
Non-Nested Hierarchy: The Cartogram 201
Non-Contiguous Cartograms 201
Contiguous Cartograms 202
Hierarchical Organization Chart for This Book 206

References and Further Reading Suggestions 208
Postscript 215
Index 223

Preface

In the post-pandemic era there is, and no doubt will continue to be, much disparity in the knowledge base of the world's current and future generations of collegiate students. Those whose educational opportunities were severely constricted, as unforeseen and unintended consequences of the coronavirus pandemic gripped the world from late 2019 to 2022, grew up in a world in which consistency in education was at least partially lost (Baumgaertner, 2023).

In a world rapidly changing due to natural forces and those caused by 8 billion humans, technological solutions to build resiliency and sustainability are needed and must be proposed, evaluated, and put into daily action. These solutions across all disciplines—engineering, planning, health, public safety, biology, and many more—increasingly depend on the application of mathematical principles. In addition, new fields such as geodesign and data science also rely on inquiry and problem-solving that require a firm grasp of mathematics. A student graduating into the workforce in such a world who has no thorough foundation in mathematical concepts, principles, and applications will be at a severe disadvantage.

Mathematics curricula are typically highly linear in nature, loaded with pre-requisites in order to move from one level to the next in a sequence of (possibly required) courses. Some parts of courses are more easily grasped than are others. Yet, students may pass from one level to the next without really mastering all the elements of each course. When that happens, as is frequent, the part not understood in an early course is a seed of misunderstanding around which later topics also become misunderstood. For example, a student who does not have a good grasp of how to add fractions with different denominators may move from the fifth grade to the sixth grade, only to find out in college that he/she cannot effectively make comparisons on maps using the scale of the map, stemming from an earlier lack of mastery of operations with fractions.

In this book, we select such seeds, based on years of teaching experience, and offer multiple activities, based on interactive mapping, to try to encourage better mastery of basic mathematics in an interesting and stimulating manner. Thus, you may see a topic mentioned early in this book only in an intuitive fashion, to be repeated later in a bit more systematic fashion. There are two reasons for this somewhat non-standard approach:

- Repetition from different vantage points promotes understanding that leads to mastery.
- Alignment of a linear curriculum (mathematics) with a non-linear one (mapping) leaves gaps, wrinkles, and puffiness where one fails to fit perfectly with the other. Again, different vantage points help to overcome such issues.

The intention is to motivate students to want to master these roadblocks so that they become sources of pride for them to point to and say, "I know how to do that!"—a far healthier attitude than the far too often heard comment "well, I just never was very good at math." The latter attitude is one which we have found often stems from tripping on an earlier math roadblock and never recovering from the fall.

The thread that links one chapter to the next is the underlying concept that all chapters present activities to overcome the effects of roadblocks—many of which were evident in the years before the global pandemic and which now will become even more pronounced in the years following it as students who were left behind may have extra difficulty recovering.

Introduction

Conceptual Approach and Purpose

The approach taken here is both innovative and exciting. In it, one learns mathematical tools and concepts as needed to solve real-world problems, as they arise, rather than only in a conventional, pre-established linear order. The mathematical concepts featured in this approach are:

- Often-taught concepts.
- Foundational concepts.
- Concepts associated with sources of gaps in understanding that prevent progress.
- Concepts engagingly taught with interactive maps and spatial data.

Thus, we feature samples of mathematics chosen from the foundations of mathematics itself, such as the Law of Excluded Middle or various axioms and postulates, with an emphasis on how these might work together to create proofs of known facts, for example, "a negative times a negative is a positive." We offer other samples that touch on roadblocks associated with the Distributive Law, illustrating its use in various contexts and culminating in challenges involving rapid mental calculation. Further, we select topics from broader mathematical sciences, such as statistics, to illustrate strong connection between measures of central tendency and patterns of clustering of data on maps. The careful instructor will think of other facts to suggest to students, creatively imitating, but not merely copying, our example.

A single volume, which reaches across a spectrum in the mathematical sciences, cannot delve deeply into any single element. Rather, the emphasis is on creating a buffet of approaches from which samples might be tasted as the instructor systematically works through conventional materials. The reader looking for a similar approach, in terms of sampling of academic content, need look no further than in the world of computer documentation where one reads the book on an as-needed basis, rather than beginning on page one and reading the book from cover to cover. So too, a traditional encyclopedia functions in this manner, again illustrating that this book is not a substitute for a full mathematics textbook. It is a supplement designed to reinforce the roadblocks in communication that often arise from such textbooks, eliminating gaps as students move forward (independent of educational level) in their quest for a clear understanding and expression of mathematics.

Organizational Structure of This Book

The diamond-shaped image (Figure A.1) at the top of each chapter displays the organizational structure of this work; it is replicated here for reference. The numerals inside the small squares represent chapter numbers. Note the "Maps" and the "Math" axes. As one proceeds from Chapters 1–3, the content of the grouping is straightforward, perhaps elementary, in level. In the next cluster, Chapters 4–6, the level of difficulty is moderate. Finally, in the last grouping, Chapters 7–9, the degree of difficulty is highest. Proximity of a given chapter to the axes suggests its relative weighting of maps and math content. At the top of each chapter, this image is placed in the middle of a set of three images and serves as a bridge going from the very specific single chapter number, through the book represented as a diamond with all chapter numbers, to a final real-world image that suggests some aspect of the content of that chapter. Finally, the color scheme for the graphic is based on a bivariate color scheme involving the adding and blending of colors on a map. This set of three elements at the top of each chapter is simple to grasp at first glance; however, the thoughtful reader will gain far more by reflecting on the deeper elements of it. Indeed, that is an approach taken throughout the book: careful reflection on content pays off.

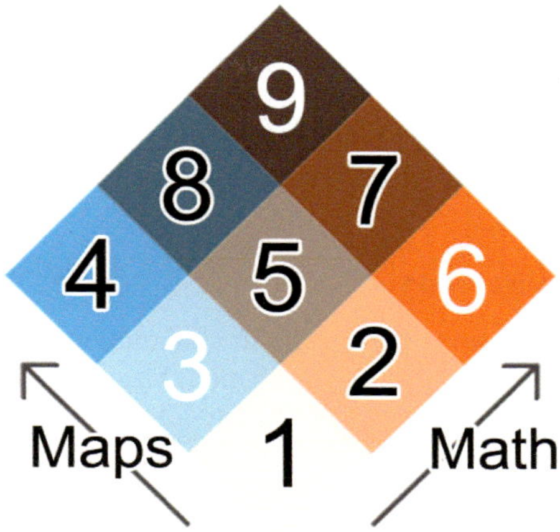

FIGURE A.1 Organizational structure of this book; numerals indicate chapter number.

Why Use Maps as a Learning Base?

From etchings in dirt and on cave walls to the 9th century BCE Akkadian language Babylonian map on a clay tablet (British Museum), to Al Idrisi's map on two plates of silver in the 1100s (Kerski, 2019), to Medieval maps carved in wood (Tuan, 1977), to the copper plates and film of the 20th century (Masterworks Fine Art Gallery, last accessed, April 2023), to the digital visualizations of the 21st century (Kerski, last accessed April 2023), maps not only help us understand *where* things are, but *why* they are where they are. They have served, and continue to serve, as a base for many real-world contexts, and they are all based on some form of mathematics.

As the modern world grows increasingly interconnected, and as change accelerates at multiple scales from local to global, maps, the geographic data behind them, and their associated mathematical foundations are essential tools for understanding human-environment interaction patterns and the planning of them for the future (see, for example, Arlinghaus and Kerski, 2013). Today, maps and math help people make smarter decisions: from rural to urban settings and across a wide range of subject domains, including social sciences such as economics and sociology, earth and environmental sciences, engineering, design, criminal justice, city and regional planning, archaeology, business, data science, and any field that uses geographic data.

Modern Mapping: GIS Concept Map

Mathematics underpins all mapping. The primary tool for all modern mapping is the Geographic Information System (GIS). Breaking apart GIS into its three letters (Kerski, 2020):

- The "G" is the fundamental map component: an underlying basemap upon which layers of information are superimposed. The basemap could be two-dimensional, three-dimensional, or even four-dimensional (with temporal components included), and it might be composed of many individual maps.
- The "I" is the Information, or table or spreadsheet, behind the map. It can be thought of as a geodatabase. The tabular data, along with associated relationships, are what gives GIS its power. A GIS is not just graphics floating around in cyberspace: it has intelligence that allows you to make wise decisions based on it. Often, GIS hinges on multiple related geodatabases.
- The "S" is the systems part that links the "G" and the "I" together—the map always works with the table, and vice-versa.

In this book, we employ a variety of software. Typically, GIS software is expensive to develop; for example, Esri (Environmental Systems Research Institute) has spent millions of dollars and years developing their ArcGIS Online mapping capability. The software we choose naturally reflects patterns of market share and availability of software. The goal is that the reader is seldom required to use a password or to need an account (except possibly a free one if it is desired to save work): software usage is free of charge.

Modern Mapping: GIS Sample Maps

Next, we illustrate how the GIS approach works in conjunction with interactive use of GIS software, directly from this book. Open the linked web map (https://www.arcgis.com/apps/mapviewer/index.html?webmap=f3a47dd8234a4de1a3dc8413651c4e80) of soil chemistry in a field in North Dakota (Figure A.2).

The "G" part is the point data of a variety of different chemical variables in the field (pH, calcium, nitrate, zinc, and others). The "I" part is the table behind the data. To view the "I" part, click on the "soilph" local hosted feature layer>ellipses>open table>you will notice that 113 records exist in the table: one record for each point on the map. The "G" and the "I" are linked so that selecting or filtering the data in the table selects or filters features on the map, and vice versa.

For added material to explore the relationship between the G, I, and S parts of a GIS, open the linked story map (https://storymaps.arcgis.com/stories/f130ea8dbc0349c9b79f36b9d934f975): The story map contains a spreadsheet. Note the latitude and longitude values in the spreadsheet along with a description, and a scenic ranking from 1 to 10. Choose a place that is meaningful to you. Enter a latitude, longitude, title, your ranking of the scenery of that place from 1 to 10, a description, and a photo URL for it (if you want to include a photo). You are entering into the "I" part of GIS. Wait a few minutes and that point with the attributes will appear on the ArcGIS Online web map below the spreadsheet; the "S" part will have been activated.

How to Use This Book: Major Web Mapping and Visualization Tools

The focus of this section is to provide a centralized reference location to illustrate usage components of major web mapping and visualization tools used in this book so that in each activity, space is devoted only to the activity itself and not to redundant explanation of how to use such tools. In addition to providing usage guidance, these brief tutorials will illustrate why interactive mapping is useful to foster spatial thinking in mathematics.

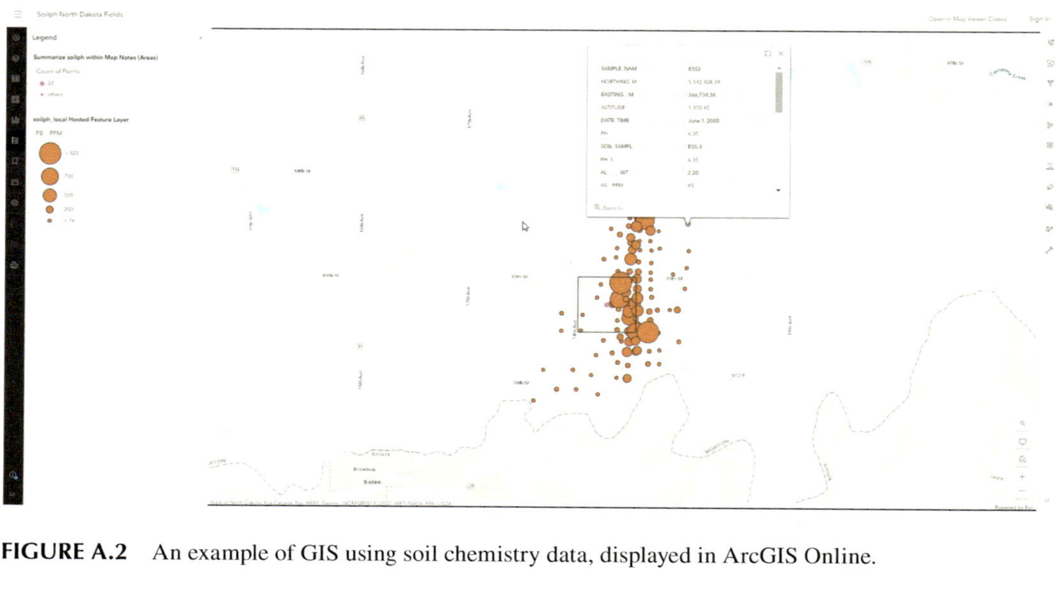

FIGURE A.2 An example of GIS using soil chemistry data, displayed in ArcGIS Online.

These tutorials will be brief because:

- The mapping tools are intuitive and long tutorials are not needed.
- We are targeting easy to use and for the most part web-based tools with no sign-in required maps and apps.
- The Graphical User Interfaces (GUIs) in these mapping tools will change in the future; thus, any screen shots are kept to a minimum and should be treated as a snapshot in time; any that appear are likely to change a bit in the future.
- We advise readers to "embrace discovery"—the intent is not to become a "mapping guru." Rather, the focus is on developing mathematical expertise and confidence, from solving real-world problems with real-world data using tools used in the workplace.

ArcGIS Online Map Viewer

ArcGIS Online is a web-based GIS from Esri Inc. ArcGIS Online includes the Map Viewer, a modern mapmaking tool where you can quickly create, symbolize, and share maps and web mapping applications to help you discover spatial and temporal patterns from a wide variety of data. To start ArcGIS Online, access the URL https://www.arcgis.com/,>click on Map>add layers, using the+sign to the left of the map, from your own device (such as a spreadsheet) or from online content from ArcGIS Online (Esri), the ArcGIS Living Atlas of the World (Esri), or from a multitude of spatial data portals (Figure A.3). These data sets are from national statistics, mapping, and science agencies (such as the United Nations (UN) Environment Programme and National Aeronautics and Space Administration (NASA)), local and regional governments, nonprofit organizations, academic institutions, and private companies. The data may be in the form of a spreadsheet, a GIS vector file, an image file, or other formats, and may cover a specific theme such as ecoregions in North America or population characteristics for census tracts for a city. The activities in this book will use maps that have already been created for you to use. Therefore, you will seldom need to start ArcGIS Online from "scratch," add your own layers, and build your own maps, but the capabilities are there for you to do so.

FIGURE A.3 The default view of the ArcGIS Online Map Viewer.

If a URL, specifically in ArcGIS Online, does not work (which might happen occasionally given the dynamic nature of the web), consider the following strategies or some natural variant thereof. The workflow is usually, if a URL in ArcGIS Online does not pull up a map, or a layer is missing in a map:

- Go to https://www.arcgis.com/>search on the title of the map.
- Or search for another layer with a similar name, such as "Human Development Index" instead of "HDI."

The map in the center of the ArcGIS Online interface is interactive; you can navigate seamlessly from east to west and at any scale. You can change the scale by zooming in or out with the + and – tools, or by drawing a rectangle with the shift key with your mouse or touchpad and dragging a box across the map. The map will zoom in to the area covered by the box. To the left of the map are important tools that you will use often, to add data to the map, to open a table of data, to change the basemap, to create and open charts, to view the map legend, to create spatial bookmarks of places on the map, to save and open maps, to access map properties, to share maps, to create web mapping applications, to print, and for more information. When a specific map layer is selected on the left, the tools on the right become activated for that layer. These tools include accessing the properties of the data layer, changing the symbols (style) and the classification of the data layer, adding visual effects, adding a sketch, searching for locations, and measuring length and area, and other tools. A home button places you in the default map view, restoring the initial map's layers and extent.

All of the activities using ArcGIS Online are accessible without a sign-in or log in required; you should have had no problem accessing and interacting with the map shown in Figure A.3. However, once you leave the map after you have completed the activity, your work disappears and is not saved. To save your work, you will need an ArcGIS Online account, available for free for your primary/secondary school via https://www.esri.com/schools. If you are in a college or university, chances are, your institution already has a subscription to ArcGIS Online, and you just need to obtain a log in to that subscription; see https://www.esri.com/en-us/industries/higher-education/overview for more. If you are outside of a formal education institution, you can still obtain an ArcGIS Online subscription via the developers' site on https://developers.arcgis.com, or via a personal use license https://www.esri.com/en-us/arcgis/products/arcgis-for-personal-use/buy.

ArcGIS Online 3D Scene Viewer

The ArcGIS 3D scene viewer is part of ArcGIS Online from Esri Inc. Access it by the URL https://www.arcgis.com/ and click>Scene. Like the 2D map viewer, ArcGIS 3D scene viewer presents several options to the left and right of the 3D scene (Figure A.4). The home tool resets the map to its default view, the + and – signs allow for changing the scale, and below that is a tool to either pan the map or rotate the map in 3D. To the right of the scene are tools to search for places, to access data layers and the map legend, to change the basemap, to change the time of day or time of year and thus the sun's illumination, to change the weather on the scene, to measure distances and areas, to create a horizontal slice of the data, and to create an elevation profile. Sharing options and settings are also available. Below is a 3D scene using the Terrain with Labels basemap zoomed and panned to look inside the crater of Mount St. Helens in Washington state, USA, from northeast to southwest (Figure A.4).

FIGURE A.4 The ArcGIS 3D Scene Viewer.

MapMaker Interactive

As the name implies, MapMaker Interactive is a web-based interactive mapping tool, created by National Geographic. It contains basemaps, data layers, and annotation tools that allow the interactive maps to be marked on. The annotation tools such as points, lines, and polygons allow the educator to highlight areas, themes, or patterns of interest on the map. Maps can be shared as URL links or exported to graphics or PDF formats.

MapMaker (Figure A.5), accessed via https://mapmaker.nationalgeographic.org/, can be signed into (allowing maps to be saved and accessed later), or simply to create a map during the active web session. Under Create a Map, the user is presented with an interactive simple basemap, which can be changed under Map Settings, a small set of annotation tools above the map, and layers to be added. Once added, most map layers include a legend and popups that expose information about each map's feature.

FIGURE A.5 The MapMaker interactive from national geographic.

Gapminder

Gapminder is a tool by an independent group of scholars that allows for data, largely by country and largely about population, to be analyzed in graphs and maps, over space and time. The Gapminder data tools are part of a larger set of information pages that include interactive quizzes about the status of major world variables. Once the Gapminder tools are accessed, (https://www.gapminder.org/tools/), the user is presented with a set of visualization options (maps, trends, ranks, ages, and more), along with animation tools to "play" the maps and countries to select, filter, and highlight. Below the country list is a set of expandable menus with the example of imports as percentage of Gross Domestic Product (GDP) (Figure A.6).

One of the strengths of Gapminder is its impactful bubble and trends charts, allowing two variables to be compared across time and against each other (Figure A.7).

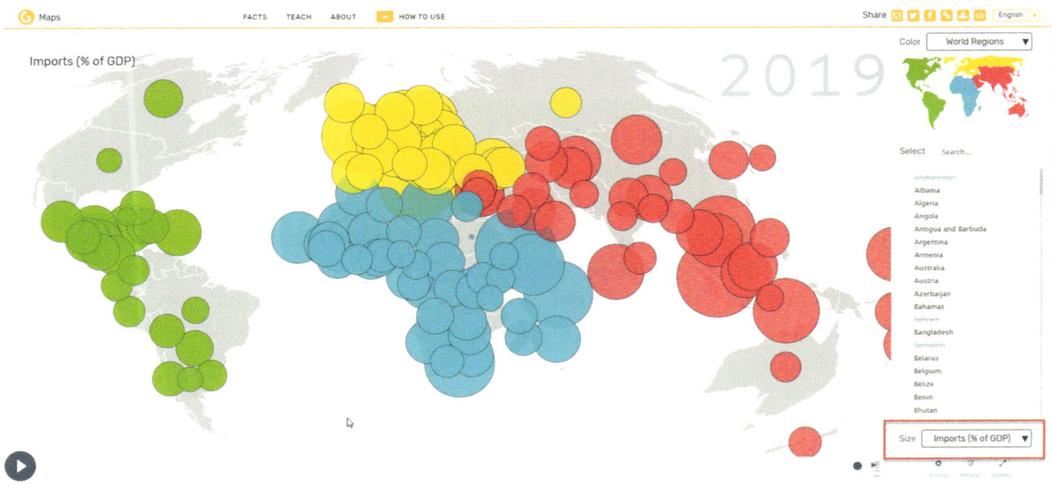

FIGURE A.6 Analyzing and visualizing variables using Gapminder.

FIGURE A.7 Comparing change over space and time using Gapminder.

FIGURE A.8 Cartogram of global population in 2018. Source: https://ourworldindata.org/world-population-cartogram Max Roser, CC BY 4.0 <https://creativecommons.org/licenses/by/4.0>, via Wikimedia Commons.

Cartograms: Bouncy Maps

Cartograms size a polygon of a geographic or administrative area based on the amount of a variable in that area; they present a very different view from most maps. Figure A.8 shows an example of a cartogram: the countries are sized based on the variable of population. A map in which a value other than land area is used to size landmasses. Here, population is used as that value. The countries are adjacent to each other in ways that we expect; there are no gaps between countries that are next to each other. Such cartograms are called contiguous cartograms. Later, we will return to consider this topic in a bit more detail; please keep this basic cartogram in mind.

BouncyMaps is a tool that allows for a wide variety of data, by country, to be visualized as cartograms. The cartograms in BouncyMaps are non-contiguous cartograms: Countries are broken apart by spaces, so their shape is maintained, but their size is adjusted based on a variable, such as road traffic deaths or refugees by country of origin (Figure A.9). To use, access https://www.bouncymaps.com, browse and select a variable, and click the "bouncy map" tab under the map. Hovering over a country displays a popup with more information. Spreadsheets of each variable can be downloaded, and a metadata page provides information about the data sources used. Most variables are for world countries, but selected variables for USA states can also be displayed as BouncyMaps cartograms.

The EPA EnviroAtlas

The EnviroAtlas, or Environmental Atlas, is an interactive mapping tool that allows for a wide variety of map layers to be visualized and queried. It was created by the US Environmental Protection Agency and its focus is on data for the USA, by state, watershed, county, water body, and other geographic and administrative areas. To access, go to > https://www.epa.gov/enviroatlas > Interactive Applications > Interactive Map > Launch the Map > Explore. The map interface (shown in Figure A.10 for amphibian species maximum richness) allows for zooming and panning, for map layers to be added or subtracted, and for tools such as elevation profiles or a raindrop's flow path to the nearest water feature.

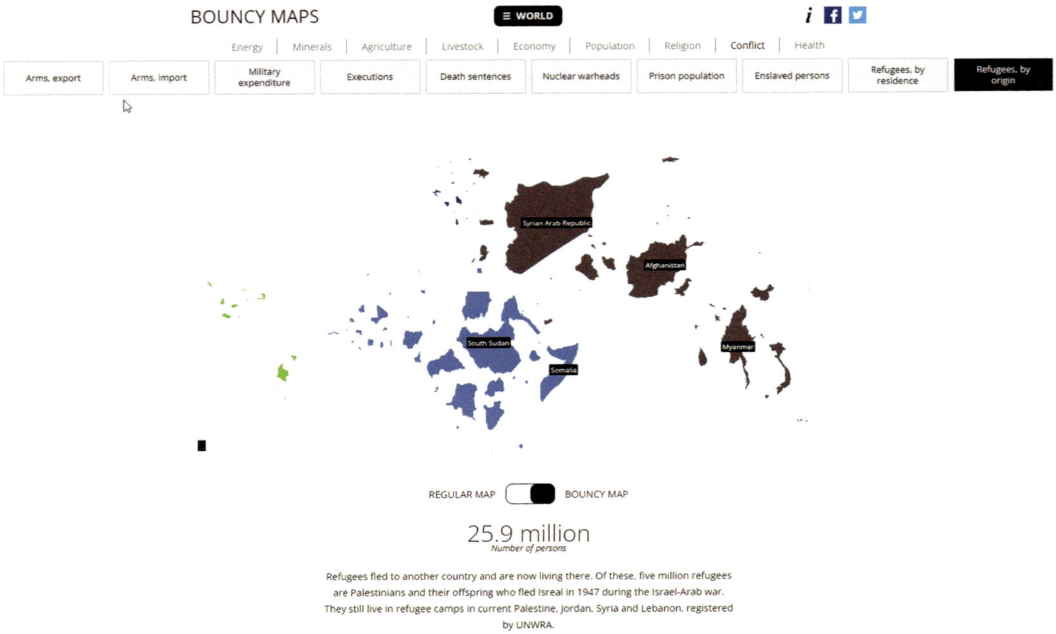

FIGURE A.9 Comparing the amount of a single variable among world countries using Bouncy Maps.

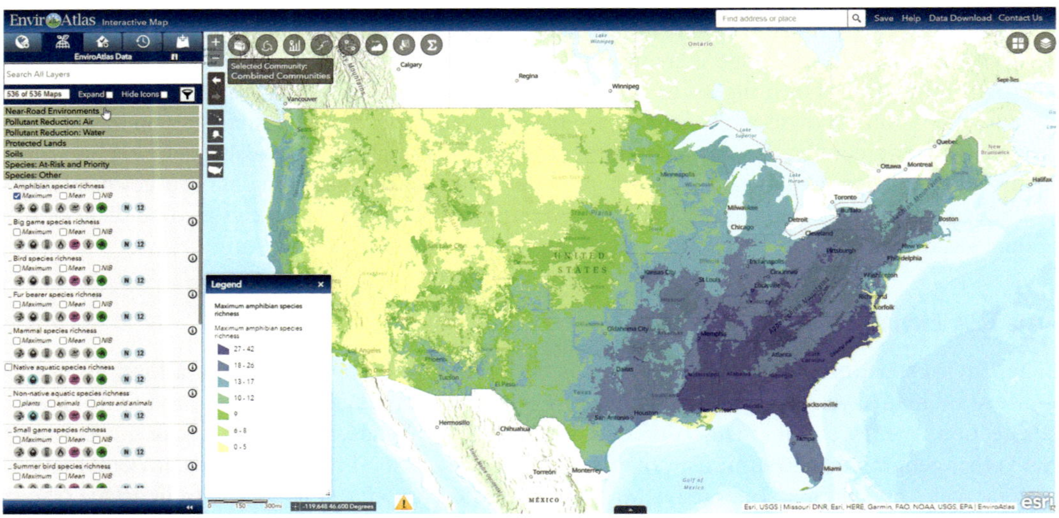

FIGURE A.10 Analyzing and mapping environmental variables using the EPA EnviroAtlas.

HOW TO USE THIS BOOK: GENERAL STRATEGY

Tools were chosen that require no sign in or logging on, that are easily operable on a phone, tablet, or laptop, in an ordinary web browser with a modest internet connection. Visualization for some activities works well on a small screen; for others, a larger screen works better. Tools and data were chosen that cover a wide variety of scales, themes, time periods, and issues, to heighten interest, foster inquiry, and promote an awareness of the connection and relevance that mathematics has for 21st-century issues. Resources are provided for those who wish to dig into the tools, data sets, and problems beyond this book.

The content can be used as lecture and presentation material by the instructor. Or it can be assigned as readings for the student. In any event, however, the content is designed to be taught *in tandem with* the activities. Such teaching may take place in any number of conventionally available interactive opportunities as well as in online and in-class discussion forums, and even embedded in video and other presentation formats such as Microsoft Sway, Prezi, ArcGIS Instant Apps, and ArcGIS Story Maps.

Each activity invites you to reinforce a specific mathematics topic either directly through more mathematics investigation or interactively using mapping tools such as those highlighted in this section. The thoughtful reader will also reflect on the fact that these tools allow you ample opportunity to keep investigating beyond the activity steps themselves—and we encourage you to do so!

ANSWERING QUESTIONS

The activities are infused with many questions; aligned with our book's inquiry-driven purpose, many questions go beyond the standard "what is the correct answer" questions. The questions encourage exploration and experimentation with the tools and data. Throughout the interactive activities, and in the foundational text, there will be questions posed to which answers are not provided. Examples include open-ended questions such as "In your judgment, which type of map classification provides the clearest view of the spatial patterns of this variable?" If instructors are assessing these questions, our intention is for instructors to look for thoughtful responses and evidence of student critical and spatial thinking. In not offering answers to everything, we also encourage instructors to be creative, knowing that creative activity from an instructor fosters creative activity from students, which then leads them to proficiency and the consequent pursuit of lifelong learning. Indeed, interactive maps, and the GIS behind them, were invented precisely for inquiry and solving problems…as was mathematics!

Authors

Sandra L. Arlinghaus, PhD, is a mathematical geographer by training who believes that pure mathematics, particularly geometry, is the key to the understanding and the creation of both spatial analysis and spatial synthesis. She holds a PhD in theoretical geography (University of Michigan, Ann Arbor) and other degrees and advanced education in mathematics (from Vassar College, the University of Chicago, the University of Toronto, and Wayne State University). She holds an MA in geography from Wayne State University, an AB in mathematics from Vassar College, and a high school diploma from the University of Chicago, Laboratory Schools. She has published over 400 books and articles, many in peer-reviewed publications. She continues innovative approaches in publication as creator of *Solstice: An Electronic Journal of Geography and Mathematics*, perhaps the world's first online peer-reviewed publication (1990–). She, together with William C. Arlinghaus and Frank Harary, created the first eBook (born digital) in 2002. One of her (and William C.'s) earliest jobs involved teaching basic mathematics in the innovative, broad program at Ohio State University in the late 1970s under the direction of program leadership from Joan Leitzel (Demana and Leitzel, 1990), Bert Waits (Demana and Waits, 2018), and Frank Demana (Demana and Waits, 2018; Demana and Leitzel, 1990). While there, she developed innovative classroom methods (Arlinghaus, 1986, p. 141) for teaching basic mathematics as derived from geographical principles.

Joseph J. Kerski, PhD, is a geographer, educator, and GIS professional who has served in four major sectors of society: government (as geographer and cartographer with NOAA, the US Census Bureau, and USGS), academia (as instructor at the University of Denver, the University of Minnesota, Sinte Gleska University on the Rosebud Sioux Reservation, Harrisburg Area Community College, and others), private industry (as education manager at Esri Inc., an international company that creates and supports GIS software and solutions), and nonprofit organizations (as the president of the National Council for Geographic Education, active in the Society for Conservation GIS, and roles in other organizations). His career focus is the implementation and effectiveness of geotechnologies across all learning domains. Dr. Kerski holds three degrees, all in geography, and has worked extensively with those in career and technology education, earth and environmental science, history, and mathematics. He has authored or co-authored ten books, including *Creating a Smarter Campus, Interpreting Our World: 100 Discoveries That Revolutionized Geography* (Kerski, 2016), *Spatial Mathematics: Theory and Practice through Mapping* (Arlinghaus and Kerski, 2013), and *The GIS Guide to Public Domain Data* (Kerski and Clark, 2012), 75 chapters and articles, 1,500 lessons, 6,000 videos, and three blogs and podcasts (Thinking Spatially, GeoInspirations, and Spatial Reserves).

William C. Arlinghaus, PhD, believes that mathematics is a critical component in becoming an educated person: "All too often, students think about academic subjects as they relate to career goals rather than as they relate to becoming a well-rounded, constructive, and productive member of society. Mathematics teaches logic and clear thought patterns. The role of any school is to educate, rather than merely to train, the future leaders of society." Dr. Arlinghaus holds a PhD in pure mathematics (*Algebraic Graph Theory: Automorphism Groups of Graphs* (Arlinghaus, 1985)) and is Professor Emeritus of Mathematics and Computer Science, Lawrence Technological University, Southfield, Michigan. He has published over 50 books and articles in the USA and abroad. He has taught all levels of undergraduate university mathematics and computer science. He has extensive administrative experience, including service as Department Chair, from which he chose to reward faculty members who took innovative leads in basic mathematics courses, and appointed Ruth Favro to head up the interaction he promoted with the Michigan Mathematics Prize Competition (MMPC). He and his wife, Sandra, donated a substantial resource to reward performance on the MMPC, and he implemented the program at the University of Detroit Jesuit High School as he volunteered to teach interested students after school and develop resources for their regular teachers. On a broader scale, he was the sole creator of a new schedule for an entire university (LTU) switching from quarters to semesters. Finally, he has enjoyed collaborating with a number of scholars (including Sandra), particularly with Frank Harary (1969), who pointed him toward his dissertation problem, and with Neal Brand (Bittinger, Brand, and Quintanilla, 2005) on research and teaching mathematics projects.

Acknowledgments

The authors extend their great thanks to the large team required to produce this work in various formats, from traditional typeset to contemporary electronic. Once again, it was our pleasure to work with two outstanding, masterful CRC Press, Taylor & Francis Group editors: Senior Editor Irma Shagla Britton and Production Editor Ed Curtis. Their personal attention to this project, coupled with their great skill, led to efficiency in communication and persistence in careful production.

The authors extend their thanks not only to them but also to their staff. To Irma's staff, including Editorial Assistants Chelsea Reeves and Shannon Welch. To Ed's staff, particularly to the thoughtful and careful efforts of Sathya Devi, our Project Manager with codeMantra. Sathya guided us through the various twists and turns of the production maze especially with regard to the fine typesetting crew.

To all of them, and to others whom we do not know, we say Thank You!

Sandra L. Arlinghaus
Joseph J. Kerski
William C. Arlinghaus

1 Classifying Numbers and the Distributive Law

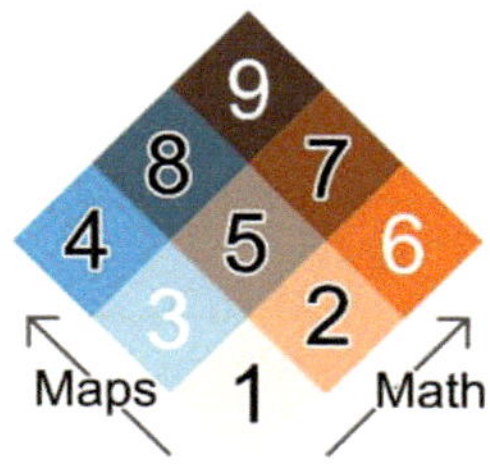

The line of images above is a visual abstract of this chapter designed to foster spatial thinking. From the chapter numeral, to the book structure, to the real world, the reader is offered gentle guidance to develop spatial intuition about what might be coming. Those thoughts are then reinforced with a detailed text outline of chapter content below. The images and outline forge as an abstract of chapter content.

CHAPTER OUTLINE

Classification
- Conceptual View
- Intuitive View: Interactive Mapping Activity

Classifying Numbers
- Natural Numbers
- Whole Numbers
- Integers
- Rational Numbers
- Irrational Numbers
- Real Numbers
- Nested Hierarchy
- Infinity
- The Number Line: Extent and Direction
- Different Classes of Numbers Positioned on the Number Line
- Absolute Value
- $a-b=a+(-b)$
- Prime Numbers

Operations
- Law of Excluded Middle
- Operations with Numbers
- Operations with Maps

Order
- Order of Mathematical Operations
- Order of Mapping Operations
- Legend: Layers and Organization

DOI: 10.1201/9781003305613-1

The Distributive Law
Terms and Factors
Parentheses and the Distributive Law
Geography and Thinking Spatially in Everyday Life: Enlarging Your World!

CLASSIFICATION

Conceptual View

Grouping similar items seems to be a natural human trait as we classify them using a single criterion or multiple criteria. Often, that action facilitates the replication of results, the organization of our lives, or clear communication with others. We might sort our clothing into summer/winter items, followed by shirts/slacks, as a simple strategy to create an organized closet. Mathematics, mapping, biology, and most other disciplines benefit from some form of classification to make organized sense of that field of academic study and of the associated real world surrounding it.

In this chapter, we focus on classification as it relates to mathematics and to mapping. Thus, we encourage the reader to think outside the box: to extend what he/she is learning into far-flung realms of inquiry and to do so creatively. Often, that underlying spark that ignites such extension comes from root principles, of which classification is an important example. Thus, we urge readers to keep such foundational concepts in their minds as they move forward in various exciting abstract and real worlds. Where do you see classification in the world around you? Is there anywhere that you do not see it? How does the concept of chaos relate to that of classification? Look words up in the dictionary and compare (look for similarities) and contrast (look for differences) their meanings.

Intuitive View: Interactive Mapping Activity

To stretch your mind a bit, consider the Human Development Index (HDI) Map that we provide through a link below (Figure 1.1). Even though you may have no experience with digital mapping at this point, dive right in. You will develop intuition and skill in navigating the software and you will see, broadly at least, some relationship to mathematics as we prompt you with questions while you move forward with the interactive mapping activity. A screen shot (static image) is shown in

FIGURE 1.1 Human Development Index data, mapped with ArcGIS from Esri.

Figure 1.1 that serves as the opening screen for the activities attached to this link: https://www.arcgis.com/apps/mapviewer/index.html?webmap=2c9f37a511a749adb8a820c3e9a9b299

The link above shows one of the maps that the reader can interact with (zoom, pan, change symbology (the process of choosing symbols to represent associated data), change classification, change variables mapped); find these tools, try them, and see what they do to the map. Do not be afraid; you cannot hurt the map. You might see mathematical ideas spring up, related to some of your earlier experiences with mathematics: creating graphs or charts, determining the position of points in the plane, counting numbers of dots, grouping dots into sets of elements (by country, for example). What else do you see? Talk about the map and learn what other readers are seeing that you might not. Maps are intended as ways to share information; begin with that idea and keep it in mind, now and in your later work.

Once you have some experience with the interface, consider the materials shown (and linked) in Figure 1.2. Click on the HDI map layer to the left of the map>using the ellipsis (...)>open the table behind the map. Note the large quantity of data that you have at your fingertips, for most of the world's countries, going back to 1980. Each row in the table represents one country: one row represents one country as a one-to-one association of content—have you heard of that idea in previous mathematics courses?

To the right of the map>Change Style ("style" is a software-specific word for symbology)>Add Field>Select a year of interest to you>Counts and Amounts (size), and>Done. Adjust the size of the points so that they are smaller, and the map is not quite so crowded.

What patterns do you notice in the world map of HDI for a given year? Next>Change the symbology to Counts and Amounts (color). Use Options>change the size of each point to something larger and better able to be distinguished. In your opinion, which map style communicates more clearly? Next, under Counts and Amounts (color)>Classify data. Use one of the classification methods and adjust the number of classes to see what happens as a sneak preview (Figure 1.3). Which classification method do you think most clearly communicates the global pattern of HDI?

In Figures 1.2 and 1.3, as readers interact with the Human Development Indicators map shown, they should notice that as they take a closer look at the map, more information can be discovered. Patterns at one scale may not be evident at a different scale, and vice versa. You will be able to change the symbology (colors, symbols, shading), classification methods, and more. Learn to be comfortable with elements of this particular mapping example. Reference will be made to it

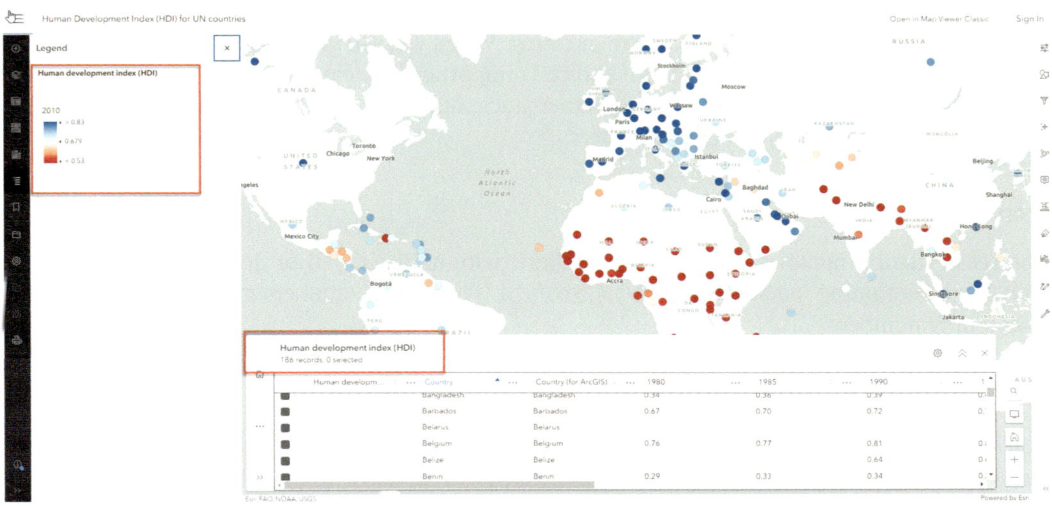

FIGURE 1.2 Human Development Indicators are mapped by quantity and color. Human Development Index data, mapped with ArcGIS from Esri.

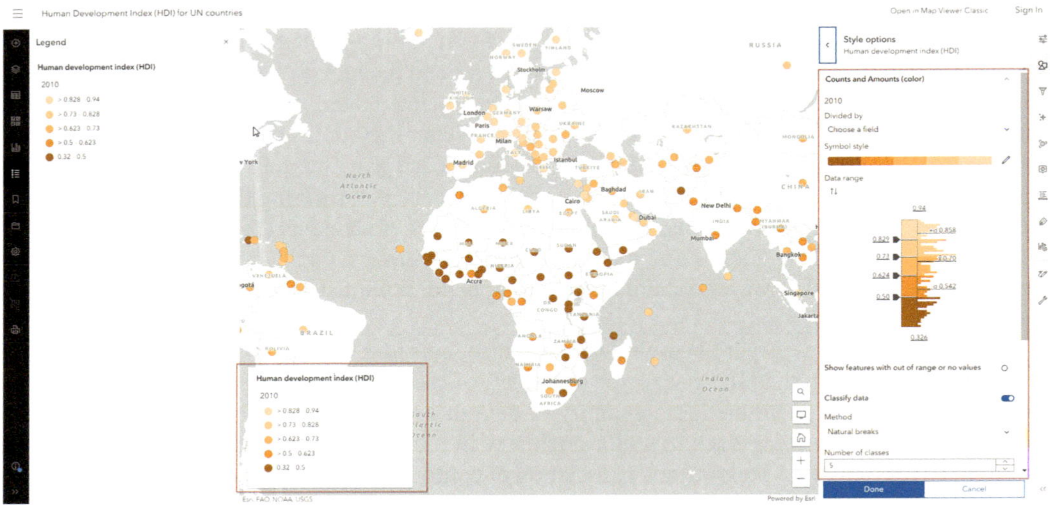

FIGURE 1.3 Human Development Indicators are mapped by natural breaks with 5 categories. Human Development Index data, mapped with ArcGIS from Esri.

elsewhere in relation to different mathematical ideas. You will gain confidence in interacting with it as you work with it; we introduce it here as a beginning. Consider that maps are powerful shapers of people's opinions, so therefore there are ethical considerations bound up in the choices of symbols and colors on maps. Think about the material of the next section, associated with classifying numbers, in relation to these maps. Reflect on the broad idea of how classification relates to mapping and math, alike.

CLASSIFYING NUMBERS

In mathematics, abstract items are typically classified in abstract space. Numbers and numerals: Are these two words interchangeable? Do they mean the same thing? Do some research on the internet to find out the answer and to understand what the difference might be. "Number" is quantitative in meaning while "numeral" is qualitative in meaning. When you use the words in your speech or writing, try to choose the correct one. Word choice is important in clear and effective communication, not only in natural language but also in mathematics; otherwise, confusion may enter the picture. We look next at classifying types of numbers.

Natural Numbers

The set of natural numbers is the set of numbers, with no fractional or decimal parts, beginning with 1 and continuing indefinitely: {1,2,3,4,5,6,7,...}. The natural numbers might also be referred to as counting numbers.

Whole Numbers

The set of whole numbers is the set of numbers, with no fractional or decimal parts, beginning with 0 and continuing indefinitely: {0,1,2,3,4,5,6,7,...}. The set of whole numbers contains the set of natural numbers.

Make sure you are as confident of your knowledge of the multiplication table (through 12) as you are of the letters of the alphabet. You need to be certain and quick; and to know not only what

9 * 7 is, but also what 63/7 is. Have competitions with your friends. What numbers are perfect squares? What is the square root of a number? When is the square root a whole number?

Integers

The set of integers is a set of numbers with no fractional or decimal parts. There is no beginning element. Integers may be positive, negative, or zero: {…–3, –2, –1, 0, 1, 2, 3…}. The set of integers contains the set of whole numbers and therefore also the set of natural numbers. Please note the correct spelling: integer. The adjectival form of the word is: integral. Then, move on a bit and consider more broadly the ideas of "positive" and "negative."

Think about positive and negative direction in finding one's way in the real world and on a map. Look at the GPS in your car. Do right and left on the map correspond to right and left in the real world? What are the default settings for the map? Is there more than one way to orient the map? Is there a setting so that north is always at the top? What does that mean when you are heading south? What is the relation in the real world to the ideas of positive and negative when associated with the cardinal points (N, S, E, W) on a compass?

Rational Numbers

A rational number is a number that can be written as the division of one integer by another—or as a "ratio." Note that the word "rational" contains the word "ratio." Between any two rational numbers there is always another one: 3/8 lies between ¼ and ½, for example. Thus, writing them out in a string, as with the previous sets, is not a particularly useful way to visualize them. Integers are rational numbers: 3 = 3/1, for example.

Irrational Numbers

An irrational number is one that cannot be written as a ratio: not rational. Name some irrational numbers: pi, the square root of 2.

Real Numbers

The set of real numbers is the set of all the sets of numbers above.

Nested Hierarchy

The sequence of sets, natural numbers, whole numbers, rational numbers, real numbers, forms a *nested hierarchy*: each successive set in the sequence fully contains the previous one. The concept of a nested hierarchy is one that comes up in many different contexts both within, and outside of, mathematics. In mapping, polygons used to display US Census Bureau data often form a nested hierarchy; Census Blocks are nested within Census Block Groups, Census Block Groups are nested within Census Tracts. Take a look at this link to see more: https://courseguides.trincoll.edu/Intro_Census_Data/census_geography. Challenges exist when areas do not nest within each other. As one example, zip code boundaries do not nest with census geography. Another example is that school district polygons typically do not nest within Census polygons; thus, it is generally difficult to use Census data in association with school data, an unfortunate situation. It is a challenge, however, that the reader of materials about school district data should be aware of and prepared to deal with: to ask pertinent questions regarding interpretation of any data associating school population data with Census data. How then should the data—whether on population or anything else—in non-nested areas be allocated to the non-nested areas?

The concept of **nested hierarchy** is a root concept linking mathematics and mapping; it arises often and it is useful to note it when it does in order to make appropriate decisions in moving forward.

Infinity

The concept of infinity is a broad one that has mathematical meaning as well as philosophical meaning, religious meaning, and a host of others. Some scholars devote their entire careers to the study of infinity. Here, we keep it simple. We refer to infinity as simply having the capability to continue the counting process forever. Each of the sets in the previous section is infinite. In all of them, if we claim to have reached the biggest number of the set, we can just add one to get an even larger number. In addition, in the set of rational numbers, for example, we can keep subdividing an interval between two selected rational numbers. Between any two elements of the set of rational numbers there is always another rational number. Is the same true for the set of irrational numbers and the real numbers? Why or why not?

The Number Line: Extent and Direction

A straight line has infinite extent in two directions; its full extent cannot be represented on any sheet of paper. It has no beginning point. A ray has infinite extent in one direction; its full extent cannot be represented on any sheet of paper. It has a starting point which may or may not be included. A line segment has two endpoints which may or may not be included; it can be represented on some sheet of paper. One number line that is important in mapping is the bar scale. It permits measurement of distance on a map: https://pro.arcgis.com/en/pro-app/latest/help/layouts/scale-bars.htm.

Different Classes of Numbers Positioned on the Number Line

We note these geometric characteristics of "lines" so that when we use lines to visualize number classes, we match their fundamental geometric characteristics to their fundamental numerical characteristics. Thus:

- Use a ray to represent the natural numbers. Label the starting point of the ray with the numeral 1 and continue to move away from the starting point spacing successive numerals evenly along the visible extent of the ray, which extends infinitely in one direction.
- Use a ray to represent the whole numbers. Label the starting point of the ray with the numeral 0 and continue to move away from the starting point spacing successive numerals evenly along the visible extent of the ray.
- Use a line to represent the integers. Label a central point as 0 and move away from zero to the left, in the negative direction, spacing successive numerals evenly along the visible extent of the line, and to the right, in the positive direction, spacing successive numerals evenly along the visible extent of the line.
- Rational numbers can be represented along a line. But, perhaps of particular interest is to represent a part of the set using a line segment. Choose any two rational numbers and connect them with a line segment and indicate selected intermediate values, using even or uneven spacing as desired. Remember: between any two rational numbers there is always another one.
- Irrational numbers. These numbers are not rational. They can be placed on the number line for the rational numbers only in approximate form. For example, pi is an irrational number; one approximation to its numerical position is 22/7 which might also be used as an

approximation to its geometric position on the number line. The value of 3.1416 is another commonly used value to estimate pi. What is critical to note, for both the numerical and geometric form of irrational numbers is that they are NOT rational; thus, any rational characterization of them is, necessarily, only an approximation. There is no "best" rational approximation for any irrational number; if there were, it would be the rational number itself (and it is not)

- Real numbers. Include the rational numbers and the irrational numbers.

Absolute Value

The absolute value of a number may be viewed as its distance from the zero point on a number line. The absolute value of 2 is 2, as it is 2 units from a zero starting point on a number line. The absolute value of –3 is 3 as it is 3 units from a zero starting point on a number line. Notation for absolute value is two vertical bars: $|-4|=4$, for example. Absolute value is a concept that arises throughout mathematics and mathematical sciences. It seems trivial to state what it is, but its application can be non-intuitive to students who are not comfortable with it.

$$a - b = a + (-b)$$

Use number line analysis to demonstrate this fact; look at using "chips" to illustrate the process: https://www.youtube.com/watch?v=YB6zcw-2Ses. Some take this fact as a definition for subtraction. Multiplication can also be viewed along a number line: https://youtu.be/PLDfl6daajo. The number line and related ideas can be quite helpful in manipulation of various algebraic expressions.

Prime Numbers

A whole number that has no divisors other than itself and 1 is called a prime number. Look at the first few whole numbers: 2 is prime; 3 is prime; 4 is not prime because 4 equals 2 * 2; 5 is prime; 6 is not prime because 6=2 * 3; 7 is prime; 8 is not prime because 8=2 * 4; 9 is not prime because 9=3 * 3; 10 is not prime because 10=2 * 5. As the numbers get larger, it becomes more difficult to tell if a number is a prime number.

Is 91 a prime number? How many numbers do you have to test to find out? What is the relation of the square root of 91 to the process of testing numbers to be prime numbers? Note that the square root of 91 is less than 10; thus, we do not need to test for whole number divisors of 91 beyond 9. The link: https://deepblue.lib.umich.edu/handle/2027.42/58264 offers insight into the fascinating world of prime numbers. One method of classifying numbers, as to whether they are prime, is animated.

Using Numbers with Maps to Analyze Demographics

In the following activity, you have an opportunity to use numbers with maps to analyze demographics—the study of human populations. You will use a tool called Gapminder, created to help people understand the health, income, life expectancy, and other aspects of the global community, with most variables provided by country. First, access the bubble chart of population and life expectancy, here https://www.gapminder.org/tools/#$model$markers$bubble$encoding$trail$data$filter$markers$dza=2022;;;;;;;&chart-type=bubbles&url=v1 (Figure 1.4 shows a screenshot). The X and Y axes may look like integers, but clicking on specific country shows that most of the numbers are rational numbers, such as Algeria, shown here, which in 2021 was shown with a life expectancy of 76.7 and a mean annual income of $11,200. Note not only is life expectancy and income shown, but the population of the country is also indicated by the size of the bubble, or circle. Click on some of the largest ones to see China and India. Considering the information about numbers that

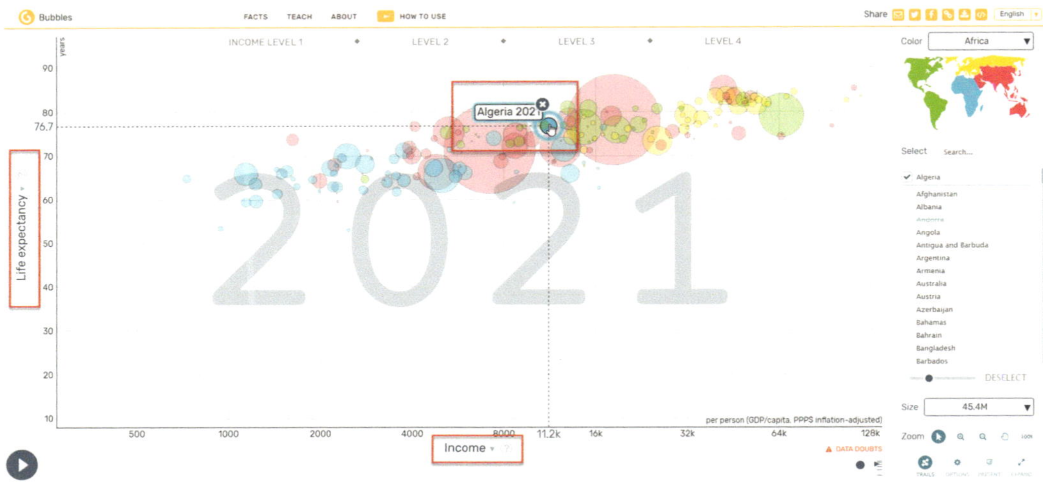

FIGURE 1.4 Comparing life expectancy to income for world countries, using Gapminder. For an additional challenge, select a subset of world countries using the select tools on the right, and compare the differences among them. Source: www.gapminder.org. License: https://creativecommons.org/licenses/by/4.0/ No changes made.

you have been learning about, which of the two axes can potentially go to infinity—life expectancy or income? Both, one, or neither? Consider the number lines that make up a part of each axis. Can either of these numbers be a negative? Click on the x-axis and note that this represents gross domestic product per person adjusted for differences in purchasing power. Even a number as straightforward as the y-axis—life expectancy—needs to be examined: Here, it is the average number of years a newborn child would live **if** current mortality patterns were to stay the same. Given the dynamic nature of our Earth, the mortality patterns might change. Graphing data in this way is a powerful aid in understanding how countries compare among key variables but click on the Data Doubts section to understand the limitations of comparing across countries in this way.

Next, click on Maps > to visualize world countries where the population is represented as a graduated symbol, with a larger symbol for a larger population. Next, in the Size box, click on a variable of your choice to map world countries based on another variable.

To visualize data in another way, access an animated map of selected countries and how their income changed over time (Figure 1.5). First, analyze the default four countries provided here. What type of number is income? Does it have a zero point? Can income be a negative? What type of number is time? Does time as measured by most of the world have a point at which one can consider it to be a "negative number"? Which of the four countries increased the most in the past 30 years? Can you name one reason why it increased so rapidly? How does animation aid you in understanding change over time?

To analyze data in still another way, for population by income, click https://www.gapminder.org/tools/#$chart-type=linechart&url=v1. The area under this curve represents all the world's people, and their income, symbolized by the continent in which they live. Note according to the map to the right of this graph that Australia and New Zealand are grouped under Asia. Note that in this graph, instead of mean annual income, income is represented as $ (dollars) per day. Do you find that dollars per day is a more helpful way of thinking about income than annual income? Based on this graph, which continent would you say is the wealthiest? The most populous? The poorest? You will learn about median, mean, and mode, as well, in this book, all of which can be visualized in this graph.

Last, look at a graph of the population of the world by age, https://www.gapminder.org/tools/#$chart-type=popbyage&url=v1. Note that the number of people in the world at each age is

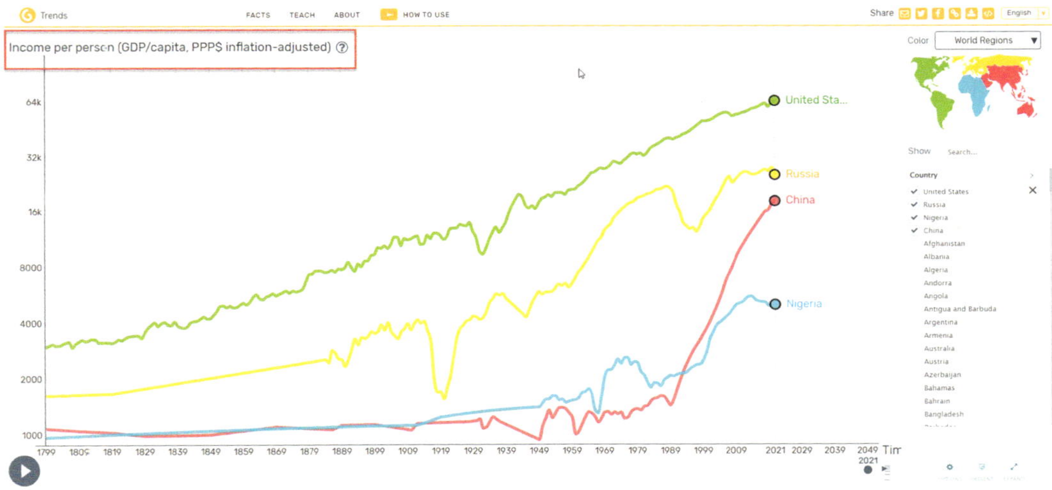

FIGURE 1.5 Comparison of income per person over time, using Gapminder. For an additional challenge, dig deeper and examine other countries besides these. Source: www.gapminder.org License: https://creativecommons.org/licenses/by/4.0/ No changes made.

shown on the x-axis, and the age of the population is shown on the y-axis. What are the factors that shape the number of people alive at each age? Will this graph's shape change over the coming years? Why, and how?

Move your touchpad or mouse up and down inside the graph. How many of the world's people are alive at **your** specific age? Are there more people alive on the planet who are 1 year older than you, or are there fewer people alive? Note that age is shown as an integer on the y-axis but in reality, is age an integer or a rational number? Like other maps and graphs you will examine in this book, there are trade-offs to be made in balancing details with clarity. There are certainly people in our world that are over 100 years old. However, how could those people over 100 years old be shown on this graph given that the scale of the x-axis is in the millions of people? Use the group function to the lower right of the chart. You will learn more about grouping and clustering of data in this book. For now, how does group aid your understanding of the pattern of the world's population by age? What information is hidden in the graph when you select group?

OPERATIONS

Law of Excluded Middle

Much of mathematics, both traditional and contemporary, is based on the Law of Excluded Middle: all answers to problems are either right, or wrong (Geach, 1980, p. 74). There is no middle value of "sort of right" or it is a "matter of opinion." This Law is foundational to most mathematics and to all of the mathematics in this book; it is a good thing—we want our bridges and buildings to be completed properly when constructed and to stand up to the test of time—"close" doesn't count!

With mapping, however, there is room for more than just the situation stated in the Law of Excluded Middle. There is the clear difference between the concepts of "precision" and of "accuracy" as two distinct ways of thinking about the concept of "error"; they are independent of each other. Accuracy refers to how close a measurement is to an accepted, true value while precision refers to how close measurements of the same item are to each other. Note that one can be very precise but not very accurate, as well as the reverse. Generally, it is best to be both. Although mapping is based on mathematical concepts, this underlying difference, between absolute and relative truth, makes it challenging to merge mathematics with mapping—it needs to be done with care.

Operations with Numbers

Familiar operations with numbers are those associated with addition, +, subtraction, –, multiplication, *, and division, /. There are others, but these are the principal ones being used in this book. There are many links to number line activities on the Internet. Consider reviewing some, as needed. Here is one link to an interactive resource (scroll down to the interactive number line and the zoomable number line): https://www.mathsisfun.com/number-line.html.

To add numbers on a number line, these examples illustrate the idea:

- Two positive numbers $3+5$. Move 3 units from 0 in the positive direction along the number line. Then, move another five along the number line, in the same direction (positive), arriving at 8. Where you land gives the correct answer.
- One negative and one positive number: $(-3)+5$: Move 3 units in the negative direction from zero (move to the left). Then, move five units in the positive direction, to the right, arriving at 2.
- One positive and one negative: $3+(-5)$: Move 3 units in the positive direction then move five units in the negative direction, arriving at –2.
- Two negative numbers: $-3+(-5)$: Move 3 units in the negative direction then move 5 units more in the negative direction, arriving at –8.

To subtract numbers on a number line, these examples illustrate the idea:

- $3-5=3+(-5)$
- $-3-5=(-3)+(-5)$
- $-3-(-5)=(-3)+(-(-5))=(-3)+5$. The double negative sign is positive; think of moving five spaces forward, then five spaces back. You have returned to your starting point. The signs are not relevant.

To multiply or divide positive and negative numbers, count the number of negative signs. If there are an odd number of negative signs, the answer is negative. If there are an even number, the answer is positive. Why is that? Relate the answer to movement along a number line.

Some expressions might have absolute value symbols within them. When that occurs, convert the expression inside absolute value symbols first, in order to remove them, and then solve as above. Thus, for example, $3+|4-6|=3+|-2|=3+2=5$.

When working with operations, it may happen that a wrong answer arises. What to do with it? Admit it is wrong and correct it; remember, that getting a wrong answer and having it pointed out is not a comment on the merits of a human being; it is simply a comment that a process has a small flaw in it. So, please ask for evaluation of the process and be appreciative if someone else is willing to try to catch small flaws you might have missed. As Strunk and White (1959 and later) have said about language use, "if you make a mistake say it loudly!" Just take error as a fact of life, try to minimize it, and when it creeps in, correct it as soon as possible, and then move on without personalizing it emotionally or trying to interpret any deeper meaning.

Operations with Maps

Operations with maps are less well-defined than they are with numbers. The focus on mapping is communications; the science of mapping, called cartography, contains some guidelines but few rules. The main guideline or rule of thumb is, "Does this map help me understand something in a deeper, richer way?" Nonetheless, operations with numbers play a critical role in looking at maps.

Interactive maps are based on underlying sets of numerical data and these sets are governed by the rules of operations with numbers.

Operations with numbers yield a single answer. It is either right, or it is wrong. The mathematics in use in this book is all based on that critical assumption. There may be equivalent answers depending on how the problem is framed: an answer might be expressed as 3, 6/2, 9/3, and an infinite number of other possibilities. These equivalent expressions all represent the same exact, single amount. They are not different answers; they are different representations of the same single exact answer.

Thus, in mapping operations, that are based solely on mathematical operations, there is a single right answer, and all others are wrong, up to equivalence. If the distance between my house and the grocery store is expressed as 1 mile, or as 5,280 feet, that difference in superficial expression has nothing to do with the amount of distance between my house and the grocery store. I am free to choose equivalent expressions but in doing so I must not distort the correct amount or the value will be wrong: 5,281 feet is close, but it is just as wrong as 6,000 feet would be. Further, my "feelings" about that distance are not relevant; if I am out of shape, and my friend is in good shape, I may feel that that distance (when walking it) is much longer than my friend does. But the distance is the same no matter how it feels when walking it.

Interactive maps may have elements that can be moved or adjusted and that are not reliant only on the sort of quantitative values expressed above. Such motion, sliding layers around, adjusting how much can be seen on a map by changing its scale, and others, are basic mapping operations that may offer room for more qualitative, or subjective, values such as "feeling"—does the map look better one way than another?

ORDER

Along with classification and operations, order is a core concept which underlies much thinking in most disciplines and in life in general.

Order of Mathematical Operations

In a mathematical expression,

- first go through and evaluate all multiplications and divisions,
- then go through and evaluate all additions and subtractions.

Order of Mapping Operations

Maps classify information that often is spatial (adjective of the word "space"). In Figure 1.6, we see an overview of Ann Arbor, Michigan, from above. In it, a layer of water features was added on top of the underlying map that comes, by default, with Google Earth. The intention is to emphasize the water features in relation to the road features. But notice: the mapped layer that contains the emphasized water lies on top of the mapped layer that contains the roads, indicating that the roads are flooded by the water feature or that the road features are in tunnels under the water. In fact, neither is the case. Instead, roads are on bridges over water. This unintended mapping consequence creates a problem; to solve the problem, we simply ensure that the layer, with the roads in it, lies on top of the layer with the water (Figure 1.7). Be careful to arrange the layers (if the software does not do it for you) to emphasize correctly that which you wish to emphasize. Think about, and observe, what you create: there are exact quantitative elements associated with the otherwise "loose" concept of layer. If you had three layers, one with point information, another with line information, and another with area information, you might stack the layers with the points on top of the lines, and the lines on top of the areas, resulting in point-line-area from the top of the layer stack to bottom of layer stack, so that all layers are simultaneously visible. Order matters!

FIGURE 1.6 Ann Arbor, Michigan. Water features are emphasized. Notice that the water covers the roads when the road and water layers appear to meet. Hydro shape file from City of Ann Arbor; mapped using Google Earth Pro with background NOAA imagery (6/30/2022).

FIGURE 1.7 Ann Arbor, Michigan. Water features and road features are emphasized. Note that the order in which the layers are presented makes a difference: roads on top of water. Hydro shape file from City of Ann Arbor; mapped using Google Earth Pro with background NOAA imagery (6/30/2022).

Legend: Layers and Organization

In parallel with the order of operations, the legend box on a map tells a story. The order in which the layers of information of a map appear in the Legend box is critical to understanding the content of the map. A change in the order can dramatically alter the meaning of the map, as we saw with roads and rivers in Ann Arbor, in Figures 1.6 and 1.7. As further noted on the following interactive link: https://pro.arcgis.com/en/pro-app/2.7/help/layouts/work-with-a-legend.htm, interactions between the legend and the map are important in various contexts.

Changes in a legend impact the associated map; so too, changes made on a map impact its legend. If a layer is removed from a map, the associated item must be removed from the legend. Four options to synchronize the behavior of a legend, based on changes made to its associated map, are listed below:

- Layer visibility: The legend displays only the visible layers in the map.
- Layer order: Legend items are reordered to match the drawing order of the layers in the map.
- New layer: A new item is added to the legend when a new layer is added to the map.
- Reference scale: The symbols in the legend change scale based on the map's reference (underlying) scale. The size of symbols in the legend matches the size of symbols on the map.

With these ideas in mind, consider the map layer activity you worked through involving legends. When breaking the conventional order, consider the impact it might have on the mapped content. For, while there is always one and only one correct answer associated with the mathematics underpinning a map, there is often more than one correct answer, and perhaps no wrong answer, associated with the mapping of content. Wrong answers can appear as in the example involving roads and rivers in Ann Arbor; but often there are multiple answers that are right. And what is a right answer for me, may not be a right answer for you. If you live in New York City and want to know about the relationship between its road and river networks, the map in Figure 1.5 which is right for me in showing those relationships for Ann Arbor, is not right for you in New York City. Furthermore, some maps are so intuitive that they do not require a legend.

To gain further insight into the concept of order, open the linked map (https://www.arcgis.com/apps/mapviewer/index.html?webmap=8c7d2d43d3ae40988c46519c6153716a). This interactive map contains three global layers: Ecoregions, hydrography (rivers and lakes), and cities. Click the layers symbol on the left side of the map to see all three layers, and then click on the world cities layer to see its symbology, as shown in Figure 1.8.

Use the eye symbol to turn each layer on and off, one at a time, to become familiar with each layer. The ecoregions layer is represented as a set of polygons: Each polygon is an area of similar landforms, climate, and vegetation. The hydro layer is represented as a set of lines and shorelines indicating rivers and lakeshores. The cities' layer is represented as a set of points: the larger the points, the greater the population for that city. The layers draw from bottom to top, with the world cities drawing first, followed by the hydro layer, and finishing with the ecoregions. Thus, the ecoregions layer obscures critical parts of the other two layers beneath it. Drag the other layers so that the map is more understandable to you. Usually, the clearest maps have the point symbols at the top, the lines underneath the points, and the polygons underneath the lines. Drag the other layers so that the world cities are at the top, followed by the hydro lines, followed by the ecoregions, respectively. When you do, the map should look like the one in Figure 1.9.

Maps are rich sources of information and require careful study to understand all that they embody. Think about layers and order of operations routinely; mapping will become much easier when you do so. And then consider what to do when you wish to break the conventional order of operations.

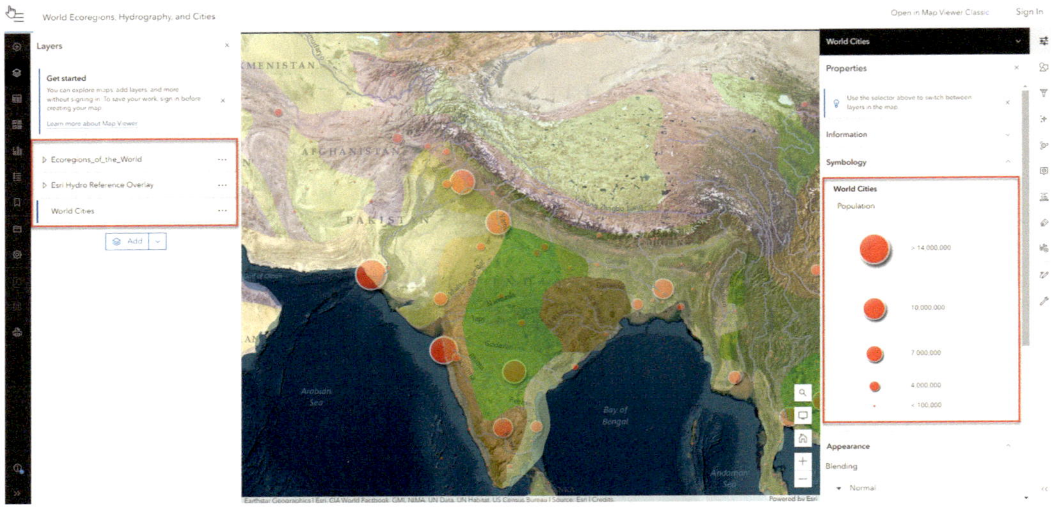

FIGURE 1.8 Screenshot of an interactive map containing three layers. Ecoregions and cities data, mapped with ArcGIS from Esri.

FIGURE 1.9 Layers arranged in order as point, line, and area, to optimize visibility of content. Ecoregions and cities data, mapped with ArcGIS from Esri.

The next section will introduce one of the most important laws in mathematics that will enable you to enrich both mathematics itself and all those things that derive from it.

THE DISTRIBUTIVE LAW

Terms and Factors

Some basic words people use to talk about parts of a mathematical expression might include the following ones. The parts of an expression linked by +signs are called "terms." The parts of an expression linked by * signs are called "factors." Make sure these words are a standard part of your vocabulary (and if you run across words whose meanings are not clear to you, ask what they mean). There is nothing wrong with not knowing; but there is a lot wrong with not asking questions! Then, of course, learn from the answer.

Parentheses and the Distributive Law

To break the standard order given by mathematical order operations, introduce parentheses.

The Distributive Law,

$$a * (b+c) = a * b + a * c$$

shows the relationship between a mathematical expression with parentheses and one without parentheses. The parentheses show how to break the conventional order of operations. Note the importance of the distributive law in linking additive and multiplicative operations. Also note that it makes sense to read it from left to right, and, from right to left. It is one of the most important laws in mathematics.

As an example, consider the following expression: $2+3 * 5 - 4 * 6$. Evaluate it using standard order of operations by first performing all multiplications from left to right: $2+15-24$, and then performing all additions and subtractions from left to right: 17–24, giving an answer of –7. Now introduce parentheses: $2+3 * (5-4) * 6 = 2+3 * 1 * 6$ (reducing it to standard order of operations) and then proceeding to evaluate: $2+18=20$. The two problems look similar; but they do not represent the same quantity. Well-placed parentheses are of critical importance; learn to be comfortable using them, note that they appear in pairs—left and right, and never ignore them.

Prove that a negative number, $-x$, times a negative number, $-y$, is a positive number. Note the reliance on the distributive law, using it in both directions around the equal sign.

$(-x)(-y) = (-x)(-y) + x(y+(-y))$ adding zero, as $(y+(-y))$, does not change the value

$= (-x)(-y) + xy + x(-y)$ multiply out using the distributive law from left to right

$= (-x)(-y) + x(-y) + xy$ rearrange terms

$= ((-x)+x)(-y) + xy$ use the distributive law on first two terms; note that the factor, $((-x)+x)$, was created by reading the distributive law from right to left, is zero

$= xy$

Look for the distributive law throughout mathematics, and elsewhere, and learn to recognize where it is being used, either directly or subtly; mathematics will become much easier for you when you do so routinely. The distributive law is a bit like a center fielder in baseball who is always in the right place to catch the flyball—never having to make a diving catch. So too, the distributive law is always there, backing you up and making sure things work right, linking operations when needed without making a grand production out of doing so. Look for the distributive law to make an appearance in the following mapping example.

In this activity, in addition to seeing the distributive law in the background, you develop skills in analyzing data over space and time. Open the linked map (https://www.arcgis.com/apps/mapviewer/index.html?webmap=f5e6469be6c34f2194758b7af02109f3) of global population.

The map opens with world population by country shown in 2015 from United Nations data. Each country's population is shown with a symbol sized according to its population; hence, a "graduated" symbol map, shown in Figure 1.10.

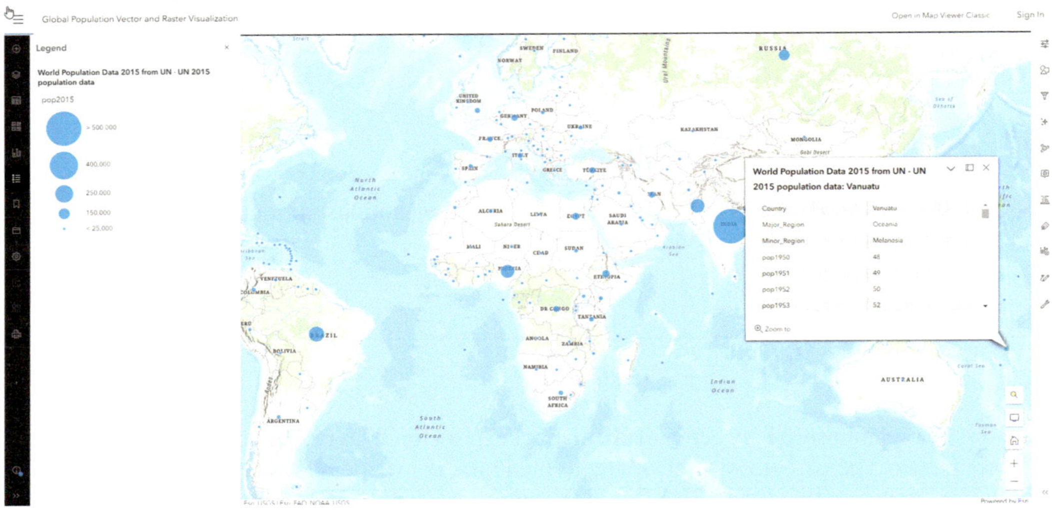

FIGURE 1.10 Graduated symbol map of the world showing population of countries. World population data, mapped with ArcGIS from Esri.

What pattern of symbols and thus country population do you notice on the map? Which is the most populous country on each continent? Which continents contain six of the largest symbols and consequently contain the six most populous countries? What is the relationship of the areal (adjectival form of the noun, "area") size of countries to their population? Do the larger countries in area tend to have more people? Yes, often they do, but not always: Can you find two large countries with a relatively small population?

Sometimes on a map and in charts, large numbers are represented in text as "in hundreds" or "in thousands" so that they can be more easily compared as shorter numerals. This qualitative practice may make the legend more readable. The visual representation of the numbers may vary but the value they represent does not. In this case, population figures are given in thousands. Click on the USA and you will note that its population in the pop-up table was 321,774 in 2015. This is in thousands; three digits have been removed. What, then was the USA's actual population that year (it was in the millions)?

The way you have been analyzing data in this map is with Earth features represented as points, lines, or polygons/areas. Attributes (oftentimes numbers), or characteristics, are assigned to each feature. Here, each country is a polygon, and a point represents the population of that country, represented by several fields of data covering the population over different periods of time.

You have been examining population for 1 year—2015. Mathematical expressions can be used to compare numbers over time and space. Population is an example of a data set that changes frequently. Visualizing numbers on maps often reveals the unexpected: in the following example, you may unexpectedly find that the population for some countries is decreasing. To compare 2010 population data with 2015 population data, you will create an expression that will show population change, as follows: To the right of the map > Styles > Choose attributes > delete the 2015 variable that is currently being shown, then > New Expression: Choose the variables below one by one by clicking on the list of variables, using the (x), and enter the parentheses and the operators (subtract, divide, and multiply) manually, as suggested in Figure 1.11.

When done, your expression should be, as shown in the top line:

```
(($feature.pop2010 − $feature.pop2015)/$feature.pop2015) * 100
= ($feature.pop2010/$feature.pop2015 - $feature.pop2015/$feature.pop2015)
  * 100
= ($feature.pop2010/$feature.pop2015 - 1) * 100
```

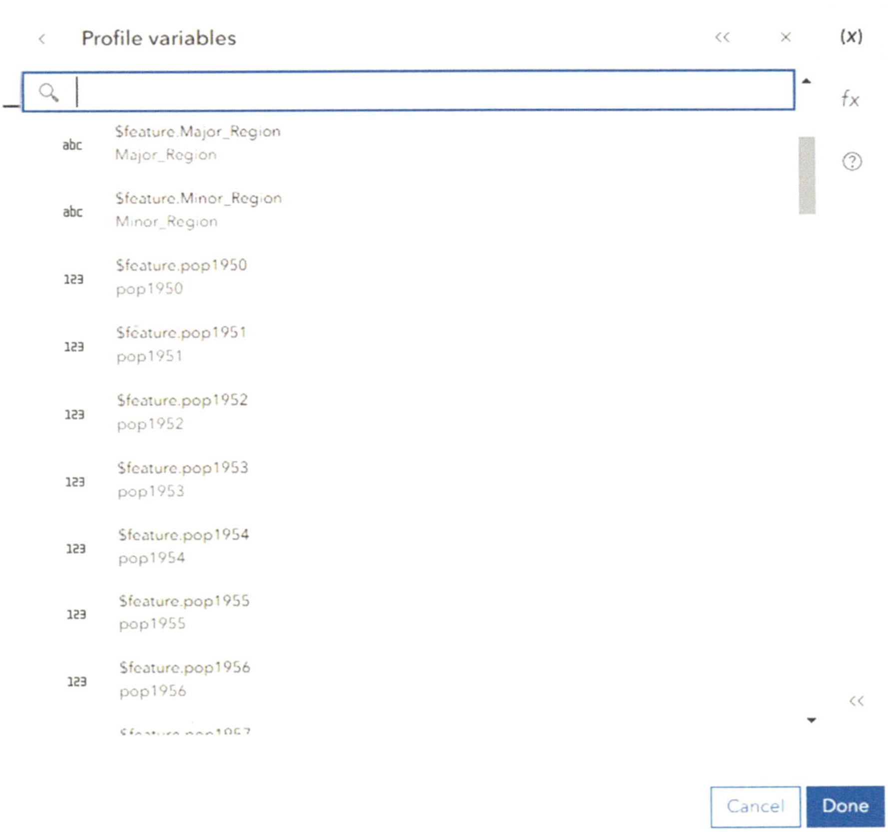

FIGURE 1.11 Building an expression from mapping variables. ArcGIS from Esri.

The two lines below are equivalent forms, based on the distributive law and are shown here to illustrate how the distributive law can work behind the scenes—where did the numeral 1 come from in the last line? Look at the structure of the expression here—long variable names may cloud general structure. The form of this set of expressions is: $((x-y)/y) * 100=(x/y-y/y) * 100=(x/y-1) * 100$, and now you see clearly where the 1 came from—the Distributive Law. This latter expression, with the numeral 1 in it, is more compact, mathematically, than the form on the top line, but is it more useful for mapping? Mapping and mathematics often agree, but not always—especially when utility is involved.

Name your expression (using the pencil and editing the title) for convenience: 2010 vs. 2015. This expression will result in countries that are increasing in population to be negative numbers, and those that are decreasing in population to be positive numbers. Why? Note that when 2010 population is greater than 2015 population, that is, when the population is decreasing, the associated rational number is greater than 1. When the 2010 population is less than the 2015 population, that is, when the population is increasing, the associated rational number is less than 1.

GEOGRAPHY AND THINKING SPATIALLY IN EVERYDAY LIFE: ENLARGING YOUR WORLD!

Geographic and spatial thinking are part of our everyday lives. Read the following article *Geography in Everyday Life* (Duke and Kerski, 2022) https://www.directionsmag.com/article/11897.

Pay particular attention to the section in the article in which the authors describe a trip they are making in order to meet each other at a conference. In that trip description, from packing of the suitcases, scheduling and taking rideshares, boarding airline flights, and deciding on the location to meet, the authors highlight how they think spatially throughout the day. At what moment in their day could you most personally identify with the authors' experience of thinking spatially? At what points when you are taking a trip do you consciously think spatially? Did the spatial implications of any of the tasks surprise you? If so, which one(s)?

Which instance of spatial thinking have you done most frequently? Which instance of spatial thinking do you feel is most connected to mathematics?

At what points during a typical day is your spatial thinking challenged? Do you have trouble remembering which direction to turn off the stairwell to get to your hotel room, or what floor to find your car in the parking garage? Do you find it challenging to orient right, left, top, and bottom on directions for assembling furniture or equipment? Take heart: Many people who are challenged by directions or other spatial tasks are very successful in science, geography, and in a whole host of fields! Be tenacious and think thoughtfully and carefully about spatial tasks, and don't be hesitant to ask for help.

The authors' success during their work trip described in the article is also partly dependent on *others* thinking spatially. Those other people, for example, had to design the numbers and letters for airplane rows that would allow the authors to find their airplane seats. Name two other instances in their day where they depended on others' thinking spatially. Think about your own activities yesterday—can you name at least two instances where your success was dependent on someone else thinking spatially? What is one argument that the authors make to advocate for spatial thinking in education and society that strikes you as the most interesting?

Note the infographic at the end of the article. How does the infographic provide a helpful way of visualizing some ways in which spatial thinking is intentionally or unintentionally used throughout a typical day that the text does not provide? Conversely, what information is provided in the article's text that is not provided in the infographic?

Take the survey linked midway through the article. The survey asks you to identify your last geography course, when you first heard about GIS, how often you use a navigation map or app, if you use an artificial intelligence enabled phone app to identify plants, clouds, bird songs, or something else in the field, if you have participated in a crowdsourcing project, and your approximate location. After you fill out the survey, consider the following: Were you surprised at how much or how little you use spatial thinking tools in day-to-day life? Keep engaged in, and enlarge upon your use of, spatial thinking as you move forward throughout the chapters of this book!

2 Fractions and Decimals

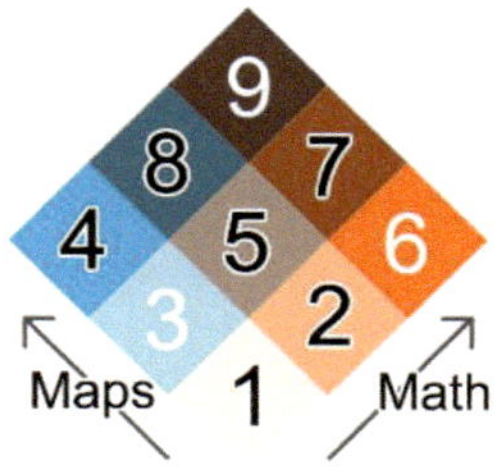

The line of images above is a visual abstract of this chapter designed to foster spatial thinking. From the chapter numeral, to the book structure, to the real world, the reader is offered gentle guidance to develop spatial intuition about what might be coming. Those thoughts are then reinforced with a detailed text outline of chapter content below. The images and outline forge as an abstract of chapter content.

CHAPTER OUTLINE

- Fractions
  - Adding and Subtracting Fractions with Like Denominators
  - Multiplying Fractions
  - The Importance of Multiplying by 1
  - Adding and Subtracting Fractions with Different Denominators
  - The Importance of Cancellation
  - Divisibility of Numbers
  - Divisibility, Prime Numbers, and Unique Factorization
  - Dividing Fractions
  - Fractions Larger than One
  - Fractions, Decimals, and Maps
  - Arithmetic with Fractions: Rapid Mental Calculation
- Exponents and Logarithms
  - Exponents
  - Logarithms
- Decimals
  - Convert Fractions to Decimals
  - Convert Decimals to Fractions
  - Adding and Subtracting Decimals
  - Multiplying Decimals
  - Dividing Decimals
  - Rounding and Estimating
    - The Eiffel Tower
    - Reflections on this Activity
    - Comparing and Measuring Decimals
- Bases Other Than 10
  - Binary
  - Octal

DOI: 10.1201/9781003305613-2

Hexadecimal
Operations
Time and Circular Measurement
Metric
English
Time
Circular Measurement
Latitude and Longitude
Latitude-Longitude Formats
Great Circles
Measurements and the Great Circle
Time and Longitude: Circling Back
Looking Ahead

FRACTIONS

Fractions are parts of whole numbers. Half, third, quarter, fifth, and so forth. Fractions can be meaningfully compared to each other. Which is larger, a half or a third? Would you rather have half a pie or a third of a pie? Learn about numbers in the abstract and keep relating them to the real world so they become your friends. Interactive mapping will provide a big assist along the way!

In a fraction, the top part is the "numerator" and the lower part, under, is called the denominator. Understanding fractions, rather than getting hung-up on non-intuitive terminology, should be the focus. Call the parts of a fraction "top" and "bottom" or "upper" and "lower"—whatever works to move forward with full focus on meaningful content.

Adding and Subtracting Fractions with Like Denominators

Fractions with like denominators are easy to add. Think about it. If you have a third of a sandwich and then get another third, what do you have? Two thirds. To add fractions with like denominators just add the top parts as you would whole numbers and keep the bottom part the same. The same idea works for subtraction. The answers may be positive or negative.

Stated generally: $a/c+b/c=(a+b)/c$

Samples:

- $1/5+2/5=3/5$
- $4/7-1/7=3/7$
- $1/7-4/7=-3/7$
- $1/5+0=1/5$.

Multiplying Fractions

It is also easy to multiply fractions. Just multiply the top parts together to make the top of a new fraction and then multiply the bottom parts together to make the bottom of a new fraction.

Stated generally: $a/b * c/d=(ac)/(bd)$

Samples:

- $1/2 * 1/2=1/4$
- $3/5 * 1/2=3/10$.

- $(-3/5) * 1/2 = -3/10$
- $3/5 * 4/9 = 12/45$
- $3/5 * 0 = 0$

The Importance of Multiplying by 1

Multiply any number by the number 1 and what is the result? Just the number you started with.

Stated generally: $a * 1 = a$

Samples:

- $27 * 1 = 27$
- $3/5 * 1 = 3/5$
- $(-2/7) * 1 = -2/7$
- $0 * 1 = 0$

A similar statement is that a number divided by itself is 1.

Stated generally: $a/a = 1$ assuming a is not zero

Samples:

- $2/2 = 1$
- $(1/2)/(1/2) = 1$
- $(-3)/(-3) = 1$
- $(-2/5)/(-2/5) = 1$

A related idea is $a/b = a * 1/b$.

These seemingly trivial observations play out in numerous important ways throughout mathematics. We consider two of them in relation to fractions. Master them and the mystery, that many feel toward fractions, will vanish.

Adding and Subtracting Fractions with Different Denominators

The puzzle of how to add fractions with different denominators has led many fine young math students astray. Generally stated:

$$a/b + c/d = (ad + bc)/bd$$

The general statements about adding fractions with like denominators, and multiplying fractions, were intuitive and followed naturally from what had already been learned with whole numbers. This general statement is not at all intuitive and usually leads to confusion, possibly culminating in full-blown math anxiety that lasts forever, until it is resolved. So, why does this general formula work?

If we can reduce the case with different denominators to the first case with like denominators, then we are done. Life once again becomes intuitive and easy. To make such a reduction, use $a * 1 = a$, or written equivalently, $a/a = 1$ (for a non-zero).

We can multiply a/b by the number 1. The form of the number 1 we choose is $d/d = 1$: $a/b * d/d = a/b$. Similarly, multiply c/d by the number 1, expressed as b/b: $c/d * b/b = c/d$.

Thus, $a/b + c/d = a/b * d/d + c/d * b/b$ (multiplying by 1, and choosing values to yield the denominator that we want)

= *ad/bd*+*cb/db* (multiplying fractions…see why you learn to do that first!)
= (*ad*+*bc*)/*bd* (using simple adding of fractions with like denominators).
Sample: 1/5+2/3=(1 * 3+2 * 5)/(3 * 5)=(3+10)/15=13/15.

When teaching, it is CRITICAL therefore that students first understand well:

- how to add simple fractions with the same denominator,
- how to multiply fractions,
- why multiplying by 1 is important,

and THEN finally learn to add fractions with different denominators.

Did you see, in the reduction of the process adding fractions with a different denominator to that of adding fractions with the same denominator, where the distributive law was hiding in the background? Look at the step involving: *ad/bd*+*cb/db*=(*ad*+*bc*)/*bd*. There are several background steps involved: *cb*=*bc* and similarly; *ad/bd*=*ad* * 1/*bd* and similarly. Thus, *ad/bd*+*cb/db*=*ad/bd*+*bc/bd*=*ad* * 1/*bd*+*bc* * 1/*bd*=(*ad*+*bc*) * 1/*bd* (using the distributive law)=(*ad*+*bc*)/*bd*. As these fundamental concepts become a natural part of your thought processes you will no longer need to disentangle them from broader thinking; however, you should be able to do so if asked—that action demonstrates clear and complete understanding.

The Importance of Cancellation

Figuring out why we add fractions with different denominators in the way that we do, was one spinoff from understanding the importance, and power, of multiplying by 1. Now we see another that makes arithmetic with fractions far more efficient to do and, again, clear to understand.

The fraction 2/4 is the same as 1/2. Just cancel 2 from the numerator and 2 from the denominator. Why does that work? We are multiplying by 1 again, written as 2/2. 1/2 * 2/2 =2/4. Cancellation simply requires finding 1s, written as fractions, inside problems.

Samples:

- 2/3 * 4/5=(2 * 4)/(3 * 5)=6/20. Both the numerator and denominator are even numbers and so we can use 1=2/2. 2/2 * 3/10=6/20. So, 6/20=3/10.
- Consider 6/15 * 45/48. Cancel first.
  - 6/15=(3 * 2)/(5 * 3)=2/5 * 3/3=2/5 *1=2/5.
  - 45/48=(15 * 3)/(16 * 3)=15/16 * 3/3=15/16 * 1=15/16.

    So, now the problem has become: 2/5 * 15/16 which is 2/5 * (3 * 5)/16; look to cancel again…cancel the 5s. So, now the problem is 2/1 * 3/16=2/1 * 3/(2 * 8) cancel the 2s. So now it is 1/1 * (3/8)=(3/8).
- Consider 6/48+10/15: cancelling, the problem becomes 1/8+2/3; convert the denominator to 24 and the problem is 3/24+16/24=19/24.

You can do these problems in your head when you use cancellation to make things more efficient! Look to cancel first before performing the multiplication or addition (or subtraction) of fractions. See why it is good to be quick with the multiplication table, both in multiplying two numbers to get an answer and in doing it the other way around, expressing a single number as a product of other numbers.

The broad idea here, as it is elsewhere, is to think about what you are doing before applying some memorized rule. Be efficient and do as much as possible in your head to make things easy. Relax; math is fun, not tedious. Challenge others to compete doing fraction problems in your heads—no calculators, no pencils. Just use the most powerful calculator in the world: your brain.

Divisibility of Numbers

Cancellation is critical in rapid mental calculation. To find common factors, it is important to know what numbers might or might not be a factor of a given number. Some types of numbers are easy to tell about. All even numbers are divisible by 2. All numbers ending in 5 are divisible by 5. All numbers ending in 0 are divisible by 10. It is easy to tell about divisibility by numbers within the multiplication table we already know. But what if we have a large number and want to know if it is divisible by 3. Is there a quick test? Yes, there is. Add the digits in the number. Is that sum divisible by 3? If so then the larger number is divisible by 3. Still not sure? Add the digits again. Keep adding them until you can say whether the result is divisible by 3.

Samples:

- Is 247,688 divisible by 3? Add the digits: $2+4+7+6+8+8=35$. Is 35 divisible by 3? No. Therefore, the original number is not divisible by 3.
- Is 986,959,981 divisible by 3? Add the digits: $9+8+6+9+5+9+9+8+1=64$. Still not sure? Add the digits in 64: $6+4=10$. No. The original number is not divisible by 3.

Why is this process for division by 3 an effective one? Suppose that you have a four-digit number, N, that is composed of digits abcd. First, observe that any number composed only of 9s is divisible by 3. Write

$$\begin{aligned} N &= 1{,}000 * a + 100 * b + 10 * c + 1 * d \\ &= (999+1)a + (99+1)b + (9+1)c + 1d \\ &= (999a + 99b + 9c) + 1(a+b+c+d) \\ &= 3(333a + 33b + 3c) + (a+b+c+d). \end{aligned}$$

Clearly the first term is divisible 3 so N is divisible by 3 if and only if the sum of the digits $a+b+c+d$ is divisible by 3. Where did you use the Distributive Law? Did you use it sometimes reading it from left to right and other times reading it from right to left? Do you think this argument will generalize for any number of digits in N? Why or why not? Can you construct a similar argument for divisibility by 9?

Divisibility, Prime Numbers, and Unique Factorization

How can the divisibility by three tests be helpful in determining if a given number is a prime number? How many numbers need to be tested to determine if a given number, N, is prime? That is, up to what value? Recall the material from Chapter 1 involving the square root of a number in relation to testing factors of a given number to test divisibility. How many factorizations does a given number have? For example, the number 32 can be written as 16 * 2; or as 8 * 4; or as 2^5. Are there other factorizations for 32? Notice that the last factorization, 2^5, is the only one that is a product of prime powers. That observation is an important one because the factorization of any integer into a product of powers of primes is unique. Thus, we can always tell how many different factorizations there are for a given number simply by looking at all possible factorizations derived from its unique prime power factorization. Hence, the importance of being able to figure out which integers are prime numbers.

Dividing Fractions

Consider dividing two fractions, $(a/b)/(c/d)$. Generally stated: $(a/b)/(c/d)=(a/b) * (d/c)$

So, it is an easy matter to divide fractions because the problem reduces to multiplying fractions which we already know how to do. Why is this true?

$(a/b)/(c/d)=(a/b) * (1/(c/d))$ and $1/(c/d)=d/c$. Thus, the result follows.

Sample: (3/15)/(4/16)=(1/5)/(1/4)=1/5 * 4/1=4/5—easy if you cancel first.

Once again, we are reminded of the following basic idea.

> If you do not know how to solve a given style of problem, try to reduce it, using your knowledge of fundamental concepts, to a style of problem that you do know how to solve.

Fractions Larger Than One

Consider the number 5/4. It is a fraction, and its value is greater than one. It means "five quarters." So, if I have five quarters of pizza, that is one full pizza plus a quarter of another pizza: 5/4=1¼. How many eighths are in 3 5/8? What is the general rule? To reduce a fraction greater than 1, such as 41/7 to a whole number and a fraction less than 1 (a mixed number), simply divide 7 into 41: 7 * 5 is 35, leaving a remainder of 6. Thus, 41/7 may also be written as 5+6/7. Similarly, to convert a mixed number to a single fraction, see what the denominator is, and multiply the whole number by that and add fractions with like denominators.

Fractions, Decimals, and Maps

- Where on a map might you expect to need to consider using fractions?
- Where on a map might you expect to need to consider using decimals?
- Is one use better in some way than another?
- Why or why not?

Fractions are important when dealing with the scale on a map. If 1 inch on a map represents 500 inches on the ground, the fraction 1/500, or 1:500, represents this idea. The concept of scale on a map is important. You have seen it come up casually in Chapter 1 as well as here. Look for it to keep coming up in all sorts of contexts. For now, give it a bit of thought, and then move forward.

Arithmetic with Fractions: Rapid Mental Calculation

Challenge friends to compete against you in rapid fraction problems. Make up some multiplication problems. Do not solve them, but instead, say what is the most efficient strategy for solving them. Make up some for which standard multiplication is the simplest and make up others for which some of the quick, in-the-head, strategies are the most efficient. Justify your answer. Then, go ahead and solve the problems. Was the method you chose the most efficient one? Did you discover, in the process of solving, that an alternate strategy might have been more efficient? Circle back as many times as needed so that you become quick and comfortable at calculations.

EXPONENTS AND LOGARITHMS

Exponents

Exponents are a compact way of representing mathematical operations. Thus, $2 * 2=2^2=4$ where 2^2 is read "2 to the 2nd power," or 2 squared for short. The 2nd power is the exponent 2.

Generally, m^n is the number m raised to the nth power. The number m is called the base, and the number n is called the exponent or power. Values of n may take various forms. Expressions with 2 in the exponent are often called "square" because of the way the area of a square is calculated. Those with 3 in the exponent are often called "cube" because of the way the volume of a cube is calculated. What about higher exponents? Read more possibilities; use your imagination (Coxeter, 1973). Generally, consider m^n:

- Suppose m is a positive integer.
  - If n is also a positive integer, then m^n means m multiplied by itself n times. Thus, x^3, read "x cubed" or "x cube" means $x * x * x$.
  - If n is a positive fraction, then m^n means the nth root of m.
  - If n is 0, then $m^n=1$
  - If n is a negative integer, then $m^n=1/m^{|n|}$. For example, $5^{(-2)}=1/25$.
- Suppose m is a fraction. Here similar rules apply but be careful how the exponent is applied: $3/5^2$ is not the same thing as $(3/5)^2$. The former is 3/25 and the latter is 9/25. See why the order of operations is important! The fraction may be larger than 1; try a few for yourself.
- Suppose m is a negative number; what are the corresponding sets of rules? What is the role of even and odd numbers? Are there exceptional cases that you don't know what to do with? What does it mean to consider the square root of negative 1? Such consideration is generally beyond the scope of core math courses as it is based on a different number system: read more (Bourbaki, 1998).
- Suppose m is 0.
  - If n is non-zero, then $m^n=0$.
  - If $n=0$ the expression 0^0 is called an indeterminate form; it has no agreed-upon value, and is generally excluded from consideration in elementary mathematics courses. Curious? Read more (Graham et al., 1989).

Logarithms

A logarithm is the exponent (power) to which a base must be raised to produce a given number: x is the logarithm of n to the base b if $b^x=n$, $x=\log_b n$. For example, $2^4=16$; therefore, 4 is the logarithm of 16 to base 2, or $4=\log_2 16$. Exponents and logarithms are inverses of each other.

DECIMALS

The word "decimal" means "based on 10." The decimal system is a number system that uses notation in which each number is expressed in base 10 by using one of the first nine whole numbers or 0 in each place and letting each place value be a power of 10. A period (in the US), known as a decimal point, separates the non-negative from the negative powers of 10. Thus, $124=1 * 10^2+2 * 10^1+4 * 10^0$. Further, $124.735=1 * 10^2+2 * 10^1+4 * 10^0+7 * 10^{-1}+3 * 10^{-2}+5 * 10^{-3}$.

Convert Fractions to Decimals

Fractions are division problems: 4/5 means 4 divided by 5, or, employing the base of 10, it is 8/10. Perform the division to get the decimal form: 8/10=0.8. That is, when 8 is divided by 10, it goes in 0 times and leaves a remainder of 8.

Convert Decimals to Fractions

Decimals include underlying powers of 10, which are fractions. For example, 0.75=7*1/10+5*1/100=70/100+5/100=75/100; cancelling out 25 yields ¾.

Adding and Subtracting Decimals

Add as you would any other column of numbers, but first line up the numbers in a column with the decimal points aligned.

Multiplying Decimals

Multiply the numbers as you would any pair of numbers. Place the decimal point after the multiplication takes place. Count the number of decimal places to the right of the decimal point in each of the two original numbers. Take this total and use that, moving from right to left, to place the decimal point that number of places from the left end of the answer. For example, multiply 47.356 * 23.2. The answer is 16,637.1914.

Dividing Decimals

To divide decimals, multiply each of the two numbers by the power of 10 that covers the longest number to the right of the decimal point. Then divide the corresponding whole numbers. The answer to the two problems is the same because they are proportional. For example, 2.3/1.67=230/167 (multiplying both top and bottom by 100). The answer is 1.37724551

Rounding and Estimating

In practical applications, it is often not convenient to have long decimal expansions as an answer. For example, in the problem above, an answer shorter than 1.37724551 might be desired. One way to shorten it is to round the numbers up: numbers greater than 5 get rounded to the next nearest integer. The last 5 digits above are 24551. Rounding up moves it to 24600. So, 1.377246 is a shorter form. Even shorter is 1.37725, or 1.3773, or 1.378. If instead, we decided to shorten the decimal by truncating the string of digits, to create an answer with three values to the right of the decimal point, the value would be 1.377. The digits that contribute to the degree of accuracy of the value are called "significant digits"; start counting digits as significant at the first non-zero digit.

It makes a difference how one chooses to shorten numbers. When providing general estimates, small differences may not be important. What is important, though, is to tell the reader how you obtained your shortened estimate—in this case, by rounding up, or by truncation.

The Eiffel Tower: Degrees of Precision and Rounding

In this activity, you further your understanding of decimals, degrees of precision, and rounding through a hands-on activity. Numbers will be mapped, studied, and measured: The numbers in this activity represent coordinates of latitude and longitude.

Open the linked web map (https://www.arcgis.com/apps/mapviewer/index.html?webmap=fce1f5d41bbf4eafbbc5757a57a01cfe) in ArcGIS Online. The map contains two layers, each made from a set of seven latitude-longitude coordinates, mapped as points. For positions to be shown on maps, they must have two coordinates: The longitude value represents the x-coordinate, and the latitude value represents the y-coordinate. The more numbers (within limits) to the right of the decimal place in the coordinate, the more precise it is, and the more precise its location will be on the map as it represents the actual feature in the real world.

On the left side of the map > Layers > Expand the two layers and make the Decimals and Precision Activity-Eiffel Tower layer the only visible layer. This layer shows seven points mapped from the following table. Each point represents the location at that specific longitude and latitude on the Earth's surface. The point labeled "six" includes six significant numbers to the right of the decimal place. This point is exactly the location of the Eiffel Tower in Paris, France. Each successive point is mapped after truncating one digit from the coordinates; thus, "five" contains five significant numbers after the decimal place, point "four" contains four significant numbers, and so on, down to "None" which contains no numbers to the right of the decimal point; thus, the location of point "None" is at 2° east longitude, 48° north latitude:

Point	Longitude	Latitude
Six	2.294524	48.858260
Five	2.29452	48.85826
Four	2.2945	48.8582
Three	2.294	48.858
Two	2.29	48.85
One	2.2	48.8
None	2	48

Zoom on the cluster of points in the northern part of the map, at and near the Eiffel Tower (Figure 2.1). Click on a few points to verify the latitude and longitude values that you are examining.

Observe the direction that the points "drift" as the numbers to the right of the decimal point are removed. What direction are the points moving? Why? How much distance exists between each point? Use the Map Tools > Measurement tool to the right of the map to measure the distance between point six and point five. Then, measure the distance between point five and point four. Do the distances increase or decrease between points as precision of the coordinates becomes less? Why? What is the distance between point six and point none?

On the left side of the map > Layers > Expand the two layers and make the Rounding Eiffel Tower layer the only visible layer. This layer shows seven points mapped from the following table. As before, each point represents the location at that specific longitude and latitude on the Earth's surface. The point labeled "six" includes six significant numbers to the right of the decimal place and is the same location as Point Six that you examined earlier. However, in this layer, the points are successively rounded, rather than truncated. For example, from points four to three, 2.2945 was rounded to 2.295 rather than truncated to 2.294. Look for similar comparisons between the truncated and the rounded values.

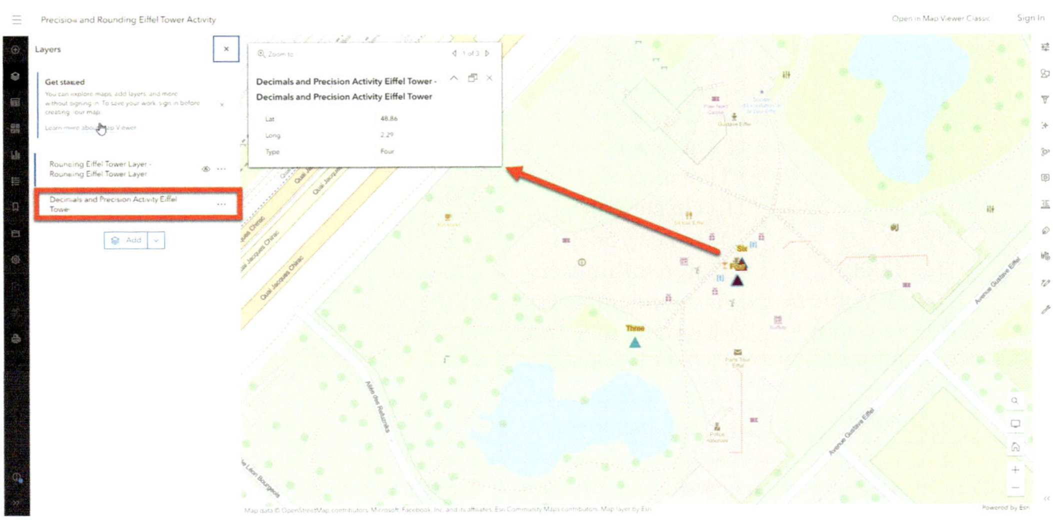

FIGURE 2.1 The Eiffel Tower location in Paris, France with a set of points with specific latitude and longitude values. ArcGIS from Esri.

Point	Longitude	Latitude
Six	2.294524	48.858260
Five	2.29452	48.85826
Four	2.2945	48.8583
Three	2.295	48.858
Two	2.29	48.86
One	2.3	48.9
None	2	49

Zoom on the cluster of points at and near the Eiffel Tower. Click on a few points to verify the latitude and longitude values that you are examining. Observe the direction that the points "drift" as the numbers to the right of the decimal point are rounded. What direction are the points moving? Why? How much distance exists between each point? Use Map Tools>Measurement to the right of the map to measure the distance between point six and point five. Then, measure the distance between point five and point four. Do the distances increase or decrease between points as precision of the coordinates becomes less? Why? What is the distance between point six and point none?

Reflections on This Activity

Turn both layers back on and zoom out until you see all of each layer's points. What cardinal (compass) direction do the rounded set of points move: north, south, east, or west? Why are the direction and the distance different for the other set of points, where digits were simply truncated, instead of being rounded?

The number of decimal places in a pair of latitude-longitude values can be of critical importance and can be used to learn about numbers, decimals, and precision. For example, consider the following chart, indicating the distance in meters between the number of degrees of longitude at the Equator:

decimal

places	degrees	distance
-------	---------	---------
0	1	111 km kilometers
1	0.1	11.1 km kilometers
2	0.01	1.11 km kilometers
3	0.001	111 m meters
4	0.0001	11.1 m meters
5	0.00001	1.11 m meters
6	0.000001	11.1 cm centimeters
7	0.0000001	1.11 cm centimeters
8	0.00000001	1.11 mm millimeters
9	0.000000001	111 μm micrometers
10	0.0000000001	11.1 μm micrometers
11	0.00000000001	1.11 μm micrometers
12	0.000000000001	111 nm nanometers
13	0.0000000000001	11.1 nm nanometers

Adding information to enhance meaning, in terms of, "how long are these units in terms of what we can really understand?", yields the following, using selected places on the left:

8	1.0 millimeter	The width of a wire making up a paper clip.
9	0.1 millimeter	The width of a strand of hair.
10	10 microns	A speck of pollen.
11	1.0 micron	A piece of cigarette smoke.
12	0.1 micron	Virus-level mapping.
13	10 nanometers	About 1/10,000 of the width of a sheet of paper, or 1/8,000 of the width of a human hair.
14	1.0 nanometer	Your fingernail grows about this far in 1 second.
15	0.1 nanometer	An atom.

This activity reinforces information about accuracy and precision together with the sometimes-subtle differences between truncation and rounding. If you were laying fiber optic cable under the Eiffel Tower, you would need a fair number of significant digits for you to accurately do your work. But not down to the sub-millimeter level: More information is not always better. Conversely, if you are studying the weather patterns of the Paris metropolitan area, just a few significant digits (so that you could differentiate between a few hundred meters) would be sufficient.

Comparing and Measuring Decimals

In the following activity, you have another opportunity to work with decimal numbers while analyzing a real-world issue on a map. Natural and synthetic fertilizers are important in aiding food crops to be grown, but their overapplication can adversely impact water quality in watersheds where they are applied. Maps can help us understand areas of concern and treatment.

Access the EPA EnviroAtlas on this link: https://enviroatlas.epa.gov/enviroatlas/interactivemap/. Then > Select the Learn option > Under Featured Collections on the left side of the map, select > Nitrogen Inputs to Watersheds > Add to Map. Your screen should look similar to that in Figure 2.2, showing an area in North Carolina.

Click on a polygon to access the amount of two types of fertilizer applied in that polygon. Note that the manure and synthetic fertilizers applied in this area are given as the number of kilograms per hectare per year. In the area identified below, how much total fertilizer is applied? This amount

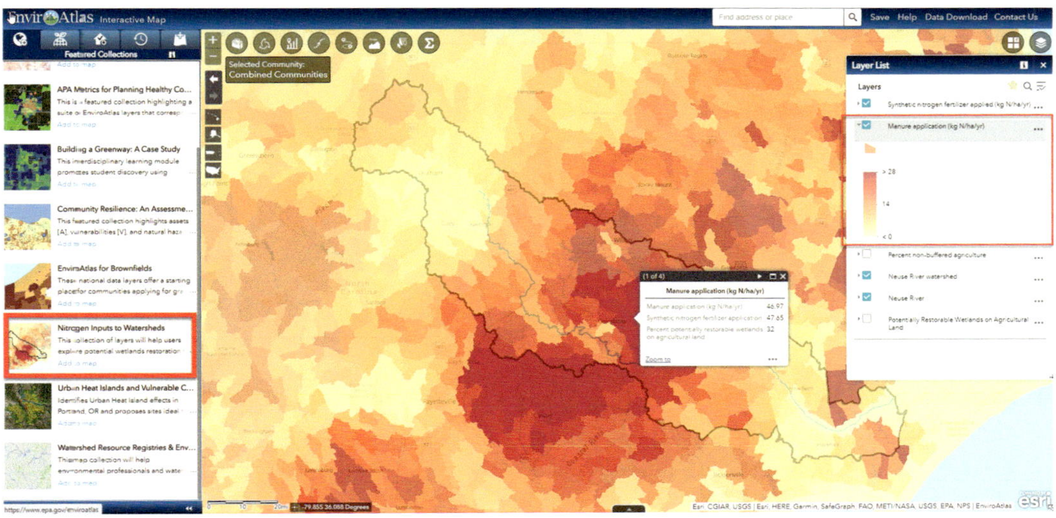

FIGURE 2.2 Analyzing fertilizer application in watersheds using the EPA EnviroAtlas.

is per hectare. There are 100 hectares in 1 km^2. How much fertilizer is applied for each square kilometer in this area? If the Neuse Watershed is 11,500 km^2, calculate the amount of fertilizer that is applied to the entire watershed. Remember that this number represents the fertilizer applied in a single year. Next, calculate the amount applied at the current rate over the next decade. That's a lot of fertilizer! Using math with maps help us plan for a more sustainable future. Next, run these same calculations in another polygon within another area in the study area of the map.

Consider the example in Figure 2.2. Note that the units in the first two figures at the location being examined provide multiple types of information, which is often the case in mapping and can be very useful because amounts, space, and time are considered. Here, it is the numbers of kilograms of nitrogen (an amount) per hectare (an area), per year (a unit of time). In the location being examined, the Manure application (kg N/ha/yr) is 53.77, the synthetic nitrogen fertilizer application is 35.46 in kg N/ha/yr), and the percent potentially restorable wetlands on agricultural land is 6%.

The material presented in this book to date has worked only with the decimal system. Next, we consider other possibilities.

BASES OTHER THAN 10

The decimal system has a base of 10. It would be possible to construct a counting system on any base; however, why bother? We already know how to use the decimal system efficiently. We have ten fingers, normally, one for each digit. A computer, however, is based on electrical switches, which are based on powers of two: "on" and "off." It is useful in today's world to have some understanding of how number systems work that are based on powers of 2. Thus, we offer the reader a glimpse into those worlds so that he/she may be able to make sensible use of online converters when a binary, hexadecimal, or other value is encountered in a piece of software.

Binary

A binary number is expressed in the base-2 numeral system using two digits, 0 and 1. Binary numbers are typically long strings of digits and are the basis for digital communication within a computer. They are often, however, cumbersome in appearance and difficult for humans to relate to. Thus, it is useful to be able to convert between binary and decimal and there are many helpful online converters available to aid in the process. For example, the number $(137)_{10}$, read "137 base 10," is a decimal number. Written as a binary number it is $(10001001)_2$. The conversion process is parallel to what we did with the decimal conversion process. Use division by the base, 2 in this case:

- Divide the number by 2.
- Get the integer result for the next iteration.
- Get the remainder for the binary digit.
- Repeat the steps until the result is equal to 0.

Thus,

137/2 = 68 with a remainder of 1—binary digit 0
68/2 = 34 with a remainder of 0—binary digit 1
34/2 = 17 with a remainder of 0—binary digit 2
17/2 = 8 with a remainder of 1—binary digit 3
8/2 = 4 with a remainder of 0—binary digit 4
4/2 = 2 with a remainder of 0—binary digit 5
2/2 = 1 with a remainder of 0—binary digit 6
1/2 = 0 with a remainder of 1—binary digit 7.

Thus, the binary form of 137 is 10001001: $(137)_{10}=(10001001)_2$.

One might also wonder, if given a binary number at the outset, how to convert it to decimal form: again, look to the forms from the decimal system. For example, $(10001001)_2=1 * 2^7+0 * 2^6+0 * 2^5+0 * 2^4+1 * 2^3+0 * 2^2+0 * 2^1+1 * 2^0=128+0+0+0+8+0+0+1=(137)_{10}$. The same sort of procedure will work for any base.

Octal

An octal number is expressed in the base-8 numeral system using eight digits, 0, 1, 2, 3, 4, 5, 6, 7. For example, the number $(137)_{10}$, written as an octal number is $(211)_8$. The conversion process is parallel to what we did with the decimal and binary conversion processes. Use division by the base, 8 in this case:

- Divide the number by 8.
- Get the integer result for the next iteration.
- Get the remainder for the octal digit.
- Repeat the steps until the result is equal to 0.

Thus,

137/8 = 17 with a remainder of 1—octal digit 0
17/8 = 2 with a remainder of 1—octal digit 1
2/8 = 0 with a remainder of 2—octal digit 2.
Thus, the octal form of 137 is 211: $(137)_{10}=(211)_8$.

One might also wonder, if given an octal number at the outset, how to convert it to decimal form: again, look to the forms from the decimal system. For example, $(211)_8=2 * 8^2+1 * 8^1+1 * 8^0=128+8+1=(137)_{10}$.

Hexadecimal

A hexadecimal number is expressed in the base-16 system using 16 alphanumeric digits, 0, 1, 2, 3, 4, 5, 6, 7, 8, 9, A, B, C, D, E, F. For example, the number $(137)_{10}$, written as a hexadecimal number is $(89)_{16}$. The conversion process is parallel to what we did with the decimal, binary, and octal conversion processes. Use division by the base, 16 in this case:

- Divide the number by 16.
- Get the integer result for the next iteration.
- Get the remainder for the hexadecimal digit.
- Repeat the steps until the result is equal to 0.

Thus,
137/16 = 8 with a remainder of 9—hexadecimal digit 0
8/16 = 0 with a remainder of 8—hexadecimal digit 1.

Thus, the hexadecimal form of 137 is 89: $(137)_{10}=(89)_{16}$.

One might also wonder, if given a hexadecimal number at the outset, how to convert it to decimal form: again, look to the forms from the decimal system. For example, $(89)_{16}=8 * 16^1+9 * 16^0=128+9=(137)_{10}$. The hexadecimal format is the most compact form typically used; that fact is an advantage when dealing with very large numbers, as computers do, often.

Operations

Binary, octal, or hexadecimal numbers can all be combined using usual operations; a parallel universe exists for each base. Because our minds tend to be trapped within the decimal world, the addition of two hexadecimal numbers (for example) can seem quite unnatural and difficult. One natural way around this difficulty is simply to convert both hex numbers to decimal format, add them as decimal numbers, and then convert the decimal result back to hexadecimal format. It is also possible, however, by thinking broadly about the process of adding decimal numbers, to add the hex numbers directly in hex format (keep your eyes open for the distributive law lurking in the background).

Consider the following example: Add $(4A6)_{16}$ and $(1B3)_{16}$. Begin at the right: 6+3 is 9, A+B is 10+11 which is 21 so 5 in hex and carry 1 (representing 16), and 4+1+1(from carrying) is 6. The answer is $(659)_{16}$. Check the work by converting it all back to decimal: $(4A6)_{16}$ and $(1B3)_{16}$ are $(1190)_{10}$ and $(435)_{10}$, respectively. Their sum is $(1625)_{10}$ which is $(659)_{16}$. The answer checks. It is important to know that this process works and that operations, as well as mathematics generally, carry over into worlds with different numerical bases. Fortunately, the computer performs these operations most of the time, behind the scenes. But when you know how to do so, you can check matters for yourself if need be.

TIME AND CIRCULAR MEASUREMENT

At a very general level, a system of measurement is composed of units and rules relating the units to each other. Contemporary systems of measurement in common use include the International System of Units (SI, the contemporary metric system), the British Imperial System, and the United States Customary System.

Metric

Initially, there were two base-units for the metric system: the meter as a unit of length and the gram as a unit of mass. Other units were derived from these two base units. Over the years, the metric system has evolved in various directions. The contemporary international standard for the metric system is the International System of Units (SI) in which there are seven base units in terms of which all others can be expressed: meter, kilogram, second, ampere, kelvin, mole, and candela.

Metric units are related by powers of ten. Thus, the metric system is compatible with the decimal number system. This fact alone, providing great convenience for understanding calculation, is often reason enough to choose the metric system, in preference to any other, as a measurement base.

English

Imperial Units, as well as US Customary Units, evolved from earlier English Units. The two systems are similar, but not identical, to each other. Units of length and area are the same: inch, foot, yard, mile, and so forth. Other units may differ: the stone and the pound are different units for the measurement of weight. Any USA resident who has bought gasoline for a car in Canada, has dealt with the difference between a US gallon and an Imperial gallon.

Time

Other forms of measurement are common in our society (Higgins, Miner, Smith, and Sullivan, 2010). The measurement of time can be viewed as a fraction: 3 and a half hours is 3.5 hours. Half an hour is 60/2 minutes, so three and a half hours is 3 hours and 30 minutes, or 3:30. Or, converting it all to minutes, it is 3 * 60+60/2 which is, using order of operations, 180+30=210 minutes.

The same principle applies to situations that are not as obvious as that one. But, it is good to begin with an obvious situation in order to check complex process against intuition. How much time has elapsed in 3 and 5/16 hours, expressed as a decimal? 5/16 of an hour is 5 divided by 16 (look at the fraction; it is more than 4/16, so just a bit more than .25). Keep a ballpark idea in mind; that way you check yourself as you go along, whether using electronic equipment or pencil and paper, so that you don't arrive at any outrageously incorrect answers. When 5 is divided by 16, the answer is 0.3125. So, 3 5/16 hours is 3.3125 hours. The same idea applies to measurement in a circle using degrees, minutes, and seconds. Convert 17 degrees, 23 minutes, and 19 seconds to decimal degrees. Express in fractions; notice that there are 60 minutes in a degree, and 60 seconds in a minute. Thus, in a degree there are 60 * 60 = 3,600 seconds.

Decimal degrees = Degrees + (Minutes/60) + (Seconds/3,600)

1. First, convert minutes and seconds to their degree equivalents. 23'/60 = 0.3833333°; 19"/3,600 = .00527778°
2. Round off for convenience and take the sum: 0.383 + 0.005 = 0.388.
3. Then, add this number to the number of degrees to obtain the answer. 17° + 0.388° = 17.388°

Circular Measurement

Time may be displayed in a digital display, on a circular face, or in other ways. Contemporary digital smart watches offer the user the opportunity to choose from among a variety of displays, with the tap and slide of a finger on the watch surface. Look at a circular display of time. When it is 1 o'clock, what is the angle between the large and small hands? It is 1/12 of 360° in the full circle, or 30°. Over the course of 1 hour, the small hand moves from the numeral 1 to the numeral 2: it moves a distance. So, the degree measure can measure distance, too, in association with its angular measure. The same is true for measurement on the sphere.

Latitude and Longitude

Maps and associated digital images are based on real-world places and are underpinned by real-world numerical measurements. The Cartesian coordinate system is the customary way of inserting coordinate axes into the plane (Figure 2.3).

Now, consider the Cartesian coordinate system if it was overlaid atop the entire Earth. It will only loosely fit on the Earth sphere. The plane does not provide an exact fit to a sphere. Nonetheless, these coordinates are based on numbers and the accurate measurement of the Earth, often represented as a sphere. Measuring the Earth accurately has been a quest for thousands of years (see the set of references on Longitude at Sea Measurement).

The activity that follows this next brief discussion of important lines on the Earth's surface will give the reader an opportunity to see, in an interactive manner, how they develop (Figure 2.4). Generally, on the Earth, the x-axis of the Cartesian coordinate system would be represented by the Equator, bisecting the distance between north and south poles, and the y-axis would be represented by the Prime Meridian, joining the north to the south pole; x-coordinates run from 0 at the Prime Meridian west across the Americas to 180° at the International Date Line in the Pacific Ocean (although the date line itself jogs on either side of 180° west so countries do not inconveniently span two different dates). The x-coordinates also run in the opposite direction, from 0 at the Prime Meridian east across Asia, Africa, the Indian Ocean, and Australia to the International Date Line. The y-coordinates run from the Equator north to the North Pole at 90° north latitude. The y-coordinates also run in the opposite direction, from the Equator south to the South Pole at

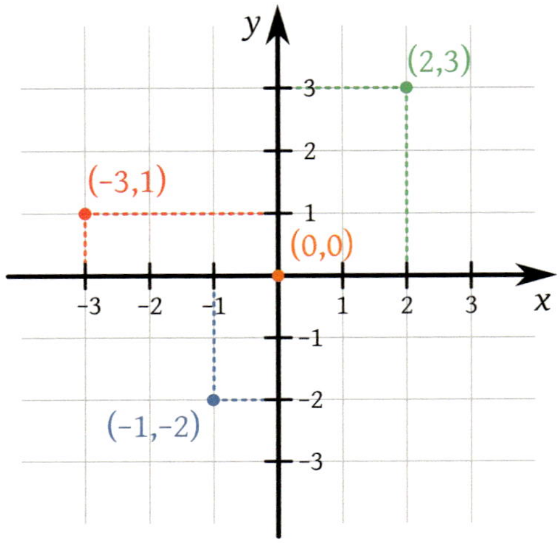

FIGURE 2.3 A small portion of the Cartesian coordinate system, showing the origin, axes (extending to infinity in both directions), and the four quadrants, with illustrative points and grid. Four quadrants: Quadrant I, upper right; Quadrant II, upper left; Quadrant III, lower left; Quadrant IV, lower right. Source: By Cartesian-coordinate-system.svg: K. Bolino derivative work: F l a n k e r (talk) – Cartesian-coordinate-system.svg, Public Domain, https://commons.wikimedia.org/w/index.php?curid=11678042

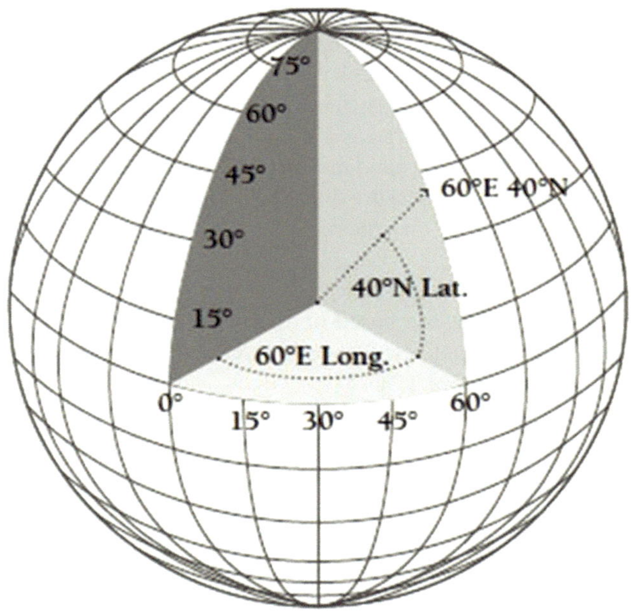

FIGURE 2.4 Latitude and longitude angles and lines of equal latitude (parallels) and equal longitude (meridians) on the Earth. Graticule mapped with ArcGIS from Esri.

90° south latitude. Why only 90° to the poles? These numbers, as "degrees" indicate, are angles—it is 90° measured from the Equator to the North Pole, for example, and the diagram shown here indicates the angles that result in a coordinate of 40° north latitude, 60° east longitude.

The x-coordinates are longitude: Anything west of the Prime Meridian, or west longitude, are left of the y-axis and therefore receive a negative number. Any locations east of the Prime Meridian, or east longitude, are to the right of the y-axis and therefore receive a positive number. The y-coordinates are latitude: Any locations north of the Equator, or north latitude, are above the x-axis and therefore receive a positive number. Any locations south of the Equator, or south latitude, are below the x-axis and therefore receive a negative number.

The Equator divides the world into two equal halves of the sphere, or two hemispheres—north and south. Similarly, the Prime meridian divides the world into two equal hemispheres—east and west. You can see that the location of the Equator was a natural one, but the Prime Meridian could have been located anywhere. Indeed, it was not until the International Meridian Conference of 1884 that an agreement was reached to set the Prime Meridian to run through the Royal Observatory in Greenwich England (Howse, 1980). This line runs north of the Observatory through England and out to the North Sea, and on to the North Pole, and south from the Observatory across the English Channel to France, through west Africa, entering the Atlantic Ocean at Ghana, and south to the South Pole.

Open the linked ArcGIS Online map (https://www.arcgis.com/apps/mapviewer/index.html?webmap=b02ef0f9af5e47eeb6b4d5da0505155d) that shows important lines such as the Equator, Prime Meridian, Tropic of Cancer, Tropic of Capricorn, and the Arctic and Antarctic Circles.

Follow the Equator around the world and note which countries and oceans it passes through. Do the same thing with the Prime Meridian. The intersection of the Equator and Prime Meridian lies off the southwest coast of a continent. Which continent is it? This location is 0° latitude and 0° longitude. Which two continents are partly in the eastern hemisphere and partly in the western hemisphere? Which continents are partly in the northern hemisphere and partly in the southern? Which continents are entirely in one hemisphere? The concept of what constitutes an island (or a mountain or island for that matter) is largely a human construct (Should Europe be considered part of Asia? Why is Australia a continent rather than an island?), but considering these questions builds your spatial and mathematical skills.

Pan to the north from the Equator and note how the 1° squares become obviously elongated rectangles, as clear trapezoids, the farther north (or south) one travels. This is because the longitude lines are converging toward the poles.

Search and find your own city or campus. Which latitude and longitude lines are you closest to? What is the distance to your nearest whole-degree latitude or longitude line?

Next, explore these lines linked (https://www.arcgis.com/home/webscene/viewer.html?webscene=037abe3b7e454a26823e05f3fa5126f1) in a 3D scene. Click on the lines to obtain the attributes for the important lines—Equator, Prime Meridian, and so on (Figure 2.5). Zoom in to see an increasingly dense network of latitude and longitude lines from 30° spacing to 1° spacing. The 3D globe may give you a better sense for how the longitude lines converge at the poles than the 2D map did. Use the Analyze>Measure Distance tool to measure the distance between longitude lines in Canada, for example, versus the distance between the same longitude lines in Brazil or other points along the Equator.

Because all of Australia lies in the southern hemisphere, its latitude values, or y values, are negative. Because all of Australia lies in the eastern hemisphere, its longitude values, or x values, are positive. In the Cartesian coordinate system diagram, all of Australia lies in Quadrant IV.

Some mapping tools allow you to enter latitude, longitude as they are commonly spoken in everyday speech ("latitude, longitude"), but in terms of map coordinates, latitude-longitude is y, x and not x, y. Other mapping tools may require you to enter the values as x, y, or longitude, latitude. One needs to pay attention to the contextual world of individual pieces of software.

Latitude-Longitude Formats

Three different formats are used for the latitude-longitude system. Thus far, you have been using decimal degree—where each degree is divided into decimals—the more decimals, the greater precision of the coordinate, for example, 40.395, –105.294 yields more precision than 40.3, –105.2. Another format for latitude and longitude is degrees, minutes, similar to the clock's hours, minutes,

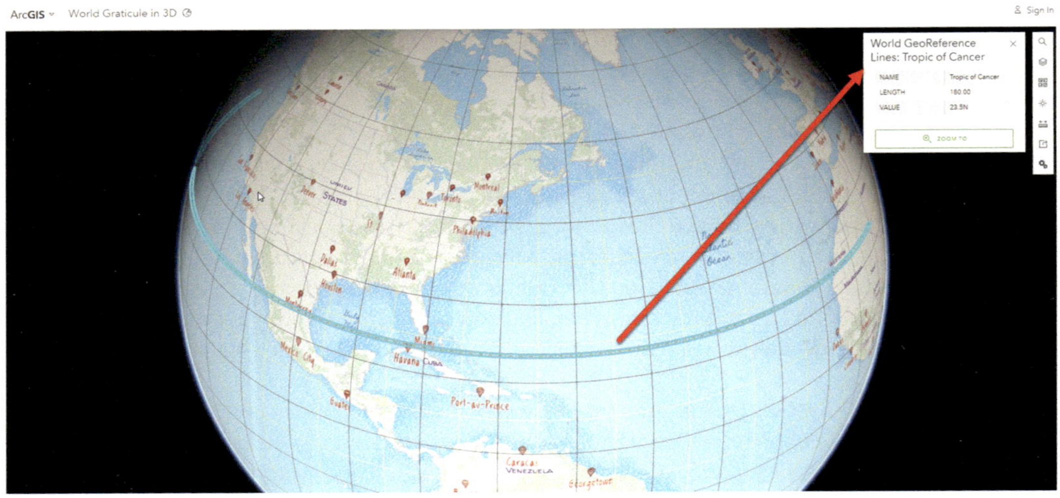

FIGURE 2.5 A 3D scene in ArcGIS online showing major latitude and longitude lines. Graticule mapped with ArcGIS from Esri.

and seconds. Recall that latitude and longitude values are really angles, and therefore, are based on 60 and 360. Similar to each hour on the clock being able to be divided into 60 minutes, and each minute can be divided into 60 seconds, each degree can be divided into 60 minutes, and each minute can be divided into 60 seconds. Therefore, 41.5° north latitude can be represented in degrees, minutes, seconds as 41 degrees 30 minutes 0 seconds. Similarly, 123.875 west longitude can be represented in degrees, minutes, seconds as 123 degrees 52 minutes 30 seconds.

Each degree of longitude at the Earth's Equator is about 111 km, or 111,000 m. Knowing this, compute the number of meters in each minute of longitude and each second of longitude at the Equator.

The third format for latitude-longitude is decimal minutes. Decimal minutes simply makes a decimal from the seconds reading. Thus, 123 degrees 52 minutes and 30 seconds in decimal minutes is represented as 123 degrees 52.5 minutes. Decimal minutes are used frequently in geocaching, the high-tech treasure hunt GPS-based game engaging millions of people.

Use the techniques for calculation that you used for measuring time to convert values from one format to another for latitude and longitude.

How many degrees, in minutes and seconds, is one-half of 1° of latitude?
How many degrees, in minutes and seconds, is ¾ of a degree of latitude?
How is 41.625 degrees north latitude represented as degrees minutes and seconds?
How is 41.625 degrees north latitude represented as decimal minutes?

Great Circles

The Equator is a circle on the surface of a sphere. It is, at 0°, the largest of the lines of latitude. It contains the center of the sphere as the center of the Equator. Other circles may be drawn on the sphere that also contain the center of the sphere as the center of those circles. These circles are called Great Circles. There is an infinity of Great Circles that may be drawn on the sphere; however, only one of those, the Equator, bisects the distance between the North and South poles. The next section presents an opportunity to interact with Great Circles on the surface of the Earth and develop skills associated with using them as measurement tools.

Measurements and the Great Circle

Access a linked interactive map (https://www.arcgis.com/apps/mapviewer/index.html) in ArcGIS Online to get started. Then > on the right side of the map > Search > search for Los Angeles CA > Add to New Sketch > search for London England > Add to your existing Sketch. You will see the two placemarkers as sketches atop the two cities. Then, > Map Tools > Measurement > measure the distance between the two cities in kilometers.

Note that the line representing the shortest distance connecting the two cities is not straight but curved (Figure 2.6). This is the shortest distance between the two points, or the Great Circle route. In Euclidean plane geometry, a straight line represents the shortest distance between two points. This line has only one length. On the surface of a sphere, however, this is replaced by the geodesic or Great Circle length. This is measured along the surface curve that exists in the plane containing both endpoints and the center of the sphere. The line here looks curved because it is drawn on top of a projected map (in this case, Web Mercator). Map projections are grounded in mathematics and are performed to represent the three-dimensional spheroidal Earth on a two-dimensional paper map or digital map. Note also that the line touches the southern tip of Greenland. Airplanes usually fly the Great Circle route to save fuel and time, and on one of those flights, you would see some ice in your view out the window on an airplane flying between Europe and North America.

Measure a location that you think is about halfway around the world from Los Angeles. You will note the rubber-banding, stretching, effect that the measurement tool has as it computes the shortest distance from Los Angeles to your desired location (Figure 2.7). Should you travel east from Los Angeles, or west, for the shortest distance?

The Great Circle routes may be more intuitive using a 3D map instead of a 2D map (Figure 2.8 instead of Figure 2.7). Access the ArcGIS 3D Scene Viewer (https://www.arcgis.com/home/webscene/viewer.html), delete the predefined scenes that you are presented with, > Analyze > Measure distance > Measure the distance again from Los Angeles to London. It will be difficult, depending on where you click on the globe, to achieve the same distance measurement that you obtained with the 2D map, but what will be clearer is that the line connecting the two cities is indeed following the shortest distance on the Earth; that is, the Great Circle route. It will also be evident that the shortest distance from your location to a location halfway around the world takes you over the North Pole (Figure 2.8)!

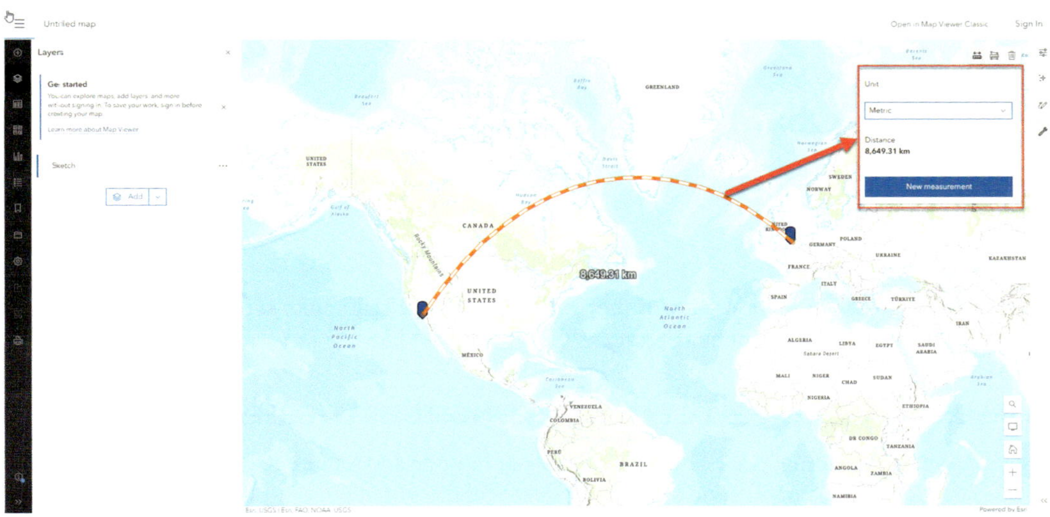

FIGURE 2.6 A great circle route displayed in the plane. Route mapped with ArcGIS from Esri.

FIGURE 2.7 A great circle route on a 3D map. Great circle mapped with ArcGIS from Esri.

FIGURE 2.8 The great circle route from Los Angeles to London over southern Greenland. Great circle mapped with ArcGIS from Esri.

Tilt and pan the globe for a view of the Great Circle route that looks even more like dragging a string across a physical 3D globe (Figure 2.9).

MEASURING DISTANCES IN LATITUDE LONGITUDE DEGREES

As you have been reading, just like hours, degrees of latitude and longitude can be divided into units of 60. Just like 1 hour can be divided into 60 minutes and 1 minute into 60 seconds, 1° of latitude or longitude on the Earth can be divided into 60 minutes, and 1 minute can be divided into 60 seconds. Remember that these units on the Earth's surface represent distance, not time.

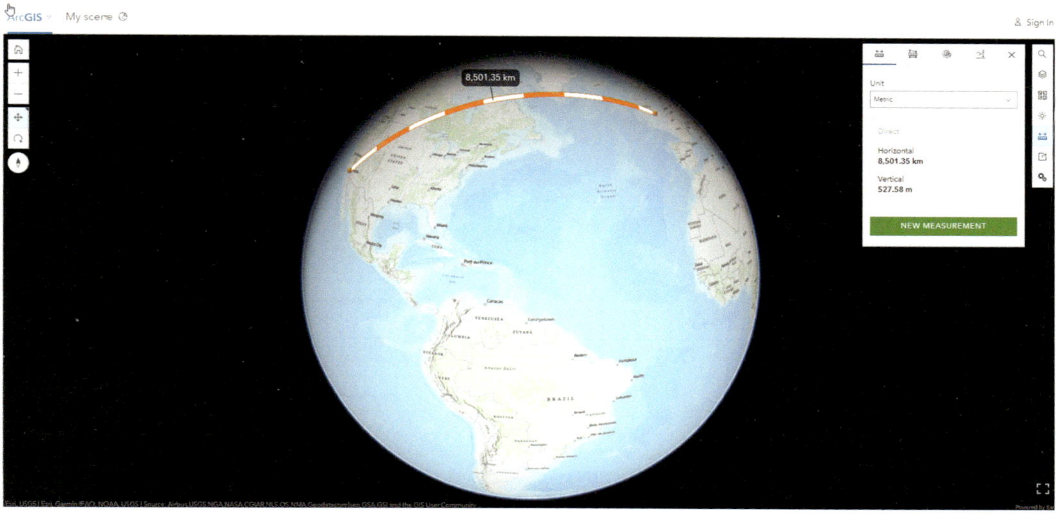

FIGURE 2.9 A tilted view of Figure 2.8 using a 3D scene viewer. Great circle mapped with ArcGIS from Esri.

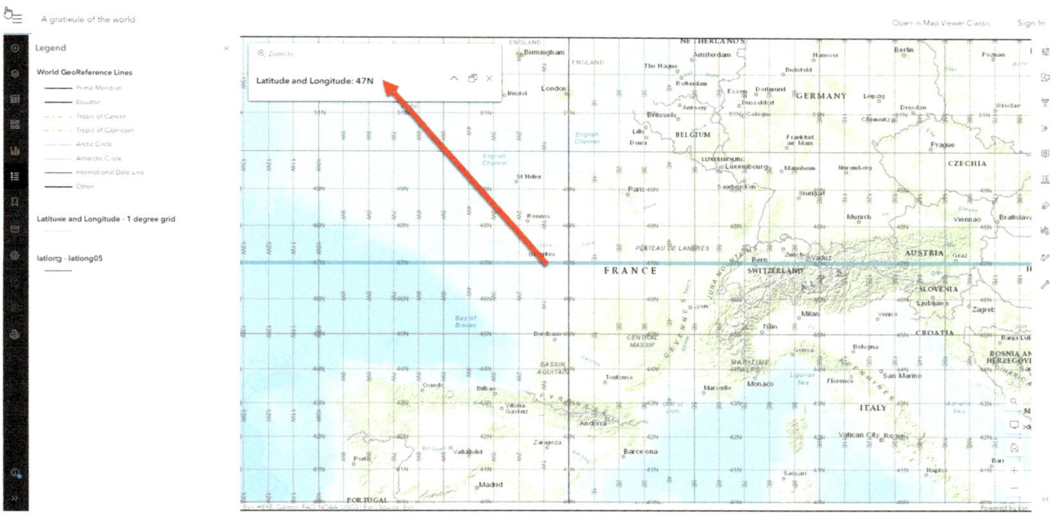

FIGURE 2.10 One degree spacing of latitude-longitude lines. Graticule mapped with ArcGIS from Esri.

Open the linked map (https://www.arcgis.com/apps/mapviewer/index.html?webmap=b02ef0f9af5e47eeb6b4d5da0505155d) and zoom to France. You will see the 1° latitude and longitude lines, Figure 2.10.

How long from north-to-south is France in degrees, minutes, and seconds? The distance tools will not help you here, as they only provide distances in miles, kilometers, and meters, and not in degrees, minutes, and seconds. You will thus need to estimate, and in so doing, practice your skills in fractions and decimals. Note that 51° north latitude is close to the northernmost extent of France, and 42° south near the border with Spain is France's southernmost extent. But not exactly! Using the zoom tools, and your powers of observation, provide the closest measurement you can in degrees and minutes (don't worry about the seconds for this activity!) for the length of France

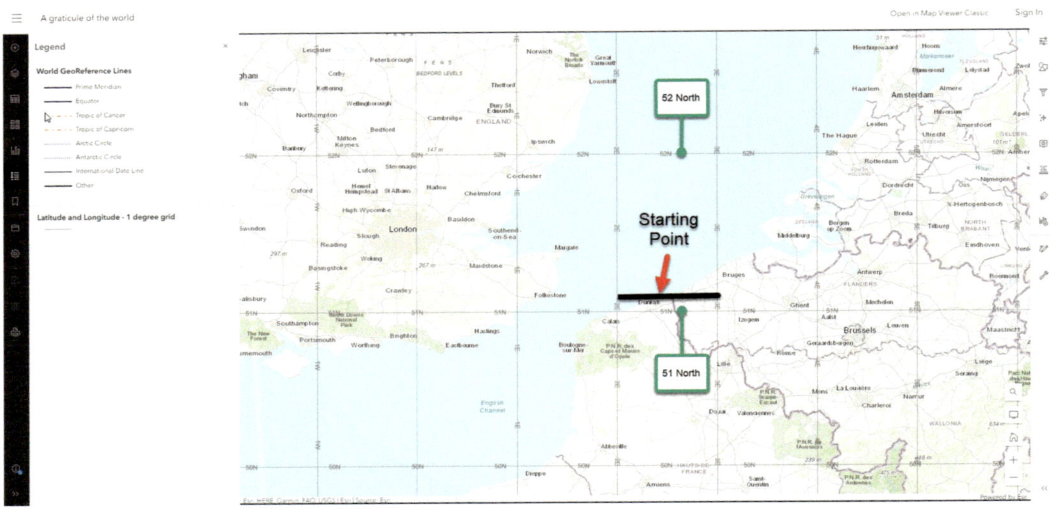

FIGURE 2.11 Northern France, closer view of Figure 2.10. Graticule mapped with ArcGIS from Esri.

from its northernmost extent to its southernmost extent. For each extent, you will need to estimate how far France extends between the full degree latitude lines. For example, the northernmost extent of France appears to be about 10% of the distance from 51° North Latitude to 52° North Latitude (Figure 2.11). That means the northernmost extent is 51° North plus 1/10 of 60 (since there are 60 minutes to 1 degree), or 6 minutes; thus, 51 Degrees 6 Minutes North.

When done, convert your degrees and minutes measurement to its equivalent number in decimal degrees; that is, converting the number of degrees and number of minutes into a fraction of a degree that corresponds to your minutes estimate. Repeat this process for the easternmost to the westernmost extent of France. Don't forget about the Brittany peninsula, that extends westward from France into the Atlantic Ocean! Ignoring this peninsula will greatly affect your measurement.

TIME AND LONGITUDE: CIRCLING BACK

The Earth makes a full rotation on its axis in 24 hours. Or, it rotates through 360° of longitude in 24 hours. Often, maps show meridians (lines of equal longitude) every 15°. Thus, each of the zones between successive meridians on such a map represents 1-hour passage of time on a clock. Do you see why time zones are roughly 15° of longitude wide? The use of the 15° zone as a time zone is not an arbitrary decision; it is one based on astronomy, and the associated mathematics, that underlie the movement or our planet in relation to the sun. However, are they always exactly that wide? Why or why not? Theory and practice may differ. Are cultural constraints, boundary issues, agricultural issues, or other issues involved in establishing the boundaries for time zones?

LOOKING AHEAD

In this Chapter, we have offered the reader an opportunity to begin to understand the deep and rich topics associated with some of the mathematics and mapping issues involved in measuring positions on the surface of the Earth. It is a topic to which we will return, after presenting some more conceptual foundations in the next chapter. When we do return to measurement issues involving position on the Earth sphere, the reader should reflect on these sections.

3 Simple Relational Measures and Measures of Central Tendency and Variation

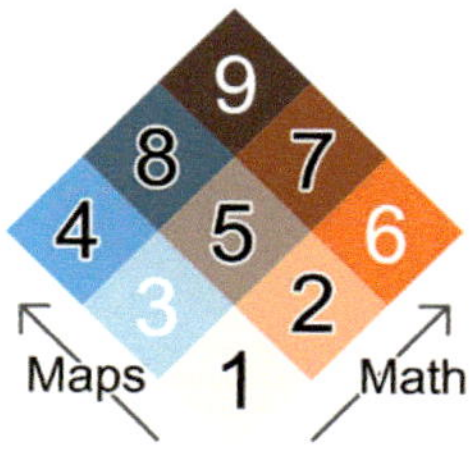

The line of images above is a visual abstract of this chapter designed to foster spatial thinking. From the chapter numeral, to the book structure, to the real world, the reader is offered gentle guidance to develop spatial intuition about what might be coming. Those thoughts are then reinforced with a detailed text outline of chapter content below. The images and outline forge as an abstract of chapter content.

CHAPTER OUTLINE

- Simple Relational Measures
  - Ratio
  - Proportion
  - Percent
  - Estimation and Rounding
  - Problems
  - Associated Interactive Mapping Activity: Mapping Urban Litter Proportions
- Math Word/Story Problems
  - General Strategy
  - Worked Problems
  - Deeper Thought Problem: Hope Village for Children
- Measures of Central Tendency
  - Mean and Related Measures
    - Standard Deviation
    - Variance
  - Median
  - Mode
  - Problem
  - Associated Interactive Mapping Activities
    - Creating and Examining a Standard Deviation Classification Map
    - Examining Distributions Using Mean Center and Standard Deviational Ellipses

DOI: 10.1201/9781003305613-3

Zebra Mussels
Geographic Centers for the State of Mississippi and the State of Virginia
Variation
Histogram
Scatter Diagram
Correlation
Positive Correlation
Negative Correlation
No Correlation
Simple Linear Regression
Associated Interactive Mapping Activity: The Histogram as a Mapping Tool
Goodness of Fit and Statistical Significance
Coefficient of Determination
z-score
Statistical Significance
p-value
Associated Interactive Mapping Activities
Examining Relationships Between Variables: Charts and Coefficients of Determination
Using Tessellations and Spatial Statistics to Examine Patterns in Numbers on Maps
Integrating Elements of This Chapter
Fibonacci Numbers and The Golden Ratio
Percentages, Estimation, and Rounding in the World of Sculpture
Questions to Think About

In this Chapter, the reader learns, through interactive mapping activities, to use mathematical tools to manage data sets and to consider how they might compare to other partial or complete data sets. The activities generate viewpoints of mathematics that may be new and different to the reader, offering fresh incentive to master the math!

SIMPLE RELATIONAL MEASURES

Ratio

A ratio is a comparison of two quantities. In a group of five people, for example, three of them might have widgets while two of them do not. The ratio of those who have widgets to those who do not is 3 to 2, or expressed in other ways, 3:2, 3/2, or 3 to 2. Reverse the phrasing: what is the ratio of those who do not have widgets to those who do? The answer is 2:3, 2/3, or 2 to 3. Notice that the fraction may be greater than or less than 1.

Proportion

A proportion is a part in comparative relation to a whole. A half a pizza is a portion of pizza that is clear in its meaning in relation to the whole pizza. How many pieces in the half? Determine equivalent ratios: 1/2; 2/4; 4/8—depending on whether the server chooses to cut the pie into 2, 4, or 8 pieces. Notice that while these three fractional amounts look "different" from each other, they are not different answers—they are equivalent answers for representing the same amount of pizza. The proportion of the whole that is being served is 1/2, independent of the number of separate pieces on a plate.

Percent

The word "percent" means "per hundred." Percentages are ratios that compare a given quantity to the number 100. The symbol % represents "percent." It may also be represented as a fraction: 7%

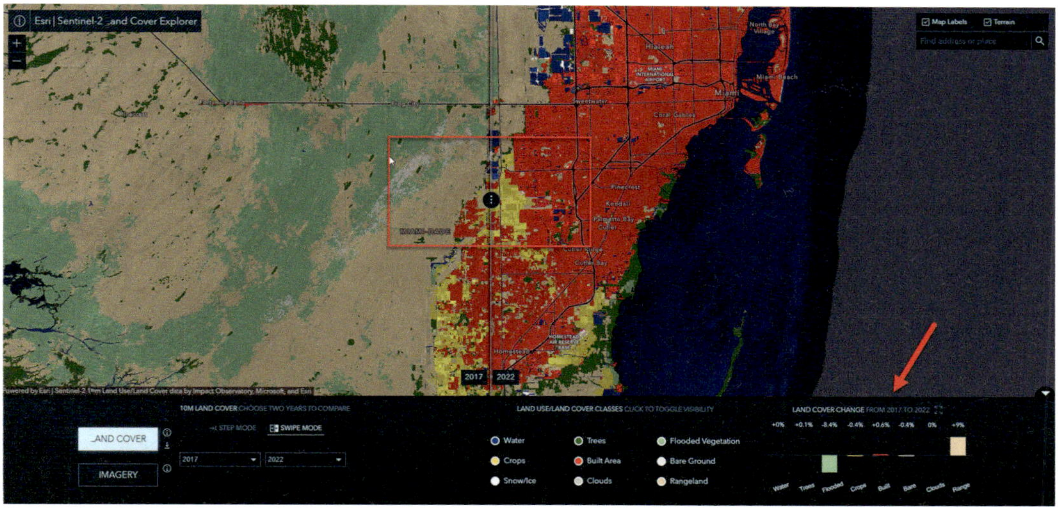

FIGURE 3.1 Visualization of change in percentages of flooded vegetation and range land from 2017 to 2022 using a swipe map, Esri.

means 7 per hundred or 7/100. Equivalently, as a decimal, 7% is 0.07. A farmer might grow corn in ¾ of a field and soybeans in ¼ of that same field: 75% of that field is dedicated to raising corn and 25% is dedicated to raising soybeans. When expressing percents be careful to use 0s correctly. Thus, 0.07 is 7%; it is NOT 0.07%. Why? Think back to expressing both as fractions: 0.07 is the same as 7 one-hundredths.

The map below shows land cover as measured by the Sentinel-2 satellite (Figure 3.1). Note the land cover totals as a percent in 2022 in the lower right. Does it surprise you that water represents 76% of the area in this image? As you have been learning, scale is important. Therefore, zoom in on a section in south Florida, and you will see the percentage of land represented as water decreases. Note how difficult it is to find any region of Florida where the land cover represented by water is 0%. Go to the URL https://livingatlas.arcgis.com/landcoverexplorer/#mapCenter=-80.445%2C25.651%2C11&mode=swipe&timeExtent=2017%2C2022 and open it to start a new investigation. Click on "swipe mode" and you will see something like Figure 3.1—the change in land cover from 2017 to 2022. By studying the bar chart, you will note how water and range land increased in percentage of the area represented by the image, but trees and flooded land both decreased. Move the swipe bar left and right to see where these changes took place. The bar chart shows that flooded land decreased almost the same amount as range land increased. The land cover on this map was derived from an algorithm applied to satellite imagery; that algorithm relied heavily on mathematics, but also reflectivity of land surfaces (physics) and is displayed in map form (geography). In addition, this map shows that maps can do more than display research results; they can also guide and create future directions for research and for planning: What is the wisest use of lands in South Florida, and how can future land use be effectively planned?

Estimation and Rounding

When calculating ratios, proportions, and percentages, a calculator is often helpful. The calculator does not make computational errors. However, that does not mean that the answer that appears on your screen is necessarily correct! You may have made a data entry error; an error in the keystrokes. So, check your answer that appears on the screen mentally using a similar ballpark problem that is easy to calculate as an estimate. For example, if you want to convert 5/8 to decimal format—you should know that 4/8 is 0.5 and 6/8 is 0.75. So, you now have bounds for your screen answer: it should be between 0.5 and 0.75. If it is not, try entering the problem again. Good estimation is important.

Also, think about the length of your answer produced by the calculator. My calculator produces an answer to seven decimal places. But using all those decimal places may be useless and indeed inappropriate; longer answers are not necessarily better answers. Think about the context of the answer. If I want to know how far it is from my house to the library, and am planning to drive my car, probably all I really care about is how many kilometers it is from one place to another. Tenths and hundredths of a kilometer probably do not matter. If I am riding my bicycle, my viewpoint may be different. Think about the context of your answer and then adjust its length to suit the situation. As was the case with latitude and longitude, so too it is with ratios, percentages, and proportions: estimation and rounding can make an important difference.

Problems—Remember, "Percent" Means "per 100"

- Express: 35% as a fraction (35/100; 7/20); 71% as a fraction (71/100); 3.47% as a fraction (347/10,000); 9% as a decimal (0.09); 74.3% as a decimal (0.743); 0.002% as a decimal (0.002/100=0.00002); 6/20 as a percent (multiply by 1; (6/20) * (5/5) * 100 or 30%); 15/70 as a percent ((15/70) * 100=21.43%).
- Suppose you score 13 out of 18 correct answers on a quiz. What percentage of the answers did you answer correctly? (13/18) * 100=72.22%. What percentage of the answers did you answer incorrectly (assuming all questions were answered)? (5/18) * 100=27.78%. Or, looking at it another way, 100% – 72.22%=27.78%.
- What is 76% of 41? $x=0.76 * 41=31.16$.
- 17 is what percentage of 47? $17=47x$; $x=17/47$; $x=0.36$ multiply by 100 or 36%.
- 3 is 40% of what number? $3=0.4x$; $x=3/0.4$; $x=7.5$.
- What percentage of 5 is 1.37? $5x=1.37$; $x=1.37/5$; $x=0.274$ multiply by 100, or 27.4%.

In the interactive mapping activity of the next section, you will have the opportunity to put this conceptual knowledge into action, as you consider the patterns, across downtown San Diego, of litter that was surveyed at several intersections in the city's downtown. You can visualize the data in many ways, including as pie charts in which each slice of pie represents a percentage of the whole with the sum of all slices adding to 100%, or as ring/donut charts which are pie charts with the center removed, thereby exposing more of the underlying map. As you work through the activity, consider how mathematics through mapping can help reveal spatial and numeric patterns, and visualizing those patterns can help people to take action.

Associated Interactive Mapping Activity: Mapping Urban Litter Proportions

POP-UP BOX

Listen to co-author, Joseph Kerski, in this video to understand what you will do in this activity (QR code from the book cover): https://www.youtube.com/watch?v=B3Pm96mokE8

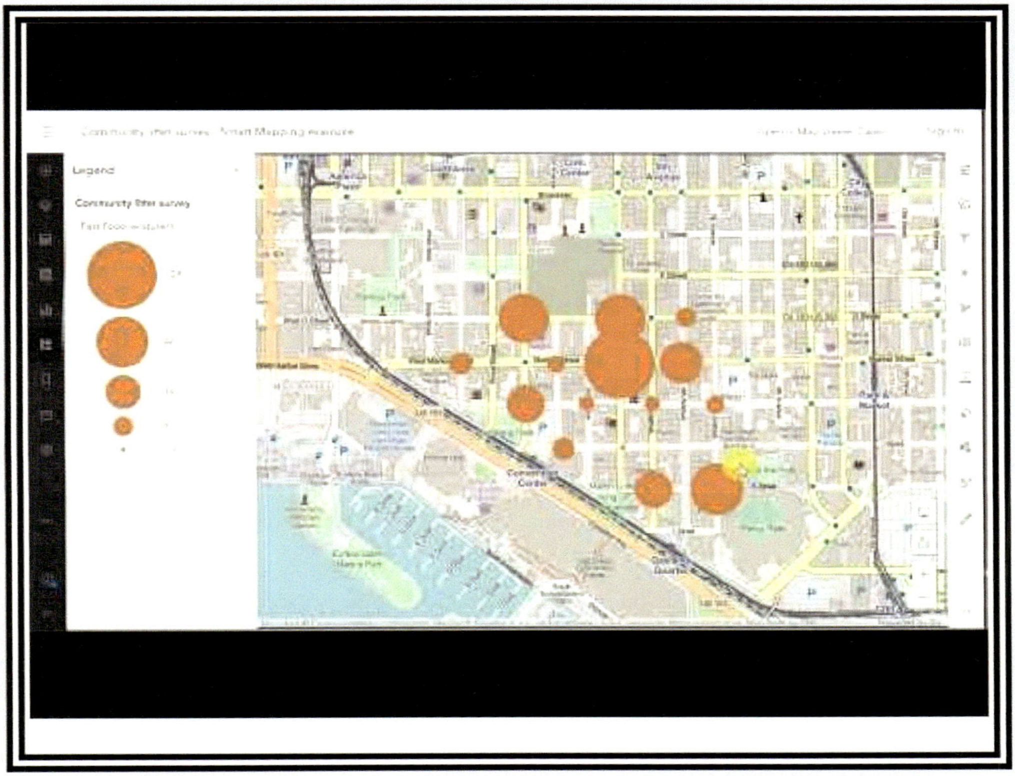

In the following activity, you will explore field data, concerning proportions of urban litter, through creating single symbol, graduated symbol, predominance, and two types of chart maps, pie charts and ring (or donut) charts. Their meaning will become clearer through the video and through hands-on work.

To begin, open the linked web map (https://www.arcgis.com/apps/mapviewer/index.html?webmap=f78dbbb432d946a58009763c68caa84f) in ArcGIS Online. The map opens with a set of litter data collected at street intersections in central San Diego, California USA (Figure 3.2). As a *single symbol map*, with a single small dot showing each data location, only the location and distribution of points where litter was found, can be seen on the map. However, the amount and the characteristics of that litter were also collected. Click on a point to view the categories associated with each piece of litter: cans, fast food wrappers, glass bottles, plastic bottles, and other plastic waste. Note also the fields—latitude and longitude—that were needed to map the points.

On the left panel>Change the base map to open street map to see the underlying city detail. Again on the left panel>Use Layers to access the community litter survey data layer>Open the table. The table shows 15 records—each record corresponds to a location in the city where litter was collected. Click on the fast-food wrappers field>Sort Descending>select the row containing 29 fast food wrappers, and note its location highlighted in cyan on the map. The same process works with the other fields; however, examining the data ***table*** will result in only a limited understanding of the spatial pattern. Instead, you will ***map*** the data to augment that understanding.

On the left panel>Layers>select the Community litter survey layer>Properties>Right Panel>Styles>Choose attributes>use the+Field button>choose Fast Food wrappers>Add.

Note that the layer is redrawn as a Counts and Amounts map style, also known as a graduated symbol map, graduated meaning scaling up or down in size. The more fast-food wrappers that were found at each location, the larger the symbol drawn. What patterns do you notice about the

FIGURE 3.2 Litter survey, San Diego, California. Mapped using ArcGIS from Esri.

number of fast-food wrappers—is one part of the city more heavily covered with these wrappers than another part?

Fast food wrappers were just one type of litter collected at each location. Therefore, you will now map *all* the types of litter to determine what kinds of patterns there might be. Right panel > Styles > Choose attributes, click + Field and choose Cans, Glass bottles, Other plastic waste, Paper waste, and Plastic bottles > Add. Now, the map redraws using the Predominant (type of greatest frequency) style. Now, two types of information are shown: The color of each point corresponds to the type of litter that was most frequently found at each location, the predominant type. The darkness of each color (its saturation) indicates the extent to which a certain type of litter is predominant: the darker the color, the more predominant the litter type. If no type is predominant, the symbol for that point would be light in tone. The size of each point is rather small. Feel free to change all the symbols to a larger circle using Style options > click on the pencil to the right of the color ramp > adjust size. What type of litter is predominant closest to the Petco Park stadium on the southeast side of the study area? How predominant is that litter type compared to the site due north of it? Think about these questions in relation to the material you have just finished that involves ratio, proportions, and percentages.

Next, change the map to show each field data collection site as a pie chart: Right panel > Styles > Charts and Size > Style options > Choose pie charts > Symbol style > Outline color > Edit > Choose a shade of black or type 000000 in the Hex # box. Close the Select color window > For Outline transparency > drag the slider down to about 40% and for Outline width, drag the slider up to 2 px > Close the Symbol style window. Reflect on what you have learned, once again, from the conceptual material in this chapter.

Mapping of this sort displays not only the type of litter at each location (from the pie chart wedges) but also the total amount collected at each location (from the size of each pie chart). What newfound knowledge do you gain about the type and amount of litter through these pie charts? How might you capture this information in terms of ratio, proportion, and percentage?

Next, change the pie charts to donut charts, or rings: Right panel > Style options > scroll down to the Shape slider > drag the Shape slider about 3/4 of the way to the Donut shape > Done. Each symbol now shows the types of litter collected as a donut chart. Next, use Effects > Drop shadow > close the effects pane. Now, each donut or ring symbol has a dark highlight around it to make it stand out on the map. See linked example here (https://www.arcgis.com/apps/mapviewer/index.html?webmap=60718b2751604ac0b70696d9df01b00f) and as shown in the static screenshot (Figure 3.3).

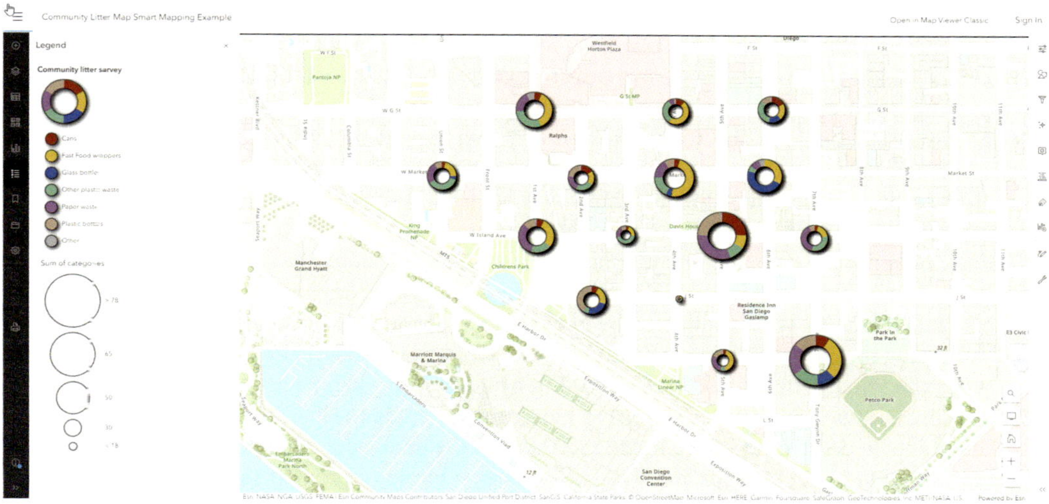

FIGURE 3.3 Ring, or donut, charts showing types of predominant collected litter, San Diego, California. Mapped using ArcGIS from Esri.

The ring or donut charts provide a great deal of information about each litter point and, again, display patterns, amounts, and the types of litter predominant among the points to be compared.

Sometimes, there is too much data in maps, making the patterns difficult to interpret. In such cases, simplifying information can aid in focusing on specific characteristics in the data and therefore the main message of the map—much as rounding off decimals can help to clarify meaning. Experiment with the Grouping tool using the charts style options. Using this tool, values below the grouping percentage are aggregated into the "Other" category to simplify each ring symbol. Setting it to 25%, for example, highlights the fast-food wrappers that dominate in many of the sampled points (as shown in the static screenshot, Figure 3.4). Notice here the use of percentages to

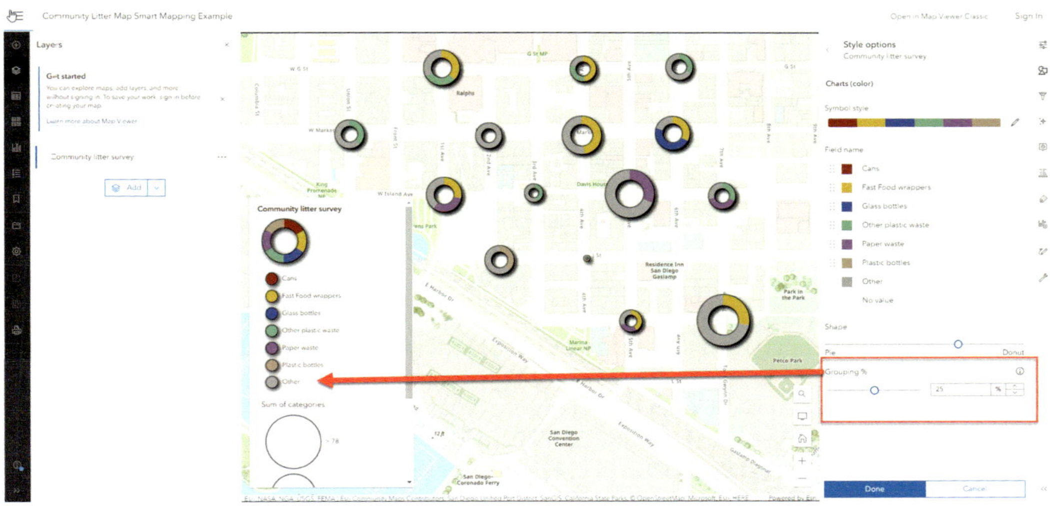

FIGURE 3.4 Grouping (aggregation) of data to simplify, and emphasize, mapped content. Mapped using ArcGIS from Esri.

make decisions to aid in understanding and in clarity of display of an otherwise overly complicated mapped pattern. Use mathematical concepts creatively and in multi-faceted ways.

You have made several types of maps in this activity. What type of map gives you the most insight? Single symbol, graduated symbol, predominance, pie chart, or donut chart? Which type of litter is most problematic in different parts of downtown San Diego? What action might such insight inspire—a community litter awareness campaign, strategically placed recycle bins, or something else? How could such action be taken, who would be the stakeholders, and how can these maps and analyses help them visualize where to focus their efforts?

Being critical of the data is a theme running throughout this book. Ask questions of this and any other type of data collected in the field: Here, what time of day, day of the week, and time of year was the data collected? Was it collected by one person or a team? How much time was spent at each point collecting data? Was the data collected in the street, the gutter, on the sidewalk, or somewhere else? Does the wind pattern influence where litter piles up? Do local activities, such as attendance at large events, influence the litter pattern? Change the basemap to satellite image and zoom in on a few data points to determine where in the intersection the data were collected. What was the shape of the actual data collection space—was it a square, a circle, or something else? How large was the space considered—1 meter, 2 meters, or something else? How would knowing these pieces of information affect your assessment?

Reflect back on the numerous times that the underlying mathematical concepts of ratio, proportion, and percentage were important. Factor them into answering questions about the litter patterns and their display, clarity, and basis for community involvement. Then, consider how the mapping activity enhanced your understanding of, and appreciation for, the importance of ratio, proportion, and percentages as mathematical tools. With a firm traditional understanding of these mathematical concepts, as fostered by mastering simple problems, enlarged by an interactive understanding using the mapping of associated spatial pattern, the reader should feel enthusiastic about seeing more mathematical tools and learning how they, too, might emerge in novel interactive contexts. To that end, we move forward in the remainder of this chapter to explore topics that deal with data and related real-world issues and their expression in prose, notation, and maps.

MATH WORD/STORY PROBLEMS

General Strategy

Success with word problems rests on success with language. Students who have reasonable mathematical ability can often just look at a word problem and an answer naturally comes to them. The usual attitude there is, understandably, "well, I don't need to show any work, I just know the answer." Indeed. However, there will come a time when the problems become difficult enough that the answer does not just appear obvious. For those situations, having a system for solving word problems is critical. And, that system rests on translating the word problem from a natural human language (such as English) into mathematical notation (another language). The following system works well.

- Find the word "is" or some other form of the verb To Be that is the dominant verb in the sentence (note: English does "count" in math class; math, English, and other core subjects transcend disciplinary boundaries). Replace "is" with an equal sign.
- What is to the left of "is" in the sentence may be translated into mathematical notation to the left of the equal sign and similarly for what is to the right. In complicated word problems, these steps may be repeated if there are sub-sentences to the left and right of the equal sign. In languages other than English, different strategies, based on variation in word order may apply. Nonetheless, the general idea of translation is universal; how it is executed may vary.
- Some words often translate to specific notation: "what number" or just "what" translates to an unknown variable such as "x"; the word "of" translates to multiplication.

Worked Problems

- If gas costs \$5.11 cents a gallon, and Bill's car gets 21 miles per gallon, how much money will it cost Bill to drive his car 4,263 miles?
    - Answer process 1 following the format involving the verb "to be": The form of the verb to be is, "will it." So, replace that verb with =. To the right of the = sign, insert what is to the right of the verb: 4,263. Now, what is to equal 4,263? The number 4,263 is a distance. To the left of the equal sign, we have a value for cost per gallon and a value for miles per gallon. Or, cost/gallon and miles/gallon. Divide: cost/gallon/miles/gallon to get rid of the denominator of gallon. Notice the work with fractions from the last chapter. How much does it cost to drive 1 mile? 5.11/21 = 0.243333 cents. So, how much does it cost to drive 4263? Multiply the value of the cost for 1-mile times total mileage of 4,263 which is 1037.33.
    - Answer process 2: For those taking a quicker view of the problem, Bill uses 4,263/21 = 203 gallons, costing 203 × \$5.11 = \$1,037.33. Once you are confident in solving word problems, shortcuts may occur to you.
- The local train leaves the station at 5:00 pm. The express is 20% faster. What time should the express leave the station to catch the local at 7:00 pm? Solution: Suppose the local travels at M miles per hour, so that the local travels 2M miles in the 2 hours from 5 to 7.
  The speed of the express is 1.2M, so it must travel 1.2Mx in the same time as 2M, where x is the time the express travels. Thus 2M = 1.2Mx, whence x = 2/1.2 = 5/3. Thus, the express train travels 5/3 hours, so leaves the station at 5:20 pm.

Distance, Rate, Time Interactive Mapping Activity

Open this map (https://www.arcgis.com/apps/mapviewer/index.html?webmap=c97b1c3300ef4ad4be868c37d04eaa2d) showing a few roads running north, south, east, and west, and a powerline running from northwest to southeast. Use Map Tools > Measurement to measure the distance between point D on the north and point B on the south (Figure 3.5). What is the distance? Measure the distance between point D and point A, where the powerline intersects the road. What is this distance? Say that you live at point D, your friend lives at point B, and you have agreed to meet on the north-south road. You each leave your homes at the same time. You walk due south at a leisurely

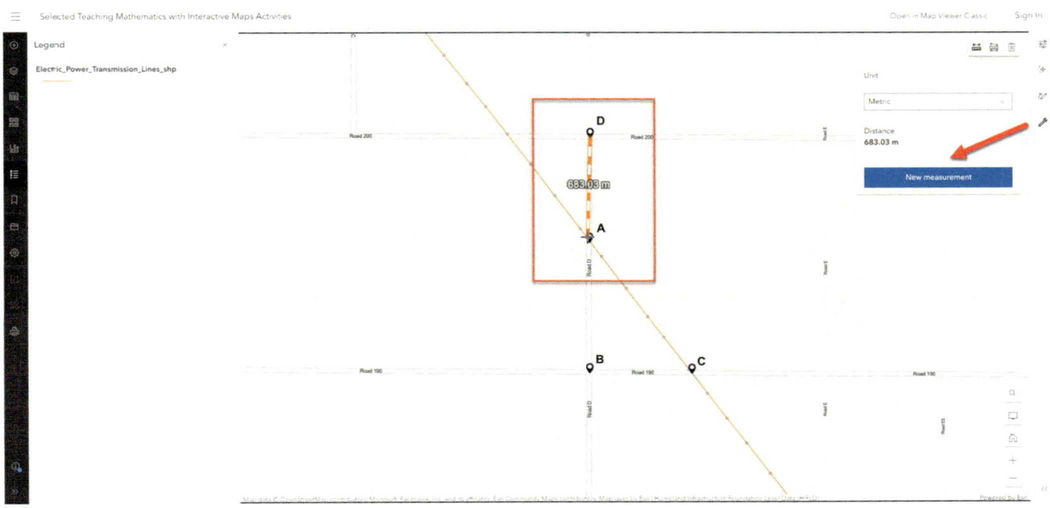

FIGURE 3.5 Using distance, rate, and time with an interactive map. Mapped using ArcGIS from Esri.

pace of 4 kph but your friend walks due north at a faster pace of 6 kph. How far from point D will you cross paths? When will you cross paths? Will you cross paths north or south of the intersection of the powerline with the road?

Deeper Thought Problem: Hope Village for Children

In traditional Word/Story Problems, there is obvious interaction between the academic disciplines of Mathematics and English (or other human language). At the obvious level, one must be able to read prose to consider solving the problem (indeed, prose is the most used form of mathematical notation). At a deeper level, we see that knowledge of English grammar is vital in solving Word/Story Problems: find the principal use of the verb To Be, and go from there, translating that "to be" form to the equals sign in an equation. Then, use the word order in the sentence to build out left and right sides of an equation on either side of the equal sign. When we were learning English grammar, we probably did not think that the verb To Be would arise as a critical element in learning to solve a whole style of mathematics problems! So, we now turn to looking around the world and seeing if there is some component in the mathematics we have been learning that might enable us to learn more about getting along in the real world.

In Meridian, Mississippi, the movie star Sela Ward created a non-profit social services venture, called Hope Village for Children, located on 30 acres of land. Its mission is: "To provide a continuum of specialized treatment programs, services, and facilities of superior quality to meet the individual needs of neglected and abused children and their families." (https://www.hopevillagems.org/about-hope-village/faqs/%20)

Hope Village posts a page of "Frequently Asked Questions" on its website. Naturally, a website that deals with services involving children is one that is likely to be, in many respects, guarded. Over the course of time, the Arlinghauses have had the opportunity to get to know both Kathy Parrish, Community Outreach Director and Terri Province, Executive Director for Hope Village. Others who have not had that fine opportunity are reliant on Internet resources and various other local resources (all of variable accuracy) to learn more about the fine work this organization does. By reading online at https://www.hopevillagems.org/about-hope-village/faqs/ and elsewhere, one might find that there are eight on-site residential cottages housing 44 children with a typical annual load of about 300 children. Thus, in the course of a decade, about 3,000 children come through the system each spending an amount of time in it as deemed suitable for the needs of the particular child.

But, how much of a contribution is 3,000 children to the community? Meridian has a population of about 30,000. So, the contribution is about a 10% service rate over a decade. In a city of 300,000 that same contribution level would be at the 1% level, and in a city of 3,000,000 it would be at the 0.1% level. Proportions—here using population, are important in analysis!

In addition to considering the helpful FAQ already available online, and the observations about population density, there are other percentages that I might wish to know about, especially those involving educational attainment of the children, that were not covered in the FAQ. For example:

- What percent of their children are they able to track after they leave Hope Village? This statistic is important so that I have some understanding of the reliability of their statements about the performance of their children in the adult world.
- What percentage of Hope Village children graduate from High School? From Community College? From a 4-year college?
- The State of Mississippi has recently been cited number one state in the country as the most improved State in reading scores on tests taken at the end of the third grade (https://www.nytimes.com/2022/10/06/education/learning/mississippi-schools-literacy.html). Do those scores for Hope Village Children compare well with the State-wide averages?

Because I have a firm grasp of the meaning of percentages, I was able to supplement the information available on the FAQ and tailor it to my interests. Mastery of learning to think clearly about percentages expanded my horizons and enabled me to think outside the box. Once again, Mathematics Matters!

MEASURES OF CENTRAL TENDENCY

Statistics is a systematic study of data (Freedman, Pisani, and Purves (1978 and later editions) is one useful textbook for readers wishing to refresh). Quantitative data sets usually capture information in numbers, and the pattern may be discrete (with values clearly separated) or continuous. Each informational unit is assigned a unique, single numeric value. Multiple values may not be assigned to a single observation. Qualitative data sets are sets that are not quantitative: the content is often captured in words or other non-numeric symbols. This topic is mentioned only briefly here as it fits with the presence of both prose and notational representation of facts; it is an important one that will resurface.

Often one wants to consider what is happening at, or near, the middle of a data set. There are three measures in common use to capture what the word "middle" might represent.

MEAN AND RELATED MEASURES

The mean is the average value of a data set. Add up all the values and divide by the number of those values The mean of a sample is often denoted by the Greek letter mu, μ.

Standard Deviation

The standard deviation measures the dispersion of values in a given set about the mean of that set. The higher the value of the standard deviation, the more far-flung the dispersion; lower values show greater clustering of values about the mean. The standard deviation of a sample is often denoted by the Greek letter sigma, σ (or S), $S=\sqrt{((\Sigma(x-\mu)^2)/(n-1))}$ where μ is the sample mean and n is the number of elements in the sample. For a bit more detail, watch the presentation in the resources section at the end of the book, or peruse the Internet (focusing on resources from sites whose URLs end, .edu, for education/university). Discussions of standard deviation often center on normally distributed data and explain how scatters of data might appear in different normal distributions. An example that the instructor (https://www.youtube.com/watch?v=09kiX3p5Vek), in the video at the end of the book, used was the distribution of height in a population. It is interesting to note that when height samples are taken from a population, the distribution of the sample averages will be approximately normally distributed (given certain assumptions on the size of the samples and such), even if the population distribution, itself, is not normally distributed! Concerns of that sort led to the Central Limit Theorem—another way of looking at central tendency.

Variance

The square of the standard deviation is called the variance. Keep this fact in mind when interpreting the variance as compared to the standard deviation. A negative standard deviation will be associated with a positive variance. An absolute value for a standard deviation that is less than 1 will result in a variance that is smaller than the corresponding standard deviation. Conversely, an absolute value for a standard deviation that is greater than 1 will result in a variance that is greater than the corresponding standard deviation. When will the variance be equal to the corresponding standard deviation?

MEDIAN

The median is the middle value in a set of numbers. If there are an odd number of elements in the set, the median is the unique value within the set above which exactly half of the elements lie and below which exactly half of the elements lie. If there are an even number of elements in the set, then there are two middle values and the median lies halfway between them.

Mode

The mode is the most frequently occurring element in a data set. It may, or may not, be unique. There may be two or more modes in a data set.

Problem

One hundred people are polled as to the length of trip they would be willing to take over a weekend. Their responses are as follows:

0–50 miles 10 treat this response as 25
50–100 miles 35 treat this response as 75
100–150 miles 30 treat this response as 125
150–200 miles 11 treat this response as 175
200–250 miles 14 treat this response as 225
What is the median response? Median = 125
What is the mode? Mode = 35
What is the mean response? Mean = avg = 116
$(10\times25+35\times75+30\times125+11\times175+14\times225)/100$

Associated Interactive Mapping Activities

Creating and Examining a Standard Deviation Classification Map

Suppose you are tasked with identifying neighborhoods in the United Kingdom (UK) for a housing assistance program. You decide to begin by finding areas of higher poverty. You start by opening the linked interactive map (https://www.arcgis.com/apps/mapviewer/index.html?layers=ef8c3344e9f7497caa5be549a0003d82) of deprivation (poverty) in the UK: The deprivation is an index of seven variables, including income, employment, education, health, crime, barriers to housing and services, and living environment. The lower the number, the higher the poverty or deprivation. As always, you spend a while examining the data source: https://www.gov.uk/government/statistics/english-indices-of-deprivation-2019. In the map (Figure 3.6), the data are displayed without classification using a color ramp from green (low deprivation, values above 10) to red (high deprivation, values less than 1). There are different varieties of classification; generally, they involve a set of criteria based on the underlying attributes of the individual observations for determining which individual observations are elements of which class. Changing the classification of a data set often creates maps of very different appearances. The lack of classification in this map, together with the large number of enumeration areas in the UK, pose a challenge to the accurate interpretation of areas on the map that face poverty challenges (Figure 3.6).

Because of these challenges, you decide to classify the data. Standard deviation classification shows how much a location's attribute value varies from the mean. By emphasizing values above and below the mean, standard deviation classification reveals which locations are above or below an average value. You decide to use this classification method because it is important to your analysis to know how values relate to the mean; here, the deprivation index (Figure 3.7).

Use the Right panel > Use styles > counts and amounts color > style options > classify data > standard deviation > Done. Note that by taking the default standard deviation classification, you are given a 4-class (category) map with the mean shown (5.5) with each class covering 1 standard deviation above and below the mean (as shown). The defaults can be overridden and it is important in any technology, including mapping technologies, to know how to override the default settings.

You decide to focus your program on enumeration areas with standard deviation of below −0.5 standard deviation from the mean of the deprivation index, focusing on the red areas as shown on

FIGURE 3.6 The data are displayed without classification using a color ramp from green (low deprivation, values above 10) to red (high deprivation, values less than 1). Mapped using ArcGIS from Esri.

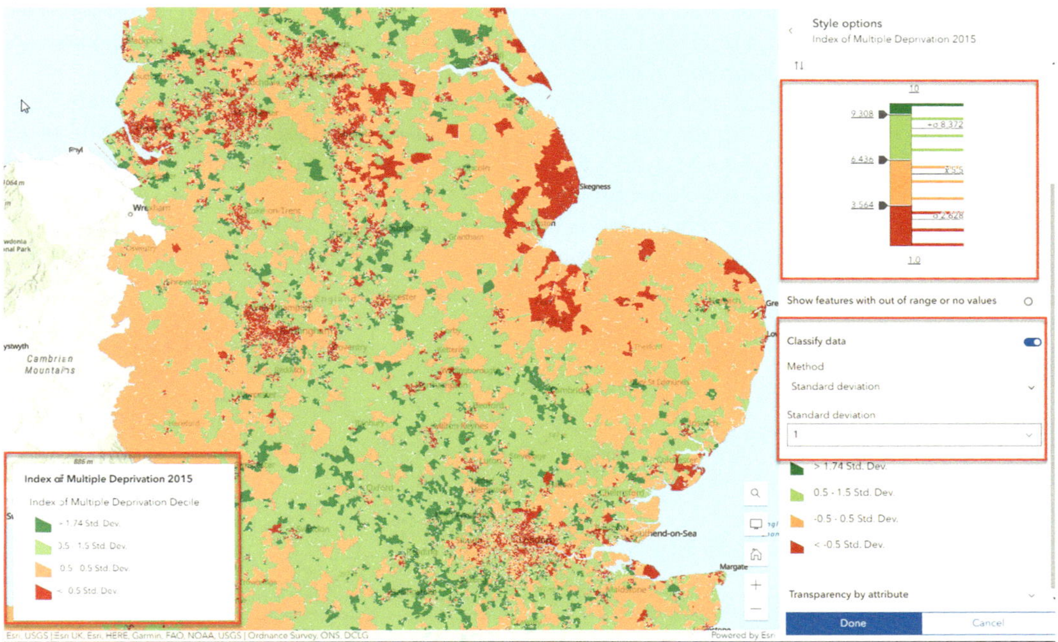

FIGURE 3.7 Data are classified using standard deviation. Mapped using ArcGIS from Esri.

the map. Experiment with a 0.5 standard deviation for each class instead of 1 standard deviation, which will result in seven categories. Which type of standard deviation map do you find more useful? Next, we look at a related mapping tool based on the concepts of mean and standard deviation that are often used in studies that involve dispersion.

Examining Distributions Using Mean Center and Standard Deviational Ellipses

Zebra Mussels

Invasive plant and animal species pose an ongoing threat to the planet's biodiversity. For your investigation of one of these invasive species, that of the zebra mussel, you decide to focus on mapping their distribution through point data and through measures of dispersion. Begin by opening the linked interactive map (https://www.arcgis.com/apps/mapviewer/index.html?webmap=460bab410f7d4d67b67072a767740342). This map shows the distribution and spread of an invasive aquatic species, zebra mussels, from 1986 into the early 2000s. Each point on the map represents a sighting of a zebra mussel: Note their heavy presence in rivers and in the Great Lakes. The zebra mussel points in lighter symbols were found in the 1980s while the darker symbolized mussels were found in the 1990s and thereafter.

Studying the map gives some indication for the spread of these mussels, but the use of spatial statistical tools will provide additional insights. Spatial statistics, as the name implies, are statistical tools applied to maps and the analytical tools behind maps. Left panel>Layers>Using the eye symbols next to the layer names, turn on the layers mean center before 1995 and mean center 2000 and later. The mean center represents the point at which all of the data would "balance" if that point were the tip of a pencil and the data were on a plane balanced on that pencil tip. As you can see, and you can verify by using the measuring distance tools on the map, the mean center migrated hundreds of kilometers from a point near the Michigan-Ontario border in 1995 to a location west of Chicago by 2000. Zebra mussels' point of origin was in central Asia; they came, attached to ships, to the USA Great Lakes. Over time, the mussels migrated from the Great Lakes to the central rivers of the USA: the Ohio, Mississippi, and Arkansas (among others). The migration of the mean center in the USA gives a good indication of changes over space and time.

Another way of understanding spatio-temporal change is through a standard deviational ellipse. As the name implies, this ellipse contains all points within 1 standard deviation spatially of the mean of the data set. Left panel>Layers>Use the eye symbols next to the layer names to turn on the layers 1 standard deviational ellipse around pre-1995 zebra mussels, and 1 standard deviational ellipse around 2,000 and later zebra mussels (Figure 3.8). The shape *and* size of the ellipses both indicate movement. In this case, the ellipses moved from being shaped more like a circle toward a

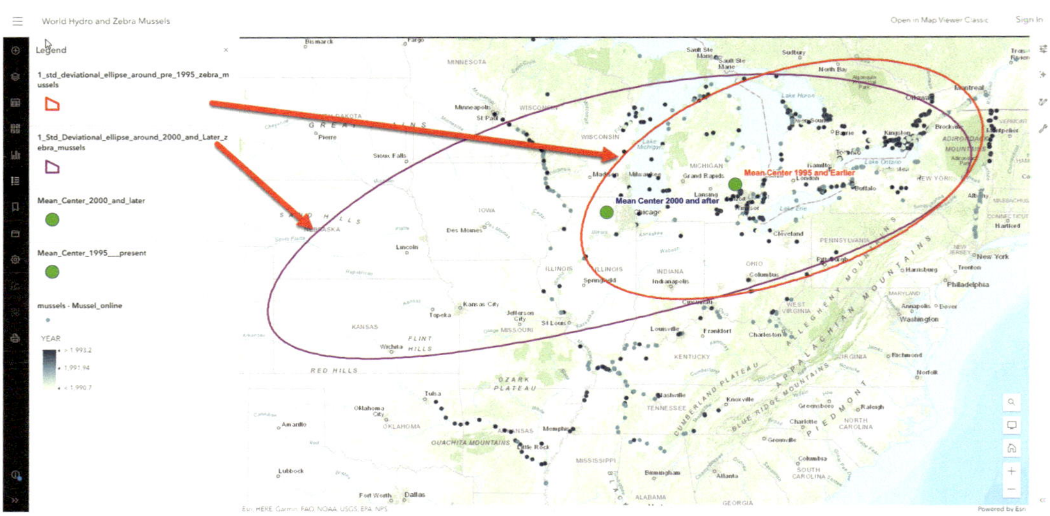

FIGURE 3.8 Map showing layers 1 standard deviational ellipse around pre-1995 zebra mussels, and 1 standard deviational ellipse around 2000 and later zebra mussels. Mapped using ArcGIS from Esri.

more elliptical shape, indicating a widening of the mussel dispersion and a migration of the invasive species to the southwest over time. Given the discussion about the mean centers, it makes sense that the mean centers are in the center of their respective ellipses.

These activities reinforce the importance of the concepts of mean and standard deviation not only for the data underlying the maps but also for their utility in determining the outcome of how the mapped data sets appear and communicate information to the reader. Often they reflect variable positions of data points over time, suggesting a view of variation in information content and, often, an interest in what is happening in the "middle" or an expression of interest in central tendency.

Geographic and Weighted Mean Centers for Population Data

In the map below (Figure 3.9), the circle with the X symbol in the middle represents the mean geographic center for the state of Nevada—the center of the state's entire area. The triangle to the south of the circle represents the weighted mean center by 2020 county population. The standard deviational ellipse of the 2020 population by county is also shown, as a blue ellipse.

The next case also shows a set of centers, but in this instance, over time, and focusing on the mean center of population. Population increase or decrease in specific areas tend to "pull" the mean center of population toward one side of the geographic area under study. In this instance, the mean center of population for the state of Nevada is examined. The geographic center is shown as a yellow star and the mean center of population for each census year as a set of red points. These red points show that the mean center of population, over time, drifts toward the south and southeast. This is due to corresponding decline in the mining industry in central Nevada and the rise in population in Clark County, of which Las Vegas is the county seat (Figure 3.10). Note the changes over time; the distance that the population center has drifted in Nevada is the greatest distance of any of the 50 states over that 120-year time span.

VARIATION

Measures of variation, in reference to a data set, describe how values in the data set vary in relation to each other. The range is the difference between the maximum and minimum values in the dataset and it offers a rough idea of the spread of the set: a small value suggests a tight, compact set while a large value suggests a broader, more spread-out, set. An outlier is a value that is significantly higher

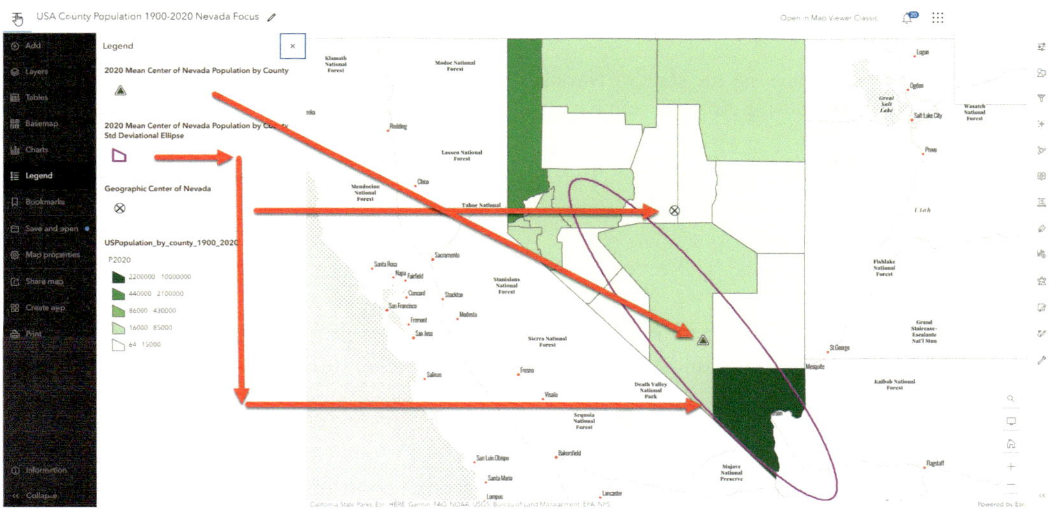

FIGURE 3.9 Geographic and mean center of population, state of Nevada. Mapped using ArcGIS from Esri.

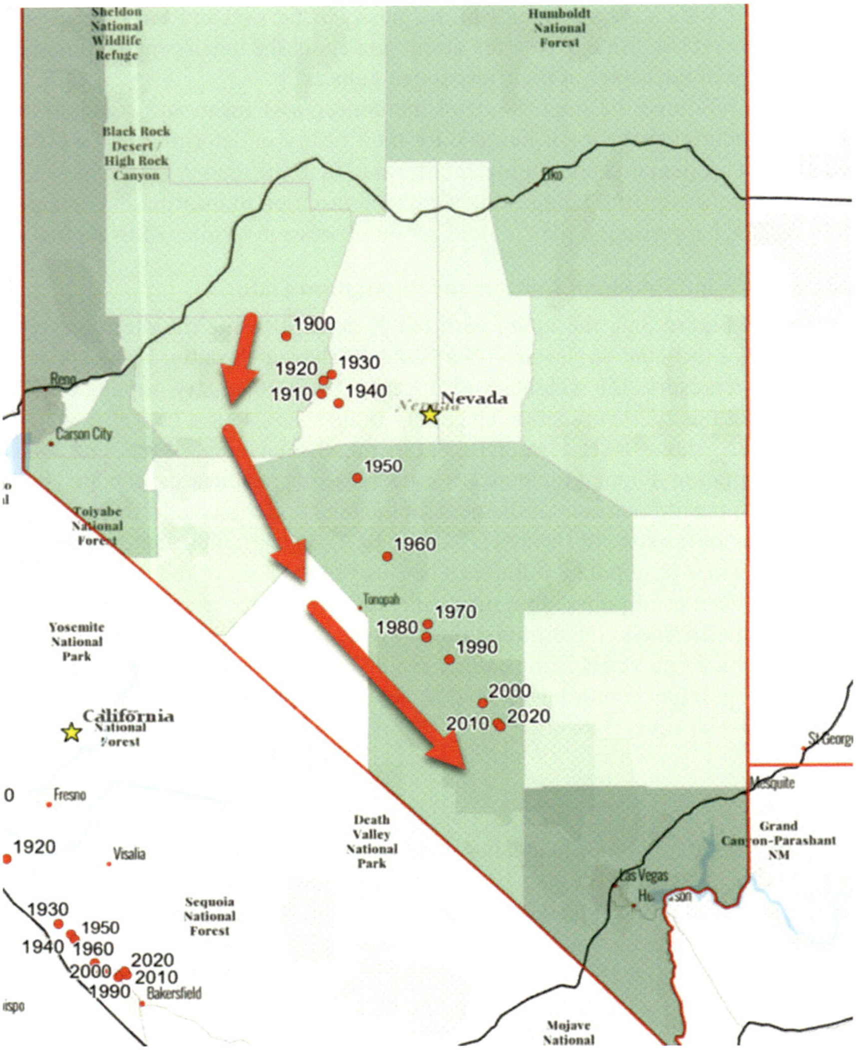

FIGURE 3.10 A map showing the state of Nevada's south and southeastern drift of the mean population center over time. Mapped using ArcGIS from Esri.

or lower than the typical values in the set. Two spatial ways to display variation involve using visual devices.

Histogram

A histogram is a visual display composed of rectangles with area proportional to the frequency of a variable and width equal to the interval between successive observations.

Scatter Diagram

A scatter diagram is a visual display composed of a Cartesian coordinate system, based on an x-axis and a y-axis, using points as ordered pairs in the xy-plane to represent individual pieces of data. When one set of data serves as x values (independent) on the x-axis, and another serves as y values

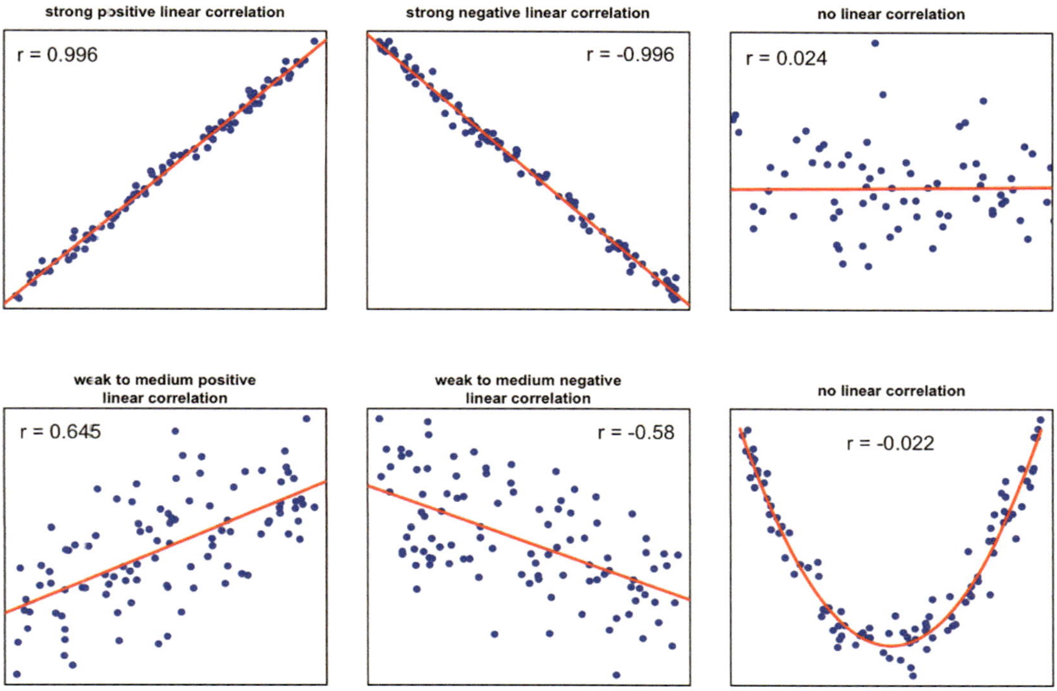

FIGURE 3.11 Figures showing scatter diagrams/plots and associated relationships: Correlation. Source: The E-Learning project SOGA was developed at the Department of Earth Sciences by Dr. Kai Hartmann (https://www.geo.fu-berlin.de/en/v/soga/legal-notice/index.html#__target_object_not_reachable, concept and content) and Dr. Joachim Krois (technical implementation and content). You may use this project freely under the Creative Commons Attribution-ShareAlike 4.0 International License (https://creativecommons.org/licenses/by-sa/4.0/). Please cite as follows: Hartmann, K., Krois, J., Waske, B. (2018): *E-Learning Project SOGA: Statistics and Geospatial Data Analysis*. Department of Earth Sciences, Freie Universitaet Berlin.

(dependent) on the y-axis, a scatter diagram may be used to see if the two variables represented on the x- and y-axes are related to each other. For example, if the scatter of dots resembles a straight line, the scatter diagram may suggest a linear relationship between the two variables (Figure 3.11). If the slope of the line suggested by the scatter is positive, then the relationship is said to be positive; if the slope is negative, the relationship is negative. Or the scatter may suggest a non-linear shape, or the scatter may suggest no relationship. The steeper the line, the stronger the relationship; the flatter the line, the weaker the relationship.

Correlation

As Figure 3.11 shows, we may look for a linear relationship in a scatter diagram by finding a line of best fit to the set of graphed ordered pairs. Simple correlation between the two data sets forming a scatter diagram is generally reflected as positive or negative (or no correlation) each measuring the amount of correlation between the two variables being graphed.

Positive Correlation

When the correlation is positive, the values in x and y rise together in unison and fall together in unison (not necessarily in a one-to-one fashion).

Negative Correlation

When the correlation is negative, the values in x and y do not rise in unison and do not fall in unison (not necessarily in a one-to-one fashion).

No Correlation

When there is no correlation, there is no discernable relationship in how the values of the two sets change in relation to each other.

The notation for simple correlation, r, called the correlation coefficient, ranges in value from +1 (for positive correlation) to 0 (for no correlation) to −1 (for negative correlation). Here we consider only simple correlation measured using Pearson's correlation coefficient. References to additional related topics are abundant on the Internet.

Causation

Correlation does not imply causation. Correlation does measure the degree to which two variables rise or fall together; however, the causes for such rises (for example) may not be related to each other. Consider the two variables, "obesity rates in the USA" and "type 2 diabetes rates in the USA." Over a decade they may both rise together. Would this topic be likely to be worth investing resources in? Consider the two variables, "obesity rates in the USA" and "ambient temperature rates in the USA." Again, over a decade these two variables might rise together. Would you rather invest valuable resources in studying the relationship between these two variables or the previous two?

Simple Linear Regression

When the line of best fit in a scatter diagram is used to predict the value of the dependent output from an arbitrary independent value as input, the line of best fit is called a simple linear regression line (Figure 3.12). That is, arbitrary input typically lies on the x-axis and the corresponding output is read off the y-axis at a position determined by the height of the line of best fit at the arbitrary x value. The 95% confidence interval, suggested in Figure 3.12, means that one can be 95% confident that a random piece of data is in the interval (within two standard deviations of the mean). For more information leading the reader beyond what is often used in mapping, consider material on the following link: https://www.westga.edu/academics/research/vrc/assets/docs/confidence_intervals_notes.pdf.

In the case of simple linear regression, a single line of best fit is used to predict a single outcome from a given input. If it is desired to predict multiple outcomes from a single given input, then multiple lines are used, and the regression analysis is termed multiple linear regression. Linear regression

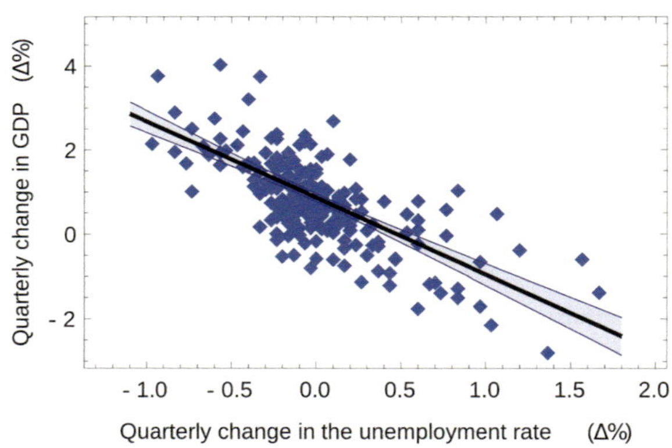

FIGURE 3.12 The US "changes in unemployment – GDP growth" simple linear regression with the 95% confidence bands. Source: https://commons.wikimedia.org/wiki/File:Okuns_law_quarterly_differences.svg

is a tool that is frequently used in the Social Sciences; we consider it here by looking at a mapping example that employs the histogram as a tool to begin to look for relationships.

Associated Interactive Mapping Activity: The Histogram as a Mapping Tool

In the following activity, you have the opportunity to develop skills in analyzing data over space and time by working different ways of visualizing numbers through mapping. In Chapter 1, we focused on graduated symbols, charts, and building expressions with strong ties back to the underlying importance of the distributive law. Here we take the results of that work and move in a different direction, related to material in this chapter. Begin by opening the linked map (https://www.arcgis.com/apps/mapviewer/index.html?webmap=f5e6469be6c34f2194758b7af02109f3) of global population.

The map opens with world population by country in 2015 with data from the United Nations (Figure 3.13). Each country's population is shown with a symbol sized according to its population; hence, it is a graduated symbol map.

Visualizing numbers as maps often reveals something unexpected: In Chapter 1, you created an expression resulting in the display of positive and negative population change on a map. You may now need to re-create that expression, and in so doing, you may find some countries where the population is actually *decreasing*. To more easily visualize this, create a two-class (or two-category) map showing the countries that are increasing vs. those that are decreasing over the 5-year time span from 2010 to 2015. After creating your expression > choose Style Options > Create a 2-class natural breaks map, and in the histogram (Figure 3.14), move the break point to be 0.0 (or type in 0 at the break point). Make the point sizes similar and larger than the default so that they will be identifiable.

Label the classes as increasing or decreasing, as appropriate, as shown in Figure 3.15.

What patterns do you notice? Which world regions contain countries where the population is decreasing? Why is the population decreasing? Decreasing population may arise from low birth rates, emigration to other countries due to political instability, economic hardship, or famine, ongoing natural hazards, or other factors.

The ability to map *change* is powerful. Our Earth is a dynamic planet, and mapping change enables more effective plans to be proposed, tested, and instituted to build resiliency and sustainability in

FIGURE 3.13 World population by country (United Nations data); a graduated symbol map. Mapped using ArcGIS from Esri.

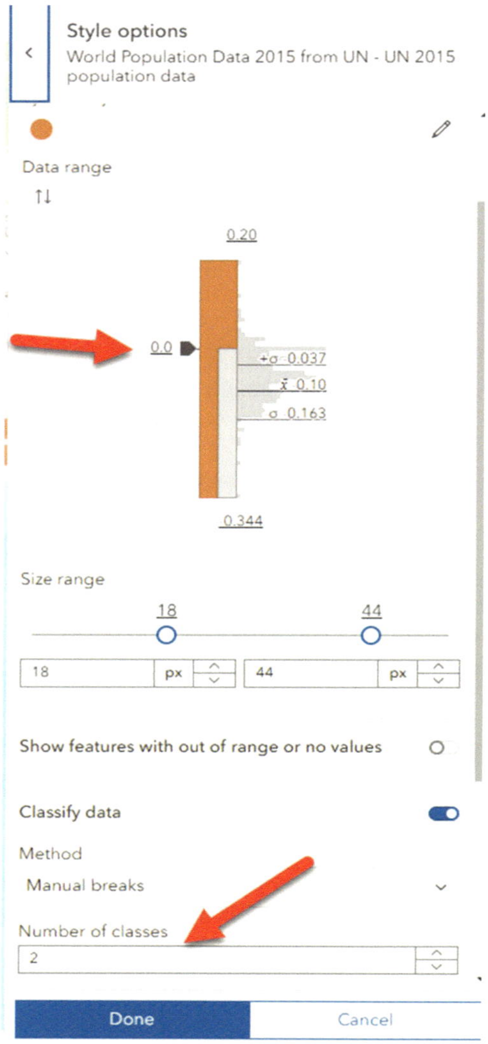

FIGURE 3.14 Working with the histogram and classification methods in the mapping process. Mapped using ArcGIS from Esri.

communities. Like all data, population data should be viewed critically: It is useful, but not perfect. Consider the population data that you have been examining. Population data is collected in specific ways around the world by national statistics agencies, in some countries on a regular basis (such as in the USA and UK, every decade) or in some countries occasionally, or in some countries not at all. In some countries, sampling and estimates are conducted, and in others, a complete census of the entire population is attempted. Furthermore, going back in time, in this case to 1950, the figures might not be as accurate as in more modern times; on the other hand, one might argue that challenges in our modern world make today's population enumeration *more* difficult than in the past. Therefore, use population data thoughtfully and carefully, always checking the metadata to find out how the data were gathered or estimated, how often it is updated, the enumeration area used, and other characteristics. Be able to justify the choices you made in terms of the data you used and the tools you used. How confident are you that the analytical tools employed are suitable to the data that you are analyzing? The next section treats some of these topics.

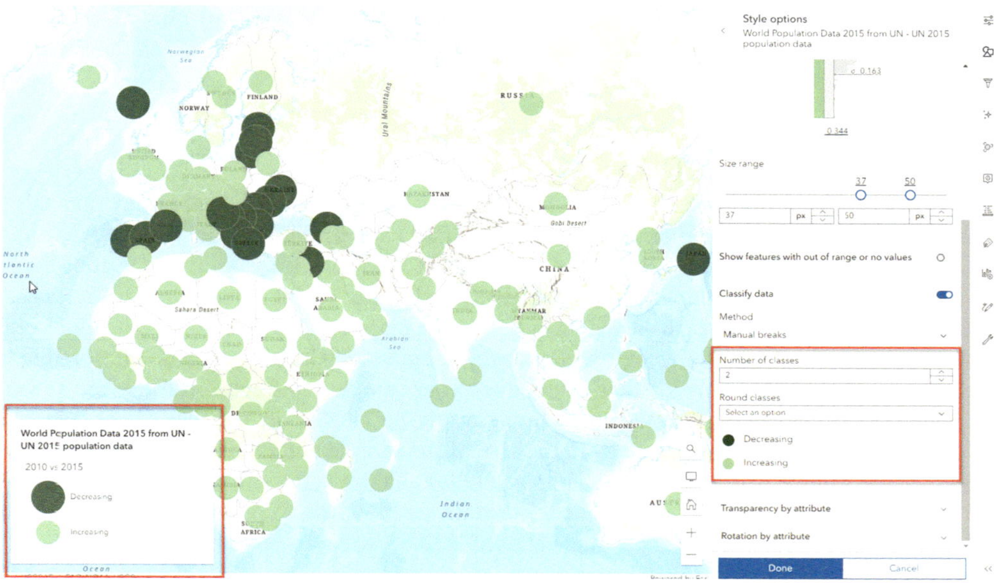

FIGURE 3.15 Histogram-generated two-class map showing decreasing and increasing population. Mapped using ArcGIS from Esri.

GOODNESS OF FIT AND STATISTICAL SIGNIFICANCE

When using models to simulate some aspects of reality, it makes sense to consider how well they fit reality. Calculations of this sort are difficult to perform; in a way, one is testing for such validity from within the system and so it is hard to know with confidence that the tests are valid.

Coefficient of Determination

It is important to know how well a regression line fits a scatter of points if we want to have any confidence that predictions made will have meaning. The correlation coefficient, r (or R), gives a good idea of goodness of fit for simple linear regression when one input value yields one output value. However, with multiple linear regression, when one input value yields multiple (more than one) output values, the meaning becomes muddled in mixing negatives with positives. To remove that difficulty, simply square the correlation coefficient so that the measure is always positive. The resultant measure, denoted as r^2 (or R^2), is called the coefficient of determination. In interpreting the goodness of fit for simple linear regression, note that an r value of 0.80 means that 80% of the variation is explained; however, the corresponding r^2-value is 0.64. In either case, the higher the value, the more variation is explained because the plotted line fits the data points better than it does for lower values. The better the fit of the line to the data, the better the regression analysis serves as a model and the greater is our confidence in such.

z-Score

A spin-off from the concept of mean is the so-called z-score that permits standardization of units that find use in mapping and elsewhere. The z-score is calculated as the difference between x, an observed value, and the mean of the sample with that difference then being divided by the standard deviation of the sample:

$$z = (x - \mu)/\sigma.$$

A z-score of +2 means that the observation lies two standard deviations above the mean, a z-score of −2 means that the observation lies two standard deviations below the mean, and a z-score of 0 means the observed value equals the mean.

Statistical Significance

When you use data, from the field or elsewhere, you will want to know whether any analysis you perform is "significant" in relation to some underlying factor or whether the results are a matter of some sort of luck or coincidence. You do not collect data for every point under consideration but instead collect it for a sample of the available possibilities: think about the San Diego activity above—we did not collect all available litter, just a reasonable sample of the available litter. The word "reasonable" is an interesting one. In any sample that does not include all possible data points, there is sampling error: to what extent does your sample represent faithfully the entire set from which it was selected. A reasonable sample might be viewed as one that has an acceptable level of sampling error which derives from the size of the sample and the variation in the entire set from which the sample is drawn. Larger samples are less likely than small ones to produce random results; variation is of particular importance. In a data set in which all points are close to the same value there is little variation and any sample chosen is likely to represent the entire set. When there is a lot of variation within the whole set, it becomes much more likely that a selected sample might not be a representation of the entire set—the greater the variation in the underlying set, the larger the sampling error. Think about what you are doing. Getting a good sample to work with will lead you to have more confidence in your results.

Generally, the process for considering statistical significance centers on two concepts:

- Stating a null hypothesis—something you try to disprove; and stating an alternative hypothesis.
- Setting a target significance level—how rare results are is the null hypothesis is true. This value is frequently expressed as a "p-value"—the lower the value, the less likely the results are due to chance. If the p-value is less than the target value, reject the null hypothesis and work with the alternative hypothesis.

Statistical significance is important for considering sampling error; however, data sets may become contaminated with other forms of error that have nothing to do with sampling. For example, some data may degrade over time, some data may get lost, some data may be derived from participants who lied, and so forth. Run tests of various kinds, but keep your eyes open, and your mind alert to the set of information you are dealing with and try to ensure that it is as free as possible from error of all sorts so that your subsequent analysis of it does not lead to immediate unintended interpretation or to long range unintended consequences.

Then, after you have run your analysis, you should wish to know how much confidence, quantitatively, you might have in the results. Your software might return a value of a 95% confidence interval for the true proportion of results. What it means is that if the same analysis were repeated many times, 95% of the constructed confidence intervals would contain the true proportion of the results. Generally, the higher the value, the greater your confidence should be.

p-Value

In a mapping context, the null hypothesis is Complete Spatial Randomness. The p-value is a measure that determines the probability that the observed spatial pattern was created by a random process. When that probability is lower, the observed pattern becomes more significant (not random)—the clustering you think you see is actually there. Software may calculate various tests of significance for you; nonetheless, it is a good idea to have a general idea of how it calculated results so you can tell if the answers are reasonable.

Associated Interactive Mapping Activities

Examining Relationships between Variables: Charts and Coefficients of Determination

In this activity, you can advance your mathematics knowledge by creating and interpreting charts and plotting coefficients of determination. Your goal is to understand the relationships between two different variables.

Open the linked interactive web map (https://www.arcgis.com/apps/mapviewer/index.html?webmap=4168c7a43f6d423dad872796af9f9c9e) in ArcGIS Online. The map opens to Colorado USA with 2 layers visible: Colorado 14ers (peaks over 14,000 feet in elevation), and ACS (American Community Survey) county level housing and income data from the US Census Bureau.

To the left of the map, use Layers>turn off the ACS layer>so that only the mountain peaks are visible. Open the table of data behind this layer to determine that 58 records exist, corresponding to the number of peaks meeting this elevation. Also, notice that a field exists named Elevation, and another field exists named Difficulty. The Difficulty values are ranked from 1 (most difficult) to 58 (least difficult). The criteria used to generate the Difficulty variable include (1) climbing duration and challenges inherent in the trek, (2) terrain stability, (3) cliff exposure, (4) presence of a trail, (5) elevation gain, and (6) total roundtrip hike distance. Thus, the Difficulty variable is really an index that incorporates six different criteria.

Is there a relationship between elevation and difficulty? To find out, create a chart and create a coefficient of determination fit on that chart. To the right of the map>Configure Chart>Add Chart>Scatter Plot>Data>X-axis: Elevation. Y-axis: Difficulty>Show Linear Trend (the coefficient of determination)>observe your chart (as shown in Figure 3.16). Hover your touchpad or mouse over some of the points on the scatter plot to observe their values (elevation and difficulty).

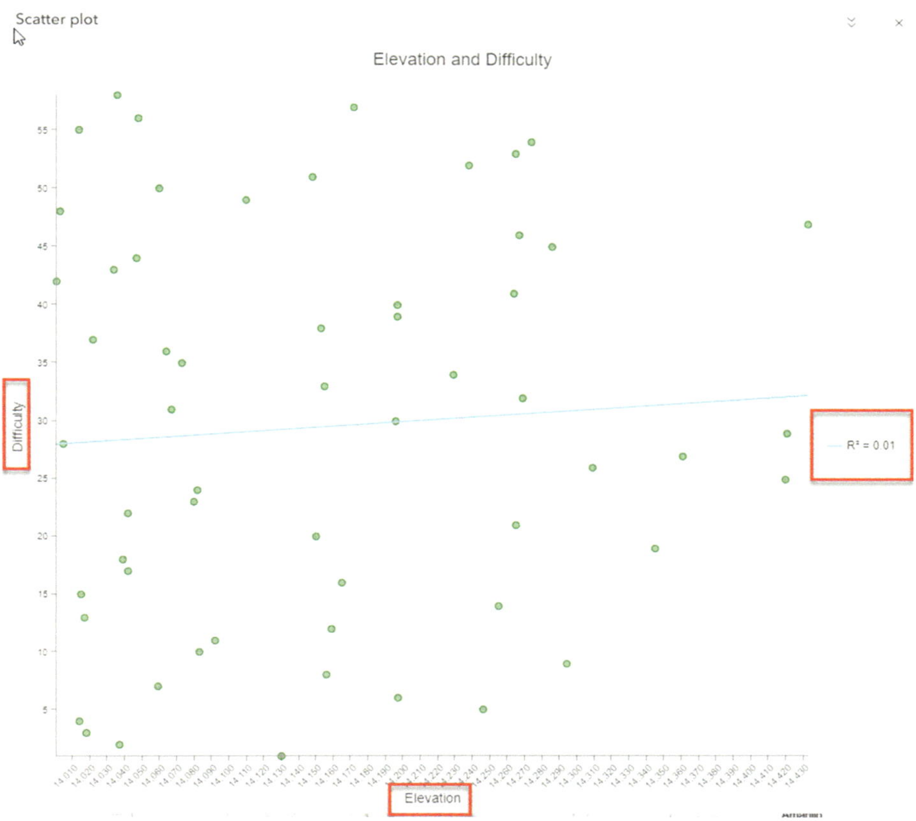

FIGURE 3.16 Scatter plot suggesting possible relationship (or lack of it) between difficulty and elevation.

Based on what you learned about the r^2 value, interpret the chart: is there a relationship between difficulty and elevation? How strong or weak is it? Why do you suppose this is the case? Why are the highest peaks not necessarily the most difficult to climb? Consider the discussion in this book about the range of data: If the range of peaks was widened to include *all* the world peaks between 10,000 and 29,000 feet (3,048–8,839 m), do you think you would see a stronger relationship between elevation and difficulty than for the Colorado peaks? Why or why not?

Next, in the same way, examine two different variables—this time, median income and median home value. Left panel > layers > Turn off the peaks layer. Turn on the layer ACS Household Income Distribution Variables > Expand > County > Configure Charts > Add Chart > Scatter Plot > X-axis: Median Household income in past 12 months. Y-axis: Median Home Value for owner-occupied units > Show linear trend (shown in a screenshot in Figure 3.17).

Based on what you learned about the r^2 value, interpret the chart: Is there a relationship between median home value and median household income? How strong or weak is it? Why do you suppose this is the case? Do you think the same relationship would exist for other states?

Note the relationship between tall mountains in Colorado atop the shaded relief base map that is in use. Traditionally, it was more expensive to build homes in the mountains and to be able to afford to live there. Use your mouse or touchpad to click on a point in the scatterplot where the home value is high, and the household income is high. Use the shift key to select additional points (or use shift-and-band an area on the scatterplot to do so). Use the "lower" symbol to minimize the

FIGURE 3.17 Scatter plot suggesting possible relation between median home value and median household income.

area occupied by the table so you can see the resulting map. Where in Colorado are these higher income and higher home value areas? Do they correspond with the most mountainous areas of the state? Conversely, do the same for the lower income and lower home value areas in the state, noting any geographic region(s) in the state in which these are concentrated. Do these areas correspond to the Great Plains region of Colorado to the east and the canyons and high plateaus of the northwest part of the state? Using the mapping techniques you have already learned, make a map showing two variables—difficulty and elevation. Then, make a map showing the two other variables—median home value and median household income. Where in the state are *both* the home value and household income the highest, and why? Reflect on how a single mapping activity can draw together so many different mathematical concepts. The more you learn and understand, the deeper you can go into understanding spatial and temporal relationships.

Using Tessellations and Spatial Statistics to Examine Patterns in Numbers on Maps

In Chapter 2, we examined some ways to group data. Grouping, or aggregating data, organizes data sets to prevent clutter of graphic symbols, and in the case of the mapping of data sets, to prevent clouding the desired transmission of information. Here, we offer an activity based on tessellation, a form of laying geometric tiles across a surface to cover it with no overlap, as an aggregating technique.

Suppose you are a crime analyst, tasked with reducing crime in Lincoln, Nebraska USA. You begin your analysis by opening the linked web map (https://www.arcgis.com/apps/mapviewer/index.html?webmap=cbe7a6bba2d24e06931ad93d8c6f123a). The map shows crime incidents in Lincoln, Nebraska, over a period of several months, the Lincoln city limit, the police districts, and the police stations (Figure 3.18 shows the mapped data). View the legend to see that the crimes have been categorized by the type of offense—larceny, assault, and auto thefts (as shown).

Left panel>Layers>Open the table for the crime data: You will see that 1,025 crimes exist in this data set, making it difficult to determine patterns and whether crimes are significantly higher or lower in certain parts of the city. You can map the crimes by day of the week using the Day_From field and determine how many crimes occurred within each police district. But even these methods may not help you determine which areas are significantly high or low in crime.

Therefore, you turn to tessellations, a repeated pattern covering a region, as another way to visualize data. Use Left Panel>Make visible the layer named Hot Spots Crime as Tessellations.

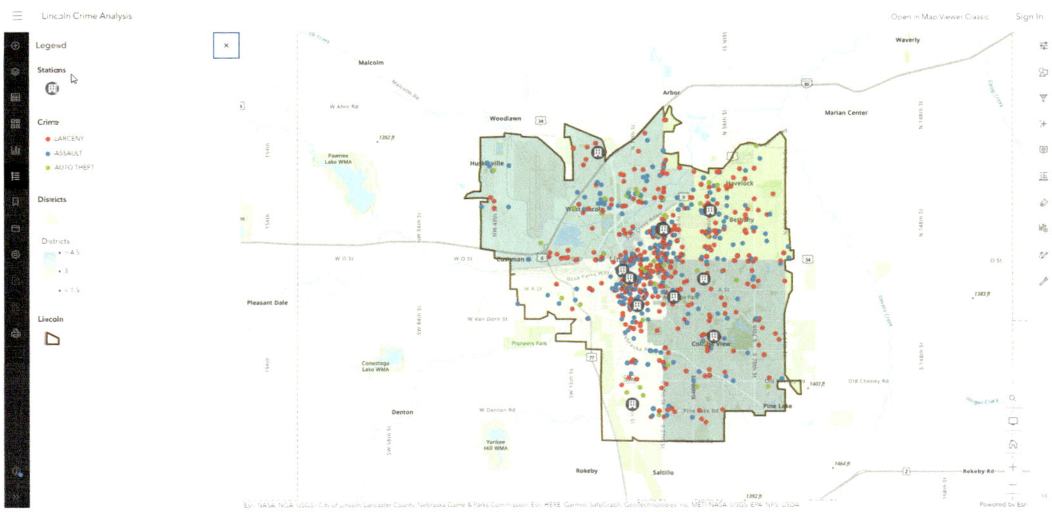

FIGURE 3.18 Crime incidents in Lincoln, Nebraska, over a period of several months. Mapped using ArcGIS from Esri.

To represent data in this layer, a tessellation was created, repeating a shape over and over until it covered the city limits. In this case, squares were used as the tessellation shape—much like a large sheet of graph paper overlaid onto the city of Lincoln.

Your organization, like all others, faces constraints: Budget, time, staff. Therefore, for the tessellations, you ran a specific set of spatial statistics tools (statistical tools that incorporate some elements of spatial observation) so that you could focus on areas in Lincoln with significantly higher crime. Specifically, you ran a hot spot analysis on the crime data, using a specific tool. This tool calculates the Getis-Ord Gi* statistic for each crime point in the dataset. The resultant z-scores and p-values indicate where features with either high or low values cluster spatially.

The Getis-Ord Gi* tool works by looking at each feature within the context of neighboring features. A feature with a high value (in this case, of crime) may be interesting (such as the numerous crimes on the east side of Lincoln at the shopping mall) but may not be a statistically significant hot spot. To be a statistically significant hot spot, a feature will have a high value and be surrounded by other features with high values as well. The local sum for a feature and its neighbors is compared proportionally to the sum of all features; when the local sum is very different from the expected local sum, and when that difference is too large to be the result of random chance, a statistically significant z-score results. Left panel>Access the legend again to discover where significant hot spots and cold spots in Lincoln occur. Clicking on any of the square tessellations reveals the hot spot with confidence levels (as shown in Figure 3.19).

Consider: What insight do you gain from the areas of significant crime activity as visualized through the tessellations over and above the crime mapped as point data? What areas of Lincoln will you focus on to reduce crime? List the specific street names as you list each of the areas. Again, the power of mapping is that you are mapping real, not imaginary, places, and grappling with very real and serious issues. What strategies will you use to reduce crime (a community awareness campaign, neighborhood watch groups, increased patrols, or something else?).

Tessellations do not have to be squares; they can be hexagons or other shapes. They can also be extruded into 3D space to aid decision-makers in seeing patterns. For example, traffic accidents of a certain county in Florida were extruded to help decision makers focus on consecutive hot spots. Investigate this idea on a linked interactive web map (https://www.arcgis.com/apps/Cascade/index.html?appid=9a27635635c940539b96fb5ef954e4d5) (screenshot in Figure 3.20).

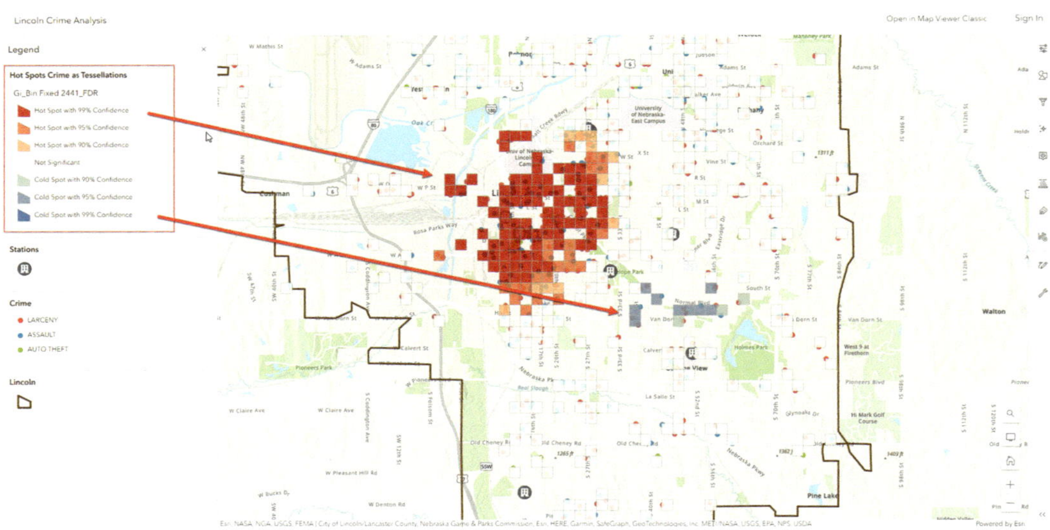

FIGURE 3.19 Clicking on any of the square tessellations reveals the hot spot with confidence levels. Mapped using ArcGIS from Esri.

FIGURE 3.20 Map based on a hexagonal tessellation with extruded values related to Florida traffic accident hot spots. Mapped using ArcGIS from Esri.

Finally, returning to the content of this chapter, consider how much this last activity on tessellations has drawn together: proportions, *z*-scores, *p*-values, confidence levels, and all the concepts that underlie each of them.

Executing an analysis is important. Reflecting on it to look at its conceptual underpinnings is even more important: for then, and only then, one has a clear thought process as a guide to different, but related (sometimes only loosely), applications that can be justified based on underlying mathematical principles.

INTEGRATING ELEMENTS OF THIS CHAPTER

To follow on the thought at the end of the last section, we offer some interesting uses of percentage and number pattern from the world of art and sculpture. They offer remarkable instances of tying the infinite to the finite and the abstract to the real.

Fibonacci Numbers and the Golden Ratio

The Fibonacci sequence is an infinite sequence of natural numbers that has inspired numerous ways to observe the real world, for centuries in fields of art, science, and more. In it, each number is equal to the sum of the preceding two: 1, 1, 2, 3, 5, 8, 13, 21, 34,…

The spiral patterns of leaves on branches of deciduous trees follow a Fibonacci pattern with both the size and position of the leaf on a twig arranged in a pattern to provide optimum sunlight access. Numerous artists have seen beauty in the Fibonacci sequence: one 20th/21st Century artist who did was David Barr. His structurist sculpture in the Sears Court of Fairlane Town Center (in Dearborn, Michigan), is an abstract imitation of the Fibonacci arrangement of leaves on a tree branch https://www.flickr.com/photos/thecaldorrainbow/41690210575.

Notice the thin blue slab near the middle of the stack of slabs of various colors. It is one unit thick. The light red ones on either side of it are two units thick. The next slabs on either side of the blue on are three units thick; then the next ones are five units thick, and then eight. Finally, the mounting column, on the bottom only (as is the trunk of a tree), is 13 units thick. The spiral pattern of the slabs, as well as their thickness, is also Fibonacci in orientation.

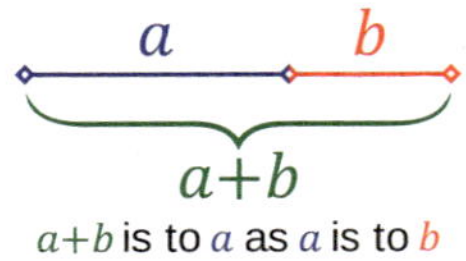

FIGURE 3.21 Line segments in the Golden ratio. Source: Public domain, Wikimedia Commons.

One way to grasp pattern that is infinite in nature is to look for what is fixed, or at least relatively fixed, within it: as a handle to hang onto. As one might imagine a sculpture spiraling out of control, think instead of what is constant about it: in this case, based on the concept of ratio. Look at the ratios of successive terms in the Fibonacci sequence: 2/1 = 2; 3/2 = 1.5; 5/3 = 1.666…; 8/5 = 1.6; 13/8 = 1.625; … 987/610 = 1.6180327…; … (the fraction 987/610 deals with the 15th and 16th Fibonacci numbers). The sequence of ratios is closing down on the "Golden ratio," φ. Two quantities a and b, with $a > b > 0$, are said to be in the **golden ratio** φ if $(a+b)/a = a/b = \varphi$; or $(a+b)$ is to a as a is to b. Figure 3.21 shows line segments in the Golden ratio.

Thus, the ratio of any two adjacent Fibonacci thicknesses in the mall column sculpture approximates the Golden ratio; that ratio thus gives an element of constant stability to an otherwise swirling suggestion of infinite appearance! Look around you—where do you see perhaps somewhat surprising evidence of the use of ratios in the real world?

Percentages, Estimation, and Rounding in the World of Sculpture

Where is the longest land border that is an international boundary between two countries? It is the land border between the United States of America and Canada. That simple, but perhaps not well-known fact, inspired Barr to create an abstract expression of that primarily east-west boundary to commemorate, simultaneously, the trajectory of the sun in the sky, via arch sculptural elements,

FIGURE 3.22 SunSweep lithograph. Source: Arlinghaus, S. and W.

and the friendly, integrative relations that have taken place across that border. Barr chose three locations along the border where one needed to cross the border to enter them from the "other side." One location was in the Pacific Northwest, at Point Roberts in the USA State of Washington, but accessible from land only from Canada; another at the Northwest Angle in Minnesota, and a third at Campobello Island in the State of Maine. The linked video (https://www.youtube.com/watch?v=CtUnTb7-5Go) in the references section at the end of this book explains more about the actual siting of the project, called SunSweep, directly from Barr and other primary sources. Figure 3.22 offers the artists abstract rendition of the project.

Now, let's think about similar projects. If Barr were to have considered commemorating other long borders, it might have been helpful to him to think about them in relation to the USA/Canada project he had completed—in terms of allocating funding resources, degree of difficulty of implementation, number of workers needed, time to complete the task, and so forth. Knowing the percentage that the border under consideration was of the USA/Canada border would give him some insight.

The USA/Canada border is 8,890 km (Source: Statistique Canada). The boundary between Russia and Kazakhstan is 6,848 km. The boundary between Argentina and Chile is 5,308 km. The Russian/Kazakh border is 6,848/8,890 * 100 = 77.0303712% as long as the USA/Canada border. The Argentine/Chilean border is 5,308/8,890 * 100 = 59.7075366% as long as the USA/Canada border. Do these numbers make sense? Perhaps I made a keystroke error on my calculator: 6,848/8,890 looks a bit like 6/8 which is 75%; 5,308/8,890 is clearly more than 50% but not near 75%. So, yes, both those calculated values seem reasonable. However, the output from my calculator gives me answers to seven decimal places; certainly, not necessary when, as we saw in the video, the whole action of site location for sculptural elements is nowhere near that exact. So, round off the values to 75% and 60%—values that might be helpful in assessing degree of difficulty of sculptural placement that comes from sheer border length! Again, think about the world around you and link your mathematical knowledge with it; even a little bit of thought linkage can go a long way!

4 Earth Measurement
Coordinate Systems and Trigonometry

The line of images above is a visual abstract of this chapter designed to foster spatial thinking. From the chapter numeral, to the book structure, to the real world, the reader is offered gentle guidance to develop spatial intuition about what might be coming. Those thoughts are then reinforced with a detailed text outline of chapter content below. The images and outline forge as an abstract of chapter content.

CHAPTER OUTLINE

Coordinate Systems in Common Use
- Cartesian Coordinates
- Earth Model: Sphere and Plane
- Earth Model Coordinates: Parallels and Meridians
- Earth Model Coordinates: Latitude and Longitude
    - The Graticule
    - Bounds of Measurement
    - Positional Measurement Conversion
- Geocoding
- Ordinal and Cardinal Directions
    - Direction
    - Bearing
    - Azimuth

Other Coordinate Systems
- Universal Transverse Mercator
- Polar Coordinates

Trigonometry: Visual Review of Functions and Applications
- Visual Review of Trigonometric Functions
    - The Basic Setup
    - Six Trigonometric Functions
    - Three Functions of θ and Three Functions of Co-θ
    - All Six Functions in a Single Animation

DOI: 10.1201/9781003305613-4

Applications: Conceptual and Real Worlds
Trigonometric Identities
Length of One Degree on the Earth Model
Measure the Height of the Arc de Triomphe in Paris, France
Measuring the Circumference of the Earth
- Eratosthenes Measurement
- Using a GPS
- Using a Smartphone App
- Using a Web Mapping Tool
- How Long Would It Take to Walk Around the Earth?
- Determining the Earth's Circumference Based on Measuring 1 Second of Longitude
- Determining the Distance Around the Earth at Your Present Latitude

Scale
Map Scale
Activity: Studying Map Scale with a USGS Topographic Map
Global and Local Scale
Activity: Comparing Sizes of Countries
Map Projection
Tissot Indicatrix
Scale and Measurement on Mars
Questions to Think About

COORDINATE SYSTEMS IN COMMON USE

In Chapter 2, we introduced simple views of coordinate systems that would enable the reader to participate in interactive mapping activities designed to emphasize some issue other than the coordinate systems themselves. Here we dive in and take a deeper look at those, and other coordinate systems, useful in mapping. There are many different coordinate systems in use in various disciplines and, abstractly, an infinite number can be created.

Teaching coordinate systems lends real-world relevance to mathematics instruction and fosters spatial thinking because coordinate systems are fundamental to all mapping and to much, but not all, mathematics. Coordinate systems anchor places and spaces to real-world positions and underpin everyday use of mapping tools enabled by modern GPS, including package delivery, rideshare apps, fitness apps, and more.

Cartesian Coordinates

A coordinate system in the plane specifies uniquely the location of a point by a pair of numerical coordinates. These coordinates are the signed distances to the point from two fixed oriented perpendicular reference lines of identical units of length (refer to Figure 2.3). Each reference line is called a coordinate axis of the system. The point where the perpendicular axes meet is referred to as the origin of the system, labelled with coordinates (0, 0). Convention often labels the horizontal axis as the x-axis and the vertical axis as the y-axis. Thus, an arbitrary point on the x-axis has coordinates $(x,0)$; one on the y-axis has coordinates $(0,y)$, and a general one in the plane is (x,y). In the 17th century, the French mathematician and philosopher, René Descartes developed this system in the plane; hence, it is often referred to as the Cartesian Coordinate system, in his honor. Much as the importance of the distributive law lay in providing a systematic linkage between additive and multiplicative operations, the importance of the Cartesian Coordinate system in the plane lay in providing a linkage between Euclidean plane geometry and algebraic concepts.

Earth Model: Sphere and Plane

One natural extension of Descartes' approach is to employ a similar strategy to specify the position of any point in spaces of higher dimensions. Consider the Earth to be modeled as a sphere. The Earth is not actually a sphere, but a sphere is a good approximation to its shape and the sphere is easy to work with using classical mathematics of Euclid and others. There are only a few logical possibilities about the relationships between the plane and the sphere (Figure 4.1). In this Figure, and in Figures 4.2 and 4.3, try to visualize these relationships in your mind, in relation to a globe. That is, learn to stretch your thinking as you read.

Great circles are the lines along which distance is measured on a sphere: they are the geodesics, or shortest distances, on the sphere. Figure 4.2 summarizes information about geodesics.

Diametrically opposed (at opposite ends of a diameter) points are called antipodal points: anti+pedes, opposite+feet, as in drilling through the center of the Earth to come out on the other side.

Earth Model Coordinates: Parallels and Meridians

To reference measurement on the Earth model in a systematic manner, introduce a coordinate system as a set of reference lines on the surface (Figure 4.3).

In the case of the Prime Meridian, it was historical consideration (International Meridian Conference of 1884), rather than mathematical consideration, that produced the uniqueness in selection (Howse, 1980). The unique line is the Prime Meridian; all others, as halves of great circles, are simply called meridians.

This reference system for the Earth is not unique; an infinite number is possible by simply varying the spacing used within sets of parallels and meridians. There is abstract similarity between this mathematical arrangement and the mathematical pattern of Cartesian coordinates in the plane.

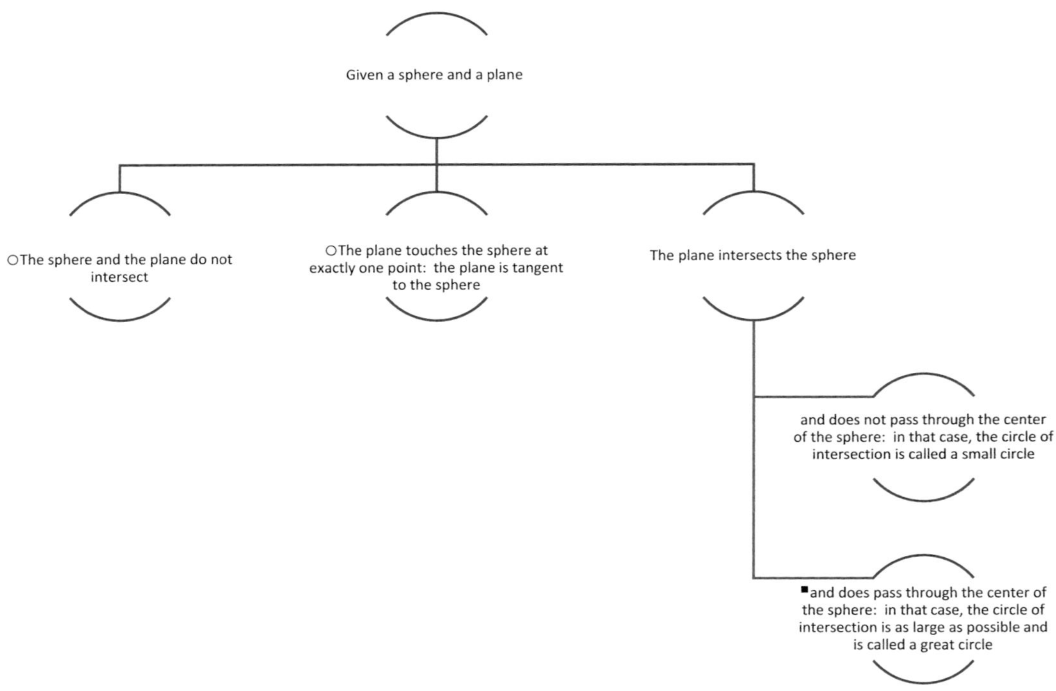

FIGURE 4.1 Sphere and plane; hierarchy of relationships.

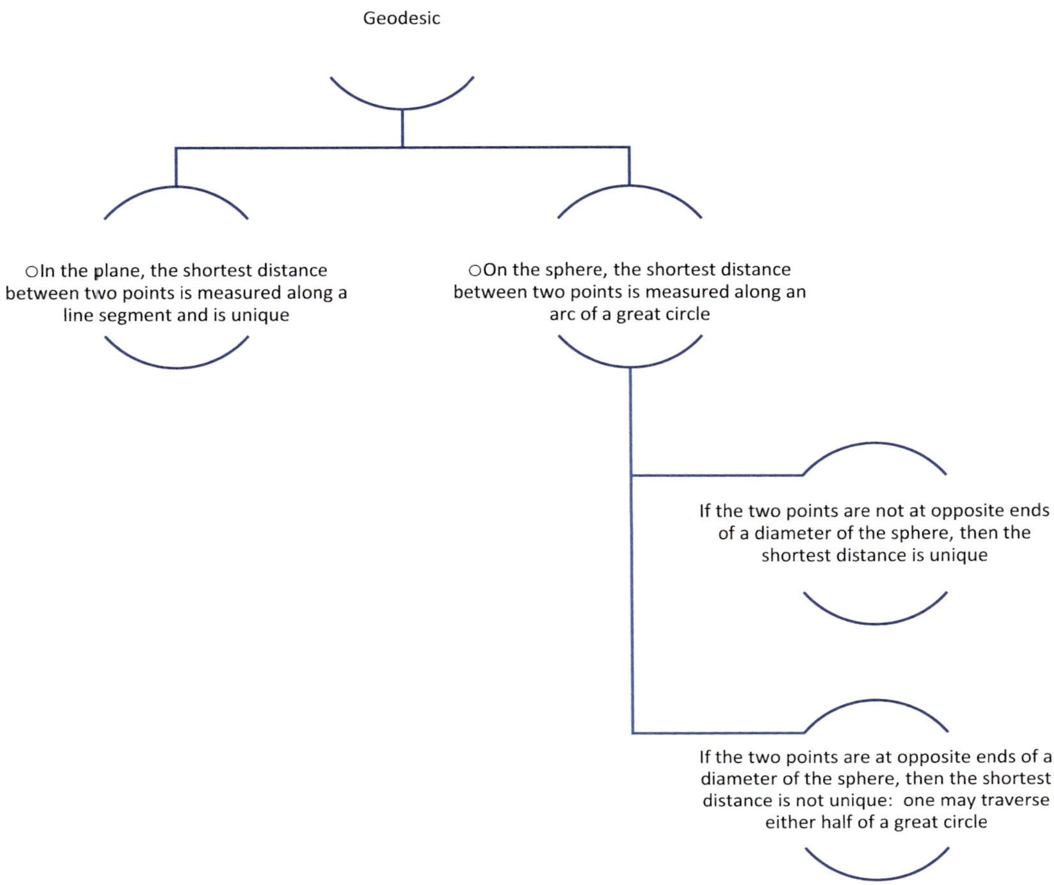

FIGURE 4.2 Great circle (geodesic on a sphere) hierarchy of relationships.

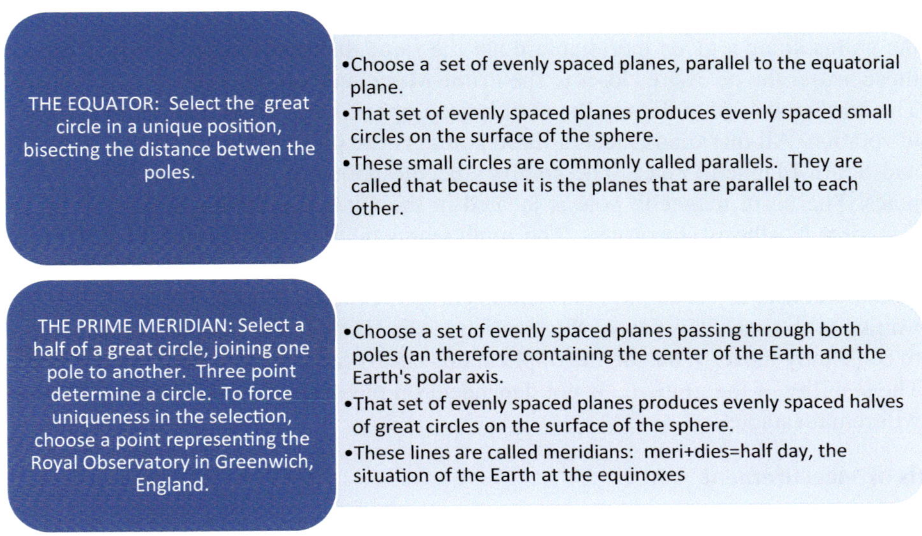

FIGURE 4.3 Axes for an Earth coordinate system.

To use this arrangement, one might describe the location of a point, P, on the Earth model as being at the 3rd parallel north of the Equator and at the 4th meridian to the west of the Prime Meridian. While this might serve to locate P according to one reference system, someone else might employ a reference system with a finer mesh (halving the distances between successive lines) and for that person, a correct description of the location of P would be at the 6th parallel north of the Equator and at the 8th meridian to the west of the Prime Meridian. Indeed, an infinite number of locally correct designations might be given for a single point: an unsatisfactory situation in terms of being able to replicate results. The problem lies in the use of a relative, rather than an absolute, locational system.

Earth Model Coordinates: Latitude and Longitude

To convert a relative parallel and meridian system to an absolute system, that is replicable, employ some commonly agreed upon measurement strategy to standardize measurement. One such method is the assumption that there are 360° of angular measure in a circle (as suggested in Figure 2.4).

- Thus, P might be described as lying 42° north of the equator, and 71° west of the Prime Meridian. The degrees north are measured along a meridian; the degrees west are measured along the Equator or along a parallel (the one at 42 north is another natural choice). The north/south angular measure is called Latitude; the east/west angular measure is called Longitude.
- The use of standard circular measure creates a designation that is unique for P; at least unique to all whose mathematics rests on having 360° in the circle.
- Parts of degrees may be noted as minutes and seconds, or as decimal degrees.
- A degree (°) of latitude or longitude can be subdivided into 60 parts called minutes (‘). Each minute can be further subdivided into 60 seconds (“).

The Graticule

What might be called a Cartesian grid in the plane is called a graticule on the sphere.

- All points along a single parallel have the same latitude; they are the same distance above or below, north of or south of, the Equator.
- All points along a single meridian have the same longitude; they are given in degrees east or west of the Prime Meridian. Because lines of longitude converge at the poles, though, the points along a given meridian are not the same distance from the Prime Meridian—those nearer the poles are closer to the Prime Meridian than those nearer the Equator.
- The north and south poles are the earth's geographic poles, located at each end of its axis of rotation. All meridians meet at these poles. These poles are not the same as the North and South Magnetic Poles. The compass needle points to either of these two magnetic poles. The north magnetic pole is located in the Queen Elizabeth Islands group, in the Canadian Northwest Territories. The south magnetic pole lies near the edge of the continent of Antarctica, off the Adélie Coast. Since the forces that generate the Earth's magnetic field are constantly changing, the field itself is in continual flux, its strength waxing and waning with time. This causes the location of the Earth's magnetic north and south poles to constantly move. What are the implications of this fact for the stability of our graticule? The stability of the graticule is not dependent on the magnetic pole: Hence, the graticule will remain stable!

Bounds of Measurement

- Latitude runs from 0° at the Equator to 90°N or 90°S at the poles because the numbers, from 0 to 90, are angles, from the Equator toward the poles.

- Longitude runs from 0° at the prime meridian to 180° east or west, halfway around the globe. Again, the numbers are angles, but this time, looking down on the Earth from a position above either the North Pole or the South Pole. The International Date Line roughly follows the 180° meridian, making a few jogs to avoid cutting through land areas (and thus keeping the residents there happily on the same day).

Remember all of this material when we come to investigate the seasons of the year and what causes them. Because mathematics is linear in its development, it is critical to both master and remember it; earlier materials keep coming up later!

Positional Measurement Conversion

In Chapter 2, we illustrated how to convert circular measure given in fractions to decimal measure. The process is the same when converting degrees of latitude and longitude to decimal degrees, as suggested in Chapter 3. It should be evident that 42 degrees 30 minutes is the same as 42.50° because 30/60 = 50/100. Current computerized mapping software often employs decimal degrees as a default; older printed maps may employ degrees, minutes, and seconds. Thus, the human mapper needs to take care to analyze the situation and make appropriate conversions prior to making measurements of position. Such conversion is simple to execute using a calculator. For example, 42 degrees 21 minutes 30 seconds converts to 42 + 21/60 + 30/3,600° = 42.358333°; powers of ten replace powers of 60.

Spacing between successive parallels or meridians might be at any level of detail; however, when circular measure describes the position of these lines, that description is unique up to agreement to use 360° in a circle. One spacing for the set of meridians that is convenient on maps of the world is to choose spacing of 15° between successive meridians. The reason for this is because the meridians converge at the ends of the polar axis, each meridian then represents the passage of 1 hour of time, given that we agreed to partition a day into 24 hours: 24 times 15 is 360. Meridians may also mark the passage of time! Think about the amount of each day in which the sun will shine, and other implications of time and position—if you know the longitude of two points you then know the east-west time vector from one location to the other. Sailors have long been aware of these connections, and have used contemporary technology, from sextant to satellite, to determine longitude while at sea. In the next activity, you have a chance to learn to use technology of today to locate position on the Earth.

Geocoding

Consider the following activity to see how coordinate systems are of critical importance in making positional information available in a variety of contexts, particularly in digital format, as a basis for software dealing with geographical information. All maps and images in digital mapping displays are based on real-world places and are underpinned by real-world coordinates. These coordinates are based on numbers and the accurate measurement of the Earth; measuring the Earth accurately has been a quest for thousands of years. Modern geographical coordinate systems are quite complex. The precise standard meanings of latitude, longitude, and altitude are currently defined by the World Geodetic System (WGS) (https://en.wikipedia.org/wiki/World_Geodetic_System), and take into account the flattening of the Earth at the poles (about 68 km or 42 miles) and many other details.

Consider the following question: Is the Australian National Capitol closer to the Prime Meridian or to the International Date Line? Is it closer to the Equator or to the South Pole? Answer these same questions for your *own* location, as well.

Access ArcGIS Online on https://www.arcgis.com/apps/mapviewer/index.html, > Search > Parliament Dr, Canberra ACT 2600, Australia (Figure 4.4). This technique of mapping locations is called "geocoding"; it was made possible because of GIS (Geography Information System) software includes a database keyed to a service containing place names, streets, and addresses. Thus, searching for the address for the Australia National Capitol Building and Parliament house in Canberra places a map of this location at your fingertips. You can do the same thing in MapQuest maps, Google maps, and other mapping services as they, too, rely on street addresses for geocoding.

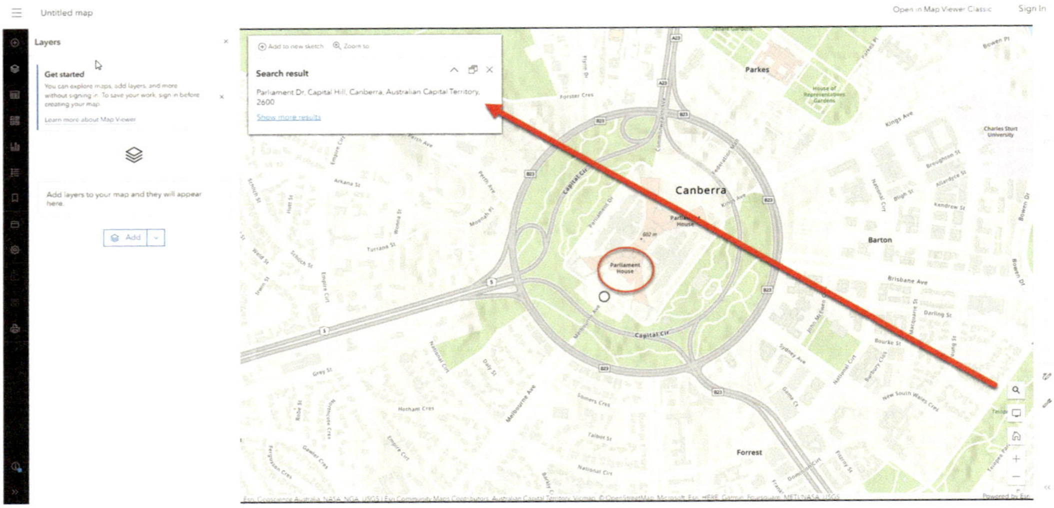

FIGURE 4.4 An ArcGIS Online map, Esri, showing the results of a street address search in Canberra Australia.

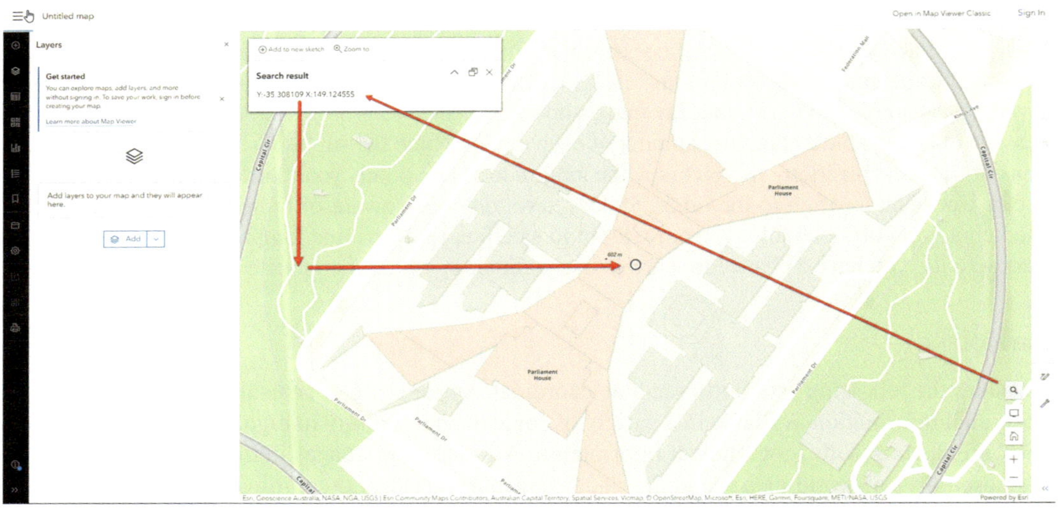

FIGURE 4.5 An ArcGIS Online map, Esri, showing the results of a latitude-longitude search in Canberra Australia.

As useful as street addresses are, they do not exist for the entire planet—not every country uses a "standard" street addressing system. Furthermore, consider places with no addresses—a water well in the middle of a field in Illinois, a dive location on the Great Barrier Reef, or a mountain peak. Even though they have no street address, these things are important. They, too, can be geocoded because they are tied to real-world coordinates that work for *any* location on the planet: latitude and longitude.

Return to your search box in either the 2D map or 3D scene, above, and enter –35.308109, 149.124555 exactly as shown with the comma and the initial negative sign. This location should be the same as the one you geocoded earlier—the Australian National Capitol Building (Figure 4.5). Because the building is in the southern hemisphere, its y-coordinate is negative. Because it is in the eastern hemisphere, its x-coordinate is positive.

Some mapping tools allow you to enter latitude, longitude as is commonly spoken in everyday speech, but in terms of map coordinates, latitude-longitude is y, x and not x, y. Other mapping tools require you to enter the values as x, y, or longitude, latitude. If you obtain a surprising result, check the tool to see if you need to reverse the order of latitude-longitude data entry.

In the next activity, the focus will be refined to illustrate both concepts and practice involving ordinal and cardinal directions.

Ordinal and Cardinal Directions

Direction

Direction is an important concept in mathematics and in mapping. Everyday speech makes use of both relative and absolute locations. You might give directions to your residence as "turn left at the stop sign, go past the public library, turn right at the bottom of the hill, go under the railroad bridge, and after the bridge I am in the second house on the left, next to the big spruce tree." Terms such as left, right, next to, under, after, and so on are given *relative to* other features on the landscape.

Access Google Maps on https://maps.google.com. Search for Ipswich Haven Marina, New Cut W, Ipswich IP3 0EA, United Kingdom. Use Directions > search for Suffolk New College, Rope Walk, Ipswich IP4 1LT, United Kingdom. Obtain the driving directions from the Marina to the College and view it on the map. How would you describe the route to someone using *relative* directions?

But absolute direction is also important in many sectors of society—the direction a ship needs to head or the direction a Lidar camera needs to be pointed to take imagery of the Earth, for example. And for phenomena that affect many people, each of whom are facing different directions, absolute direction is also important: It makes much more sense to say that the high winds will be coming from the northwest tomorrow rather than the more ambiguous, "to your left."

Absolute direction is tied to the four cardinal directions on a typical "compass rose" (Figure 4.6) of North, East, South, and West. The directions east, south, and west are at 90° intervals in the clockwise direction relative to due north. Hence, east is 90° to the right of north, south is 180°, and west is 270°. The ordinal directions (also called the intercardinal directions) are Northeast (NE), 45°, halfway between north and east, Southeast (SE), 135°, halfway between south and east, Southwest (SW), 225°, halfway between south and west, and Northwest (NW), 315°, halfway between north and west.

The intermediate direction of every set of intercardinal and cardinal direction is called a secondary intercardinal direction. These eight shortest points in the compass rose shown are: West-northwest (WNW), North-northwest (NNW), North-northeast (NNE), East-northeast (ENE), East-southeast (ESE), South-southeast (SSE), South-southwest (SSW), and West-southwest (WSW).

Answer the following question: Suppose you are flying a paraglider toward the northeast. You turn at a right angle to catch a desired wind current. Which direction are you now flying toward?

Answer the following question: Suppose you are now flying your paraglider again, this time at an angle of 180° from true north. You turn left 45° to catch a desired wind current. Which direction are you now flying toward?

As before, access Google Maps on https://maps.google.com. Again, search for Ipswich Haven Marina, New Cut W, Ipswich IP3 0EA, United Kingdom. Use Directions > search for Suffolk New College, Rope Walk, Ipswich IP4 1LT, United Kingdom. Obtain the driving directions from the Marina to the College and view it on the map. This time, how would you describe the route to someone using *absolute* directions?

Bearing

A bearing provides a direction given as the primary compass direction (north or south), degree of angle, and an east or west designation. A bearing describes a line as heading north or south and deflected some number of degrees toward the east or west. A bearing, therefore, will always have

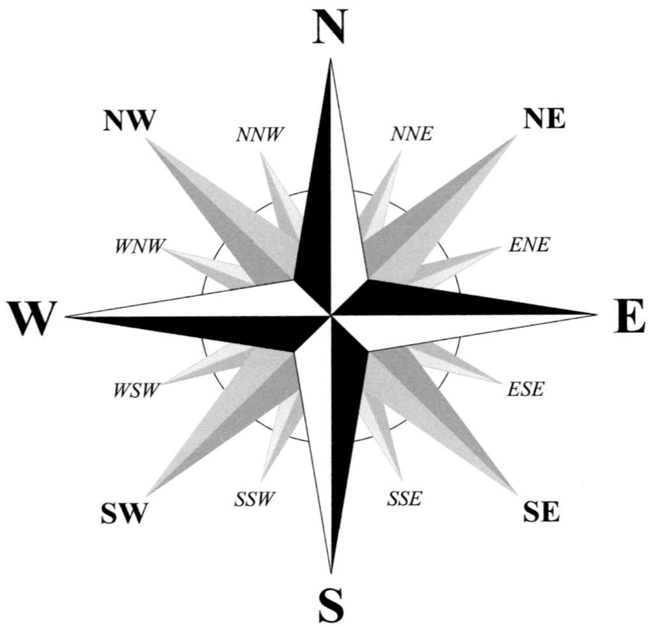

FIGURE 4.6 Compass Rose with Cardinal, intercardinal, and secondary intercardinal directions. Source: Wikimedia Commons. By Brosen~commonswiki—Own work, CC BY 2.5, https://commons.wikimedia.org/w/index.php?curid=667281

an angle less than 90°. Therefore, a person traveling to the southeast has a bearing of 45° south of east. An azimuth is the direction measured in degrees clockwise from north on the compass rose. An azimuth circle consists of 360°. Ninety degrees corresponds to east, 180° is south, 270° is west, and both 360° and 0° mark north. An azimuth of a person traveling southeast is 135°. However, as there is not a universal consensus on the difference between a bearing and an azimuth, the terms are sometimes used interchangeably.

Azimuth

Access the following IGISMAP mapping tool: https://www.igismap.com/map-tool/bearing-angle (Figure 4.7). This tool allows you to create two points and calculate an azimuth between them. Search for and zoom to Lawrence, Kansas USA>move the two placemarkers to Massachusetts Avenue in Lawrence so that they are directly north and south of each other. Verify the azimuth as adhering to the above discussion. Note that the term given here is bearing rather than azimuth, per the above discussion on the lack of consensus. Still, using the tool should confirm the above discussion on cardinal directions. Move the placemarkers to other points in Lawrence or to a route that you take from your residence to work or campus.

Questions for you to consider and answer: What direction is opposite to northeast? What direction is 45° north of WSW (west-southwest)?

Access Google Maps or ArcGIS Online to answer the following questions: What coastal city is WSW of Hyderabad, Pakistan? For a person on a bearing of 10° west from north from Tamale, Ghana, what is the first country outside of Ghana that will be entered? If you sailed on a ship heading on an azimuth of 270° from Kanazawa, Japan, what country will be the first that you will see from the bow of your ship?

Examine the real-time map of ship traffic on https://www.marinetraffic.com. Each of the thousands of ships on this map have a position, an azimuth, and a bearing that may change by the

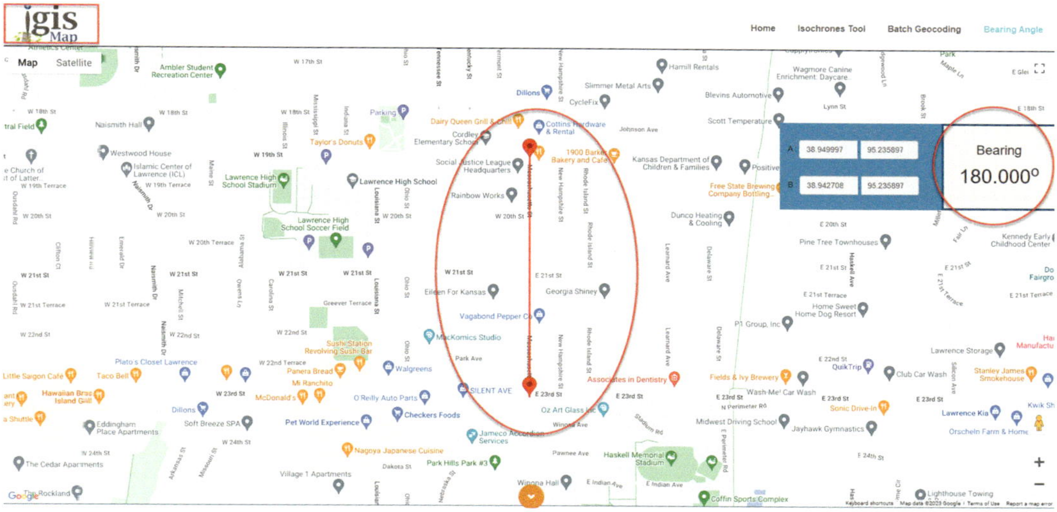

FIGURE 4.7 Obtaining the bearing from placemarkers on a map in Lawrence, Kansas. Mapped using igis.

minute! Select a few and see if you can determine their position, azimuth, and bearing. Return to the map 5 minutes later in a new tab to detect the differences in position of the ships.

OTHER COORDINATE SYSTEMS

Universal Transverse Mercator

There are hundreds of other types of coordinate systems, because there are many needs across a wide variety of sectors of society, from architecture to zoology and everything in between. You have been using the most common coordinate system—latitude and longitude. Other common ones are the state plane coordinate system (in the USA), the UK, Australian, Indian, and other national grids, and the UTM (Universal Transverse Mercator) system. Have you ever found a surveyed disk of copper or aluminum in a sidewalk or in the rocks on a mountain peak? These survey markers helped "anchor" these coordinate systems. They were used extensively over the past 200 years, though they are less needed now given advancements in GPS, GIS, and geodetic models.

No matter what system is used, it is important to use the annotation carefully: Either use the cardinal directions N, E, S, and W, or a – negative sign to designate west longitude and south latitude, but use one of these two types of annotations—omitting these annotations leaves ambiguity.

The UTM coordinate system covers most of the globe, except for the poles, dividing the world into 60 zones that are each 6° wide. These zones are numbered consecutively beginning with Zone 1, between 180° and 174° west longitude, and progressing eastward to Zone 60, between 174° and 180° east longitude. In each zone, coordinates are measured north and east in meters (thus, northings and eastings instead of latitude and longitude). The northing values are measured continuously from 0 at the Equator, in a northerly direction. To avoid negative numbers for locations south of the Equator, the Equator is assigned an arbitrary false northing value of 10,000,000 m. A central meridian through the middle of each 6° zone is assigned an easting value of 500,000 m. Grid values to the west of this central meridian are less than 500,000; to the east, more than 500,000. Since multiple zones could have a coordinate of, say, 3,500,000 m northing, 710,000 m easting, it is important to always specify the zone, and ideally, the sector of the zone indicating how far north or south in the zone the coordinate is located. Thus, a coordinate in UTM may be given as 3,500,000 m northing, 710,000 m easting, UTM Zone 14 T.

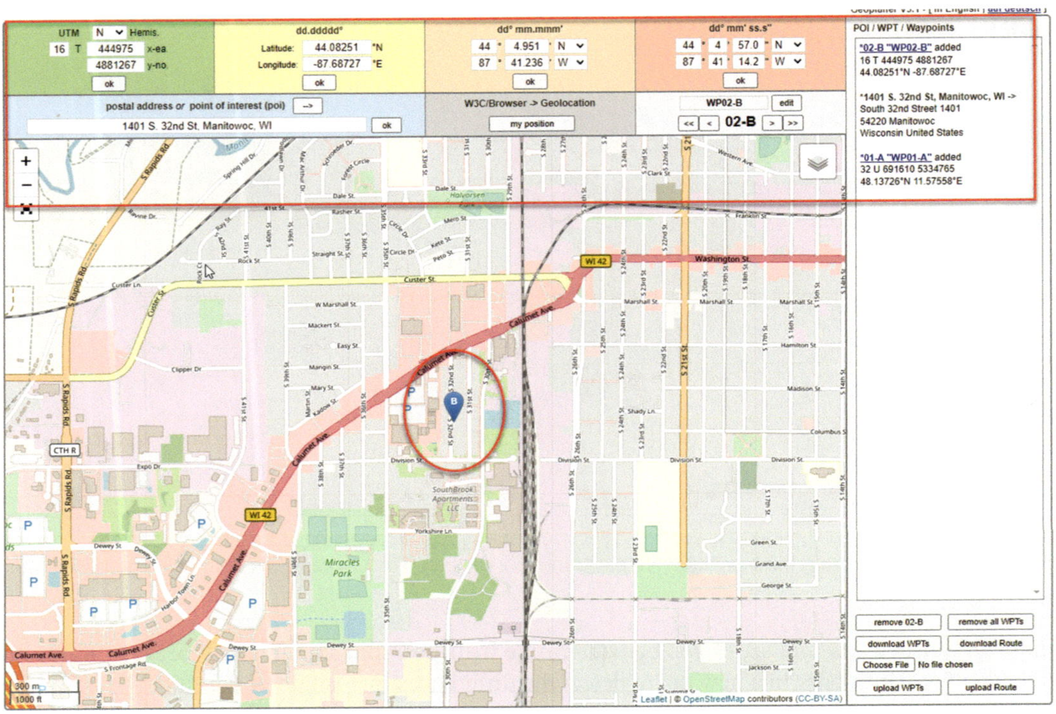

FIGURE 4.8 A map showing four different coordinate formats for a given point on the Earth. Mapped using Geoplaner.

Knowing the above information, determine the following:

Point A as plotted in a UTM coordinate system is located at 4,000,000 m northing and 550,000 m easting. Point B as plotted in the same zone is located at 4,100,000 m northing and 550,000 m easting. How far is Point B from Point A in meters? In kilometers? What direction is Point B from Point A?

Point C is at 4,200,000 m northing and 570,000 m easting. What direction is Point C from Point B? How far is Point C from Point B in meters? In kilometers?

How many meters east of the central meridian of a UTM zone is a location that is 710,000 m easting? How many kilometers east of the central meridian of a UTM zone is this same location?

Access https://geoplaner.com/—an interactive mapping tool—and enter the address 1401 S. 32nd St., Manitowoc, WI 54220 (Figure 4.8). Geoplaner shows the location of that geocoded position in UTM, decimal degree, and degrees minutes seconds (as shown).

Note the formats for this street address in UTM, decimal degree, decimal minutes, and degrees minutes seconds, respectively, across the top of the Geoplaner interface, as described in the above text. Use this same Geoplaner web mapping tool to locate your own residence or campus, or a famous landmark somewhere around the world such as the Eiffel Tower, paying attention to and comparing the values above the map.

Polar Coordinates

Another system for capturing the location of points in the plane is the polar coordinate system. In it, the coordinates of each point in the plane are determined by a distance from a reference point

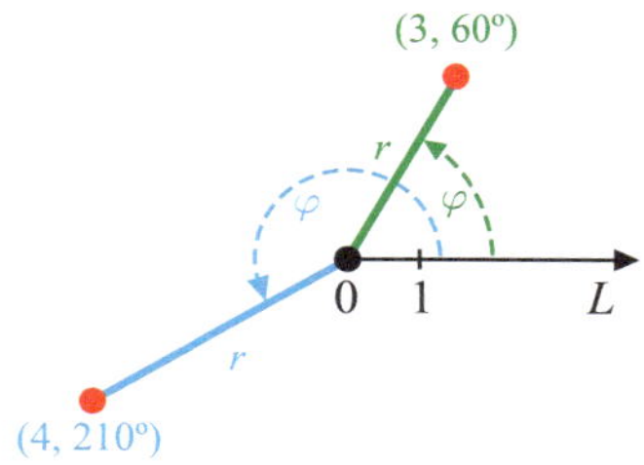

FIGURE 4.9 Points in the polar coordinate system with pole O and polar axis L. In green, the point with radial coordinate 3 and angular coordinate 60 degrees or (3, 60°). In blue, the point (4, 210°). Source Wikimedia Commons. By Monsterman222—Own work, CC BY-SA 3.0, https://commons.wikimedia.org/w/index.php?curid=33802846

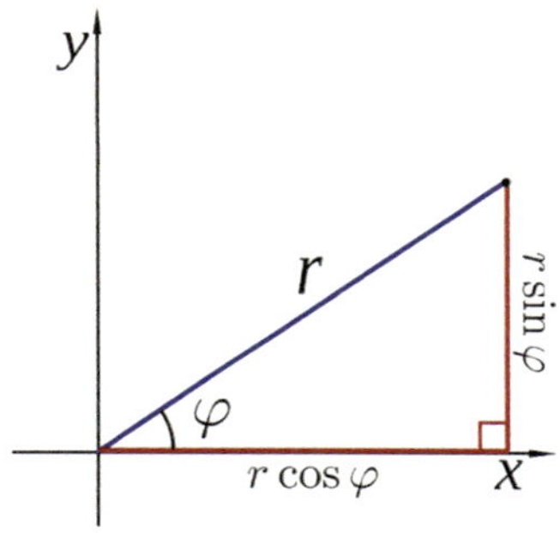

FIGURE 4.10 A diagram illustrating the relationship between polar and Cartesian coordinates. Source: No machine-readable author provided. Mets501 assumed (based on copyright claims), CC BY-SA 3.0 <https://creativecommons.org/licenses/by-sa/3.0/>, via Wikimedia Commons

and an angle from a reference direction (Figure 4.9). The reference point, serving as an origin, is called the pole, and a ray extending from the pole in the reference direction is called the polar axis. The distance of a point from the pole is called the radial coordinate of the point, and the angle a ray from the origin to the point makes with the polar axis is called the angular coordinate (or azimuth in some applications). Polar angles are expressed as degrees or radians (2π radians equals to 360°). Polar coordinates are often used to study phenomena originating from a central point.

To convert from Polar Coordinates (r, φ) to Cartesian Coordinates (x,y), use the conversion equations, $x = r \times \cos(\varphi)$, $y = r \times \sin(\varphi)$, as suggested in Figure 4.10.

The polar coordinate system may be extended to three dimensions as a cylindrical or spherical coordinate system. We confine ourselves here to coordinatization that is often found in mapping. Thus, we move on to a conceptual review of trigonometry which also sees abundant use in mapping.

TRIGONOMETRY: VISUAL REVIEW OF FUNCTIONS AND APPLICATIONS

All of our geometric analysis is based on Euclidean geometry, assuming Euclid's Parallel Postulate: given a line and a point not on the line—through that point there passes exactly one line that does not intersect the given line. Non-Euclidean geometries violate this Postulate. What does the geometry of the Earth model become in the non-Euclidean world? Stay tuned and look for it in a later chapter! Now we offer an elegant visual summary of trigonometry that takes advantage of graphical animation, as a single visual capture of the entire set of elementary trigonometric functions.

Visual Review of Trigonometric Functions

The Basic Setup

Figure 4.11 shows the basic set up, with terminology highlighted (Arlinghaus and Arlinghaus, 1989). Note the importance of language and "co-language."

Six Trigonometric Functions

Figure 4.12 shows the six common trigonometric functions cast in the basic setup of Figure 4.11.

Three Functions of θ and Three Functions of Co-θ

Figure 4.13 organizes the functions in relation to whether they are complementary.

All Six Functions in a Single Animation

Figure 4.14 shows an animation of all six functions.

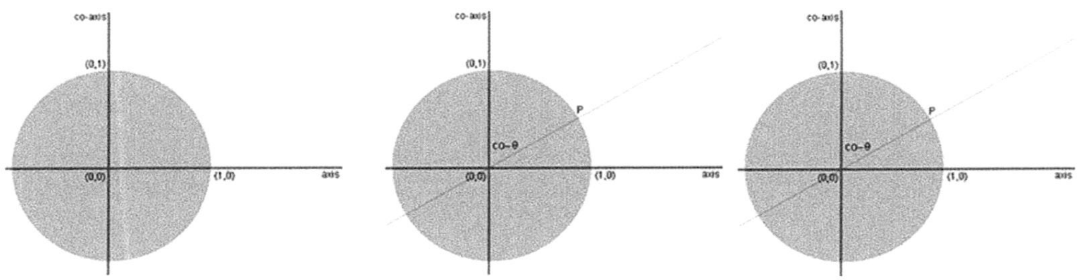

FIGURE 4.11 The Basic Setup. A: axis and co-axis, complementary and orthogonal. B: angle θ. C: complementary angle, co- θ. Source: Arlinghaus and Arlinghaus, 2005.

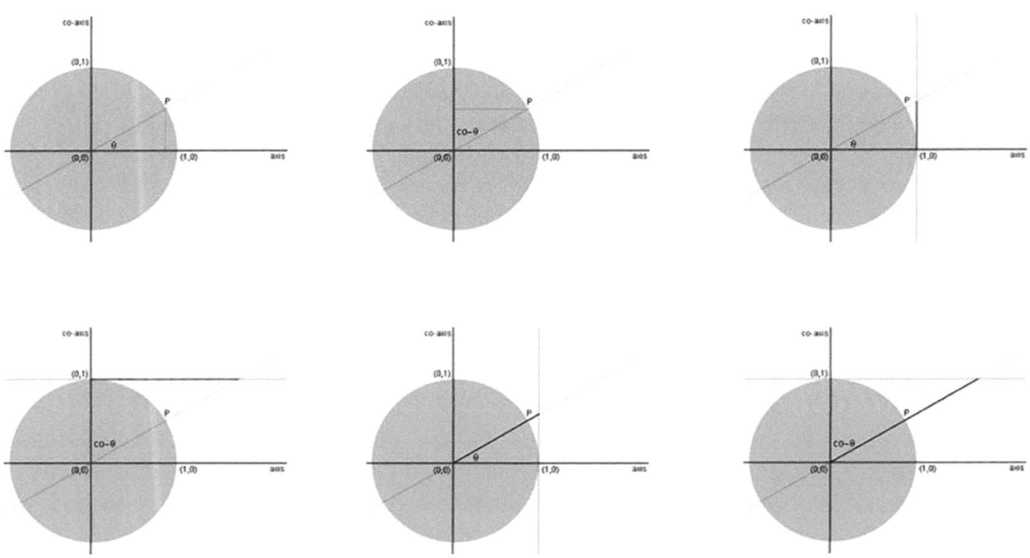

FIGURE 4.12 Six trigonometric functions. (a) Sine; (b) Tangent; (c) Secant; (d) Cosine; (e) Cotangent; (f) Cosecant. Source: Arlinghaus and Arlinghaus, 2005.

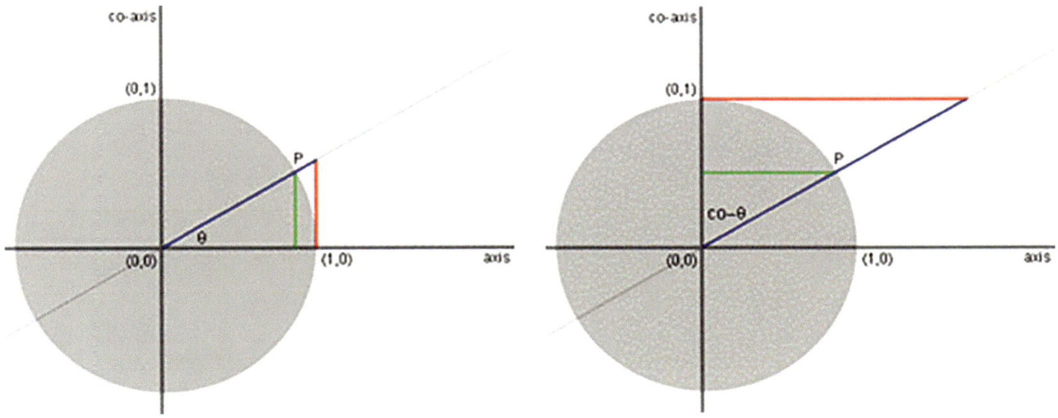

FIGURE 4.13 (a) three functions of θ. (b) Three functions of co-θ. Green, Sine and Cosine. Red, Tangent and Cotangent. Blue, Secant and Cosecant. Source: Arlinghaus and Arlinghaus, 2005.

Scan QR Code

FIGURE 4.14 All visualized trigonometric functions unified in a single animation. Green, sine and cosine; red, tangent and cotangent; blue, secant and cosecant. Source: Arlinghaus and Arlinghaus, 2005.

Measure the Height of the Arc de Triomphe in Paris, France

Access the linked ArcGIS Online 3D scene (https://www.arcgis.com/home/webscene/viewer.html?webscene=037cceb0e24440179dbd00846d2a8c4f) (Figure 4.15). The scene opens at the Arc de Triomphe in Paris. Select the analyze tool at the right>measure>measure from the top of the Arc to one of the circular streets running around its base. Which of the measurements is the hypotenuse of the triangle? What is the height of the Arc above the circular street?

Exit the measure tool>Use the search tool>search for the Eiffel Tower, zoom to it, and measure the height from the center of the nearest bridge crossing the River Seine to the top of the tower (Figure 4.16). How much taller is the Eiffel Tower than the Arc de Triomphe in meters? In percentage? How much longer is hypotenuse of the triangle for the Eiffel Tower to that of the Arc de Triomphe? In percentage? Note how different the shape of the Eiffel Tower depicted as a solid building is from what you know the Eiffel Tower looks like. Change the basemap to a satellite image base. Make sure your 3D scene has due north at the top by checking the compass on the left side of the 3D scene. Click on the Daylight/Weather tool as shown below. Scroll the time of day and time of year back and forth. How does the shadow of the tower and surrounding buildings change with the time of day? With the time of year? Why?

Next, note the direction the shadow is casting on the satellite image basemap. Based on the direction of the shadow from the satellite image (indicated by the arrow to the right of the tower in Figure 4.16) and not from the building model, at what time of day was the satellite image actually taken?

Next, use the measurement techniques you practiced above in your own community. Measure some of the buildings on your campus, and your own residence, and compare to the Paris buildings

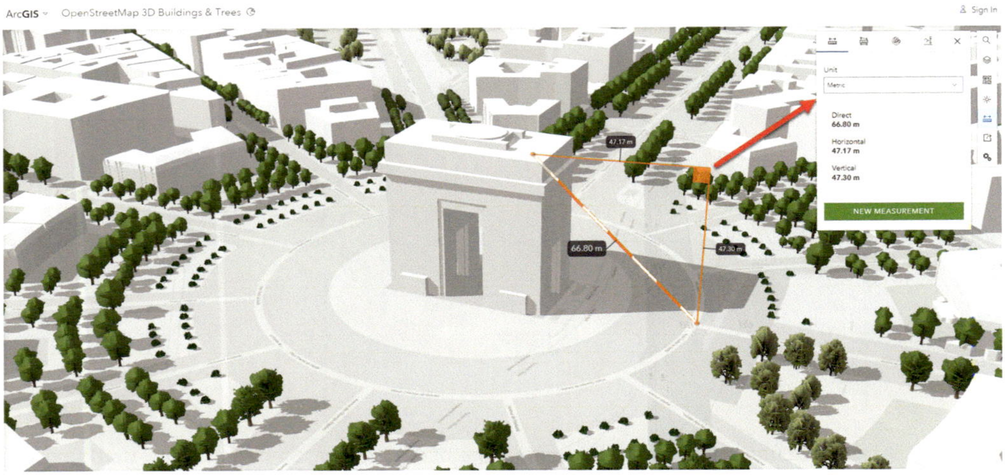

FIGURE 4.15 Measuring heights, lines, and angles on an Arc de Triomphe, Paris, France, 3D model. Mapped using ArcGIS from Esri.

FIGURE 4.16 Eiffel Tower, Paris, France; 3D model. Mapped using ArcGIS from Esri.

that you measured above. Next, measure a few of the world's tallest buildings such as the Burj Khalifa in Dubai, Merdeka 118 in Kuala Lumpur, Indonesia, and the Shanghai Tower in Shanghai, China. Compare the heights and the hypotenuse from the top to the ground, of each.

Next, we extend our capability to embrace the entire world!

Applications: Conceptual and Real Worlds

Trigonometric Identities

From the Pythagorean Theorem (which we will see again later), it follows that:

$\sin^2\theta+\cos^2\theta=1$, the radius of the unit circle measured along the secant line;
$\sec^2\theta=\tan^2\theta+1$, the radius of the unit circle measured along the horizontal axis;
$\csc^2\theta=\cot^2\theta+1$, the radius of the unit circle measured along the co-axis.

With the visualization of trigonometric functions mastered, along with the Pythagorean Theorem and good mastery of manipulating and reducing fractions, most trigonometric identities become easy to unravel within their own conceptual world.

Length of 1° on the Earth Model

Here we offer some detail in determining values for measurements to which we have made reference, but not shown derivation. With trigonometry in hand, derivation becomes straightforward.

- One degree of latitude, measured along a meridian or half of a great circle, equals approximately 69 miles (111 km). One minute is just over a mile, and 1 second is around 100 feet (a pretty precise location on a globe with a circumference of 25,000 miles). Calculation: 25,000/360=69.444.
- Next, consider the length of 1° of longitude. For example, at 42° north latitude, $r=2{,}956.882$. Thus, the circumference of the parallel at 42° north is approximately 18,578.6205 miles. Thus, the length of 1° of longitude, measured along the small circle at 42° of latitude, is 51.607 miles or 83.05 km.
- Because meridians converge at the poles, the length of a degree of longitude varies, from 69 miles at the equator to 0 at the poles (longitude becomes a point at the poles). Calculation: at latitude θ, find the radius, r, of the parallel, small circle, at that latitude. The radius, R, of the Earthsphere is $R=25{,}000/(2 * \pi)=3{,}978.8769$ miles. Thus, $\cos \theta=r/R$ (using a theorem of Euclid that alternate interior angles of parallel lines cut by a transversal are equal). Therefore, $r=R \cos \theta$. Then, the circumference of the small circle is $2r * \pi$ and the length of 1° at θ degrees of latitude is $2r * \pi / 360$. This calculation scheme is a rich source of elementary problems using geometry and trigonometry. Consider the following question: at what latitude is the length of 1° of longitude exactly half the value of 1° of longitude at the equator?

Next, we consider a global use of trigonometry that embraces global measurement!

Measuring the Circumference of the Earth

Eratosthenes Measurement

In this activity, you have the opportunity to use unit conversions and get out onto the landscape, using new technology to address a problem that mathematicians and astronomers wrestled with for millennia.

Eratosthenes (276 BCE–194 BCE) was born in Cyrene, now a part of Libya, in North Africa. After studying in Alexandria and Athens, he became the director of the Great Library in Alexandria. This Library truly lived up to its name, housing a great deal of the learned and compiled knowledge of the time.

It was at the library where Eratosthenes read about a deep vertical well near Syene (now Aswan) in southern Egypt. Once each year at noon at this well, on the day of the Summer Solstice, the bottom of the well was entirely lit up by the sun. The sun was directly overhead, its rays shining straight into the well (Figure 4.17).

Eratosthenes then placed a vertical post at Alexandria, which was almost due north of Syene, and measured the angle of its shadow on the same date and time. He made novel assumptions that (a) the Earth is round (a sphere) and that (b) the Sun was far away and by the time its rays reach the Earth, the rays are essentially parallel. With these assumptions, Eratosthenes knew from geometry that the size of the measured angle equaled the size of the angle at the earth's center between Syene and Alexandria: that is, he knew from Euclid that the alternate interior angles, formed by a transversal cutting parallel lines, are equal. Knowing also that the arc of an angle this size was about 1/50 of a circle, he then had to determine the distance between Syene and Alexandria. This was a difficult

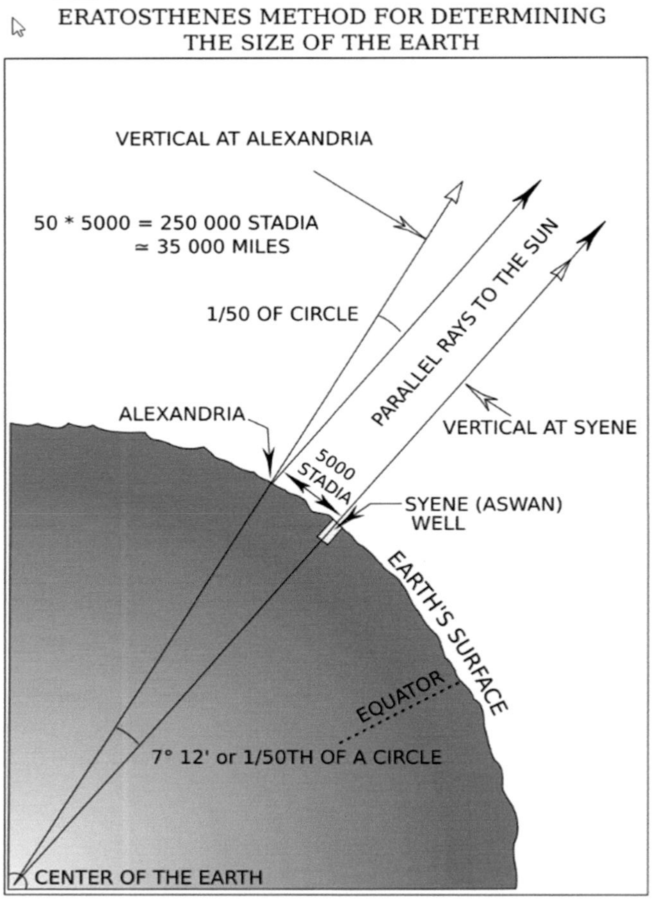

FIGURE 4.17 Geometry for Eratosthenes's Earth measurement. Source: Wikipedia, Creative Commons.

task during that time, due to the different strides of camels and human error, and despite the best efforts of the King's surveyors, required years of effort. It was finally determined to be 5,000 stadia. Eratosthenes multiplied 5,000 by 50 to find the Earth's circumference. His result, 250,000 stadia (about 46,250 km), was amazingly close to the accepted modern measurements (40,075 km around the equator and 40,008 km around the poles).

With a recreational grade GPS, a smartphone app, or a web mapping application, **you too** can emulate Eratosthenes' methods and measure the circumference of the Earth. In so doing, you are incorporating and integrating geography, mathematics, Earth Science, physics, and mapping. Like Eratosthenes, you will be bound by your own technology—the spatial accuracy of the positions provided by most recreational grade GPS and smartphones are as of this writing accurate to 2–5 m on the ground.

Using a GPS

This activity uses a recreational grade receiver such as a Garmin, one that can display coordinates in degrees, minutes, and seconds, and also in UTM. If you don't have a GPS, use the smartphone app activity below. Bring a clipboard or phone so you can take notes. You can also do this activity with a friend, with each person using a GPS receiver, one set to degrees, minutes, and seconds, and the other set to UTM.

1. Go outside with a recreational grade GPS receiver.
2. Set the GPS receiver to display coordinates in the format of degrees minutes seconds (hddd° mm' ss.s") latitude-longitude coordinates. The "H" stands for Hemisphere, ddd=degrees, mm=minutes, and ss=seconds. Set the GPS receiver's datum to WGS 84 so that you are working in an established datum (model of the Earth's shape).
3. Know how to use the GPS compass to determine which direction true north and true south lie from your present position.
4. Position yourself at a whole second of latitude (not a fraction), as measured using the degrees-minutes-second format. For example: 39 degrees, 20 minutes, and 5 seconds, rather than 39 degrees, 20 minutes, and 5.1 seconds. At this whole-second position, write down the coordinates that are showing on your GPS receiver (or mark a waypoint). This will be latitude-longitude degrees-minutes-seconds on your GPS.
5. Change your GPS receiver to display UTM. The UTM units are in meters. The units in the UTM system represent eastings ("east" relative to a Central Meridian in given UTM zone), and northings (meters north of the Equator).
6. Change the coordinates back to degrees-minutes-seconds. Using the GPS compass, walk as closely due north or due south for a full second of latitude. When a full second of latitude has been traversed, stop.
7. Write down the coordinates showing on your GPS receiver (or mark a waypoint). Again, this will be latitude-longitude degrees-minutes-seconds on one GPS.
8. Change the display to UTM and write down the UTM coordinates at your location.
9. Determine how many meters you have walked by comparing the UTM northing at the starting point to the UTM northing at the end point: ______________ meters. This is the number of meters in 1 second of the Earth's latitude.
10. Compute the Earth's circumference by using the following equations:
    1 second of latitude × 60 = 1 minute of latitude
    1 minute of latitude × 60 = 1° of latitude
    1° of latitude × 360 = the number of degrees around the Earth
    Therefore, the number of meters that you have recorded above × 60 = the number of meters in 1 minute. The number of meters in 1 minute × 60 = the number of meters in 1°. The number of meters in 1° × 360 = the number of meters around the Earth, through each of the poles.
    Fill in the following:
    The number of meters that you walked = __________
    × 60 = __________ meters in 1 minute
    × 60 = __________ meters in 1°
    × 360 = __________ meters around the Earth
    Divided by 1,000 = __________ kilometers around the Earth.
11. How close are you in kilometers to the accepted circumference of the Earth in kilometers? How close are you as a percent of the accepted circumference of the Earth in kilometers?
12. Are you closer to the accepted circumference than Eratosthenes was?
13. Name at least two reasons why your answer is not exactly the same as the accepted circumference of the Earth.
    For Example:
    The number of meters that you walked =
    4,391,181 – 4,391,150 = 31 m
    × 60 = 1,860 m in 1 minute
    × 60 = 111,600 m in 1°
    × 360 = 40,176,000 m around the Earth
    Divided by 1,000 = 40,176 km around the Earth
    Error: 40,176 – 40,008 = 168 / 40,008 = 0.004 × 100 = 0.4%.

Using a Smartphone App

Repeat the above procedures on a smartphone using an app. Most smartphones have a compass app that provides the latitude and longitude coordinates in decimal degrees. You need to find and use an app that provides the coordinates in degrees-minutes-seconds, and UTM coordinates. At the time of this writing, available apps included Mapnitude, MGRS UTM GPS, UTM Grid Ref Compass, UTM Coordinates Tool, and others. Use the procedures outlined above, only with a phone.

Using a Web Mapping Tool

You can also calculate the Earth's circumference with a web mapping tool such as ArcGIS Online. ArcGIS Online and other web GIS tools are anchored to real-world spaces, and many provide locations in several coordinate formats.

To begin, go to https://www.arcgis.com/, use>Map>to create a new map>then, to the right of the default map that draws>Map Tools>Search>Search for 14th & Grant, Blair, NE (exactly as written)>when the map zooms to this location>in the search result box, select "Add to new sketch" (Figure 4.18).

Use the Search tool on the right again>Search for 14th & Lincoln, Blair, NE (exactly as written)>in the search result box, select "Add to Sketch 1" to add to your existing sketch layer. Then, on left>Basemap>Change to OpenStreetMap. You now should see two placemarkers one city block apart, on an OpenStreetMap basemap with the second one due north of the first (Figure 4.19).

Next, to go Map Tools>Location, and place your cursor with your mouse or touchpad over the first (most southerly) point. Change the coordinate system to DMS—Degrees Minutes Seconds. Note its coordinate in degrees, minutes, and seconds. It should be close to 41 degrees, 32 minutes, 30.700104 seconds north latitude and 96 degrees, 7 minutes, and 58.166286 seconds west longitude (Figure 4.20).

Repeat the process for the second, northern, point. The longitude should be very close to the first point, since the second point is almost *due north* of the first. The latitude, however, should be close to 41 degrees, 32 minutes, and 34.215389 seconds north latitude. Determine how many seconds of latitude separate the two points. It should be around 3.55 seconds. Then, change the coordinate system to UTM, touching each point to record its UTM position. The UTM coordinate is given with the UTM zone first (here, zone 14, and the cell within the UTM zone, here, "T") followed by the easting and the northing. The easting should be similar for each point because one is due north of the other.

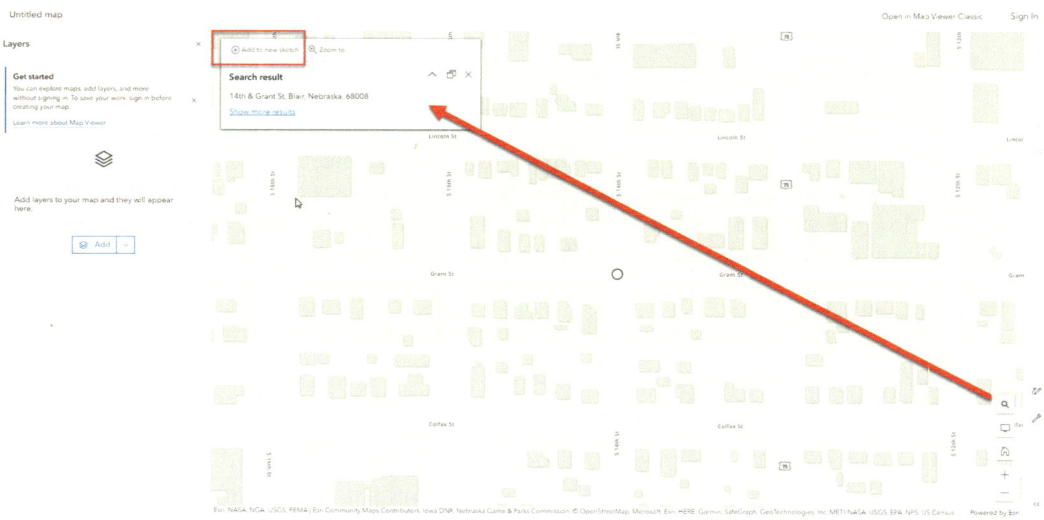

FIGURE 4.18 Adding a placemarker in ArcGIS Online, Esri, to a location in Blair, Nebraska.

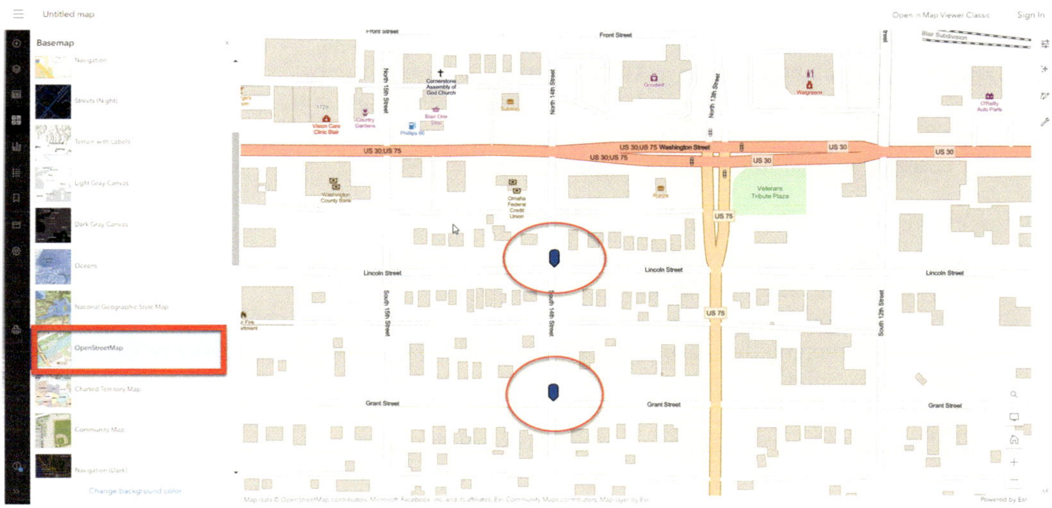

FIGURE 4.19 Adding two placemarkers to a map in Blair, Nebraska. Mapped using OpenStreetMap basemap.

FIGURE 4.20 Determining latitude-longitude coordinates in degrees, minutes, and seconds. Mapped using OpenStreetMap basemap.

But the northing is the value you are interested in here; record the northing for both points. These values are given in meters north of the Equator. Record how many meters north of the Equator each of the points is (the northing). Then, subtract the northing value on the southernmost point from the northing value on the northernmost point. This will yield the length, in meters, between the two points. It should be about 110 m. Using the value of the number of meters between the two points, and the number of seconds of latitude between each point, calculate how many meters exist in each second of latitude at this location. In other words, (the number of seconds, divided by 110 m). This value, the number of meters in each second of latitude, will be similar to the number you would have determined if you were using a GPS receiver or smartphone app in the field while pacing off the distance with your strides. Use this number, along with what you have learned about the number of seconds around the whole Earth, to calculate the circumference of the Earth. What is this value?

You could use the above activity, ArcGIS Online map, and values for your own neighborhood instead of the neighborhood in Blair, Nebraska. The most important thing to be mindful of when substituting your own location is choosing points that are aligned north-south with each other for the most accurate measurement.

How Long Would It Take to Walk around the Earth?

1. Set the GPS receiver to the screen where you can determine how fast you are moving, or use a fitness app on a smartphone.
2. How fast are you walking (in km/hour) when you are walking at a comfortable pace?
3. If you could keep up this pace and walk due north or due south from this point, and walk all the way around the Earth on this meridian, how long would it take before you arrived back at this same spot?
4. What would the date be when you arrived back at this spot?

Example:

$$\frac{40{,}176\,\text{km}}{6\ \text{km/hour}}$$

= 6,696 hours, or 279 days. If you started on New Year's Day 2024, you would arrive back here at the same location on 5 October 2024. The Earth doesn't seem to be so large after all, does it? The only problem is that there are just a few obstacles that could hamper your planned longitude journey—ice, cliffs, oceans, weather, lack of food, the fact that you need to spend time sleeping every day, and other "challenges"!

Determining the Earth's Circumference Based on Measuring 1 Second of Longitude

This variant of the above activity will help you gain further skills in measurement and calculation.

1. Using a GPS receiver or a smartphone mapping app, this time, instead of locating yourself at a whole second of latitude, position yourself at a whole second of *longitude*.
2. Record the latitude that you are standing on. Convert this latitude value to decimal degrees: Latitude = degrees + minutes/60 + seconds/3,600
3. Using an analog compass, a GPS compass, or a phone app compass, walk due east or west for a distance of exactly 1 second of **longitude**.
4. Record the distance walked in meters.
5. Use the following formula to compute the Earth's polar circumference. When doing the calculation, make sure the cosine is measured in degrees, not radians (otherwise, your final value will be a negative number!).

$$\frac{(\text{Distance walked in meters} * 360) *}{\text{Distance walked in degrees}} \quad \frac{1}{\text{cosine of latitude } \emptyset}$$

Example:

$$\frac{24 * 360 *}{.0002777} \quad \frac{1}{\cos(40)} = \frac{8{,}640 * 1}{.0002777 \quad .766} = 40{,}617{,}116\,\text{m}$$

Determining the Distance around the Earth at Your Present Latitude

Expand your skills by determining the distance around the Earth at your present latitude.

1. Determine the length in meters of 1 second of longitude using one of the methods you used above.
2. Compute:

   Distance walked in meters * 360 * (cosine of latitude)

   Distance walked in degrees

   Example:

   24 m * 360 * (cosine of latitude)

   .0002777
   = 31112711 * .766
   = 23,832,336 m, or 23,832 km

3. Why is the distance around the Earth on this latitude line **shorter** than the distance around the Earth along this **longitude** line?
4. For what rather famous line on the Earth is the distance around the Earth on a latitude line close to the distance around the Earth on a longitude line?
5. Determine the length of time it would take for you to walk around the Earth at this latitude at a comfortable walking speed (3 mph or 4.8 kph).
6. What would the date be when you arrived back at this spot? Show your work.
7. Is the time it takes to walk around the Earth at this latitude (east or west) longer or shorter than it would take to walk around the Earth on a path due north or south from any point on the planet? Why?

In the next section, we move away from the specialized, but fascinating world of GPS, back to the more general mapping scene and elements of its mathematical foundations.

SCALE

In the world of maps, two forms of measurement are critical in creating and understanding spatial displays and their analysis. One is the scale at which the map is shown and the other is the survey of spatial data and populations that underlie the content to be mapped. While the concept of measurement takes many different forms in many different disciplines, we confine ourselves to looking at these two fundamental forms needed for the effective mapping of the world's lands and its peoples (in this chapter and the next one).

The word "scale" is a common word that has multiple forms and meanings. It may be a noun; it may be a verb, transitive or intransitive. Scale has meanings in fishing, mountain climbing, and music, as well. In this chapter, we focus on its meaning in relation to maps and to statistical survey data.

MAP SCALE

As you are learning about in this book, maps are representations of reality. We live on a planet much larger than we are. Thus, we must scale the map to make it into a manageable size, in paper form (so we can fold it) or in digital form (so we can display it on a computer screen). Scale is an important

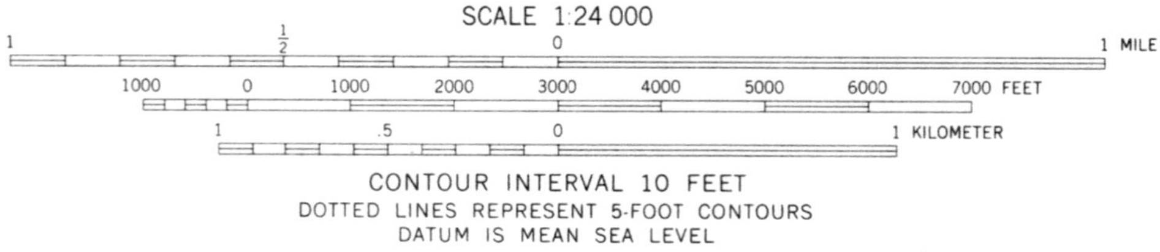

FIGURE 4.21 Map scale. Source: USGS.

concept in mathematics and in mapping. You could conceivably map a trail of leaf cutter ants, a tabletop, or a small room at 1:1 scale. But for anything larger than that, you would need a map scale that reduces the size of your map in relation to the area studied. Thus, most maps are *scaled representations of reality,* usually at a smaller scale than the real world. The scale of a map is the ratio of a distance on the map to the corresponding distance on the ground.

The Earth is often modeled by a small "generating globe" from which the points of the globe are projected into a plane to create a map. The ratio of the Earth's size to the size of the generating globe is the representative fraction (also called principal scale or nominal scale). Maps often display the representative fraction in the legend and may display it as:

- A bar scale: One advantage to using a bar scale on a paper map is that when the map is photographically reduced or enlarged, so too is the bar scale. These are geometric forms and thus do not compute in ways that numerical representations would, using greater than symbols, for example. The units can be (carefully) mixed as well: On a 1:1,000,000-scale map, 1 cm on the map equals 1,000,000 cm on the ground, or 10,000 m, or 10 km on the ground.
- A fraction. This sort of numerical representation on a map makes quantitative comparisons easy between maps, using "greater than" applied to fractions. For example, a map with scale 1:500 is a larger scale map than is a map with scale 1:1,000 because the fraction 1/500 is larger, in numerical value, than the fraction 1/1,000. Scale expressed in this manner does not remain correct under transformations of enlargement or reduction (as a bar scale does). All that happens is the text in the fraction gets enlarged or reduced as, for example, 1/1,000.

In the example shown here (Figure 4.21) there is a representative fraction (1:24,000) as well as a bar scale. For this map, one unit on the map corresponds to 24,000 units on the ground. Thus, 1 cm on the map corresponds to 24,000 cm on the ground, and 1 inch on the map corresponds to 24,000 inches on the ground. The fraction 1:24,000 is free from units; however, once a unit is chosen for one part of the fraction, the same unit must be used throughout to retain the validity of the value of the fraction.

Activity: Studying Map Scale with USGS Topographic Maps

Open a web browser and navigate to the map: Map Scale Activity Starting with USGS Topographic Maps (https://www.arcgis.com/apps/mapviewer/index.html?webmap=39b4e135a17d487c92ef9ad a99e2f7d4). The map opens with a view of the 1:250,000-scale USGS topographic base map of Bismarck, North Dakota (Figure 4.22). At this scale, 1 inch on the map equals 250,000 inches on the ground; 1 cm on the map equals 250,000 cm on the ground, and so on. Note how the cities of Bismarck and Mandan are symbolized on the map, the detail shown of Interstate Highway 94 running east and west through the cities, the detail of the Missouri River running northwest to southeast, and the detail of the fields and minor roads. This is a small-scale map series that covers the entire country. Why is this a "small scale" map? Work through the material below to find out!

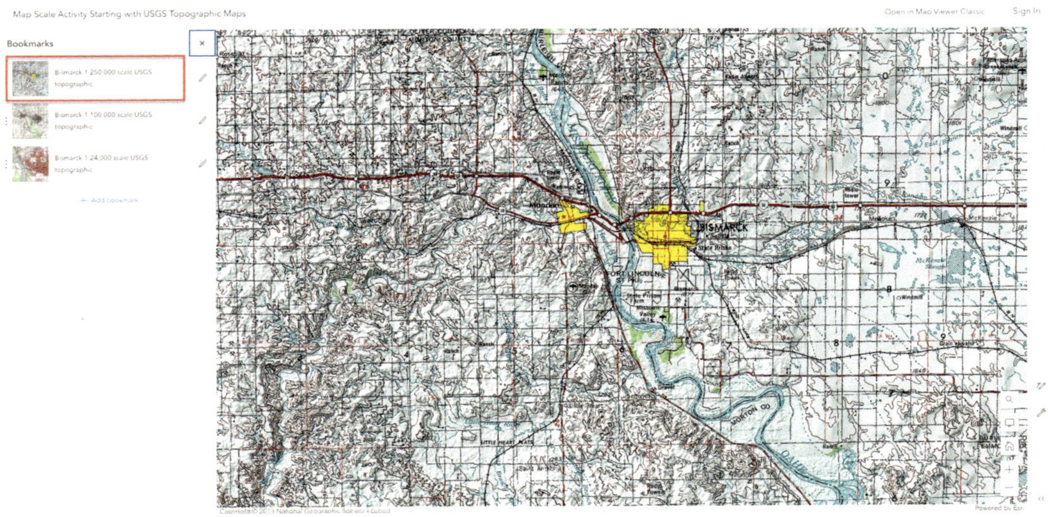

FIGURE 4.22 Bismarck, North Dakota, 1:250,000-scale USGS topographic map. Mapped using ArcGIS from Esri.

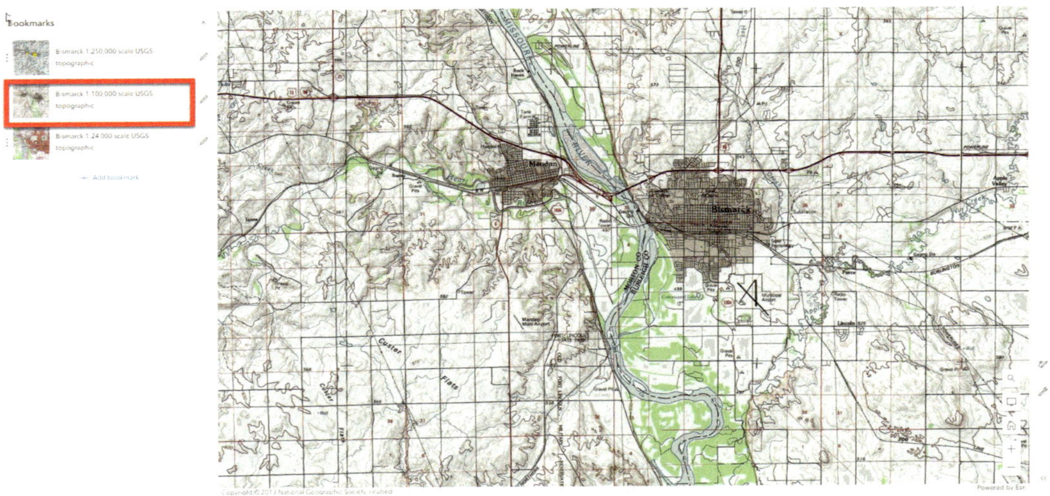

FIGURE 4.23 Bismarck, North Dakota, 1:100,000-scale USGS topographic map. Mapped using ArcGIS from Esri.

Use the bookmarks tool on the left side of the interface to change the map to the 1:100,000-scale USGS topographic base map of the cities. Again, note the details of the cities, rivers, highways, fields, and minor roads (Figure 4.23).

Use the bookmarks tool on the left side of the interface to change the map to the 1:24,000-scale USGS topographic base map of the cities. Again, note the detail of the cities, rivers, highways, fields, and minor roads. This is the largest scale USGS topographic base map covering the entire country (Figure 4.24).

Which scale of map, 1:250,000, 1:100,000, or 1:24,000, covers the most area? Which map offers the most detail? Which would you use to find the location of the state capitol building? Which map would you use to get a sense for how the two cities nest beside the Missouri River and its floodplain?

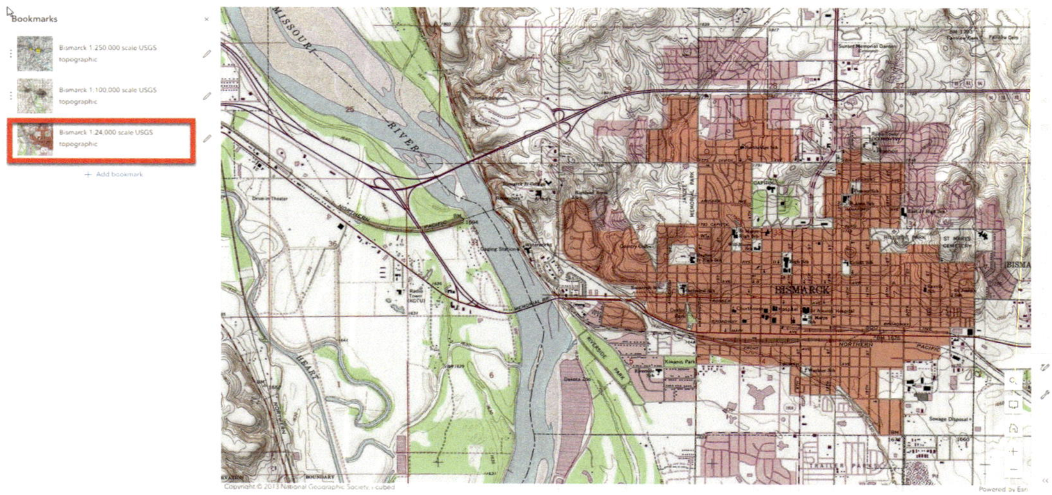

FIGURE 4.24 Bismarck, North Dakota, 1:24,000-scale USGS topographic map.

The answer to the first question is that the 1:250,000-scale map covers the most area, while the 1:24,000-scale offers the most detail. Why is the 1:250,000-scale map considered a small-scale map when 250,000 is larger than 24,000? The 1:250,000-scale map is considered a small-scale map because the number must be considered as a ratio, or fraction. Thus, one must compare 1/250,000 to 1/24,000, rather than just the denominators. Consider 1/100 of a pizza vs. 1/10 of a 12" pizza. Which is more? 1/10 is more pizza. In the same way, 1/250,000 is a smaller number than 1/24,000, and therefore, the smaller the number, the smaller the scale. 1/24,000 is a larger number and thus 1:24,000 is a larger scale map, covering less area but more detail than a smaller scale map. A globe may be at a scale of 1:20,000,000 and therefore a very small-scale representation—a large area, in the case of a globe, the whole world is shown, but because the whole world is shown, the detail that can be shown is very little. Most globes only show major cities, mountain ranges, and coastlines; sometimes they show countries, but never any detailed insides of cities.

Which of the above maps is **most** useful? This last question is subjective: Each scale of map is different, and each scale serves a purpose.

Pan and zoom the above map to another city or to your own city. Because the map uses USGS topographic maps for its basemap, the activity is only valid for the USA.

Global and Local Scale

The choice of scale impacts our perception of the size of a political, natural, or any other area, and therefore the size of a problem or situation. Scale is important to studying changes over space and over time. Scale as a straightforward linear relationship becomes complicated by the curvature of the Earth's surface, causing scale to vary across a map. Scale is fundamental to understanding the interaction between the biosphere, anthroposphere, lithosphere, cryosphere, atmosphere, and hydrosphere. The scale of your study and the scale and resolution of the data you are using impacts all mapping analysis and ultimately, how well your audience will understand what you are communicating.

Activity: Determining Map Scale

In this activity, you will determine map scale based on measurements. Open the linked ArcGIS Online map (https://www.arcgis.com/apps/mapviewer/index.html?webmap=725c686f9aff407b837d722cd114832a)

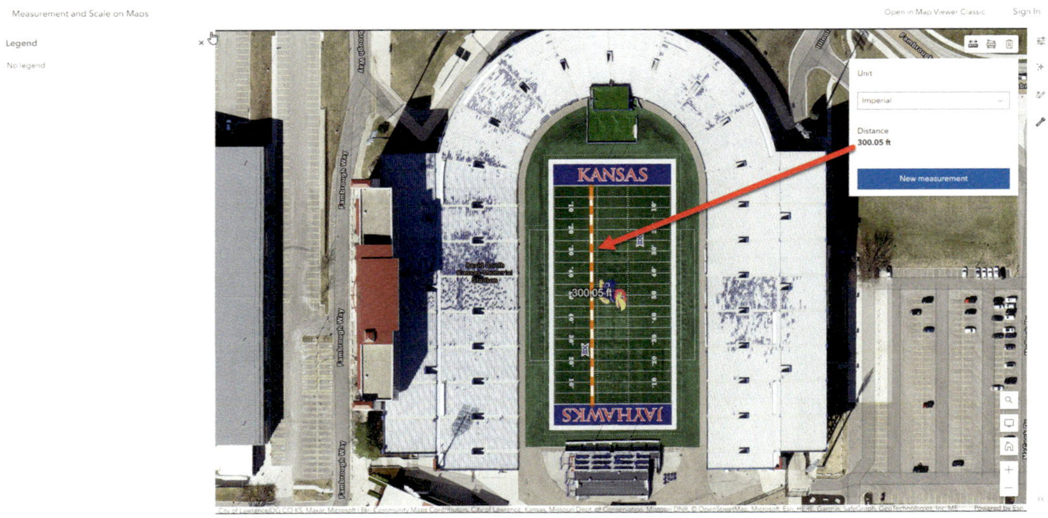

FIGURE 4.25 Measuring the length of the University of Kansas football stadium. Mapped using ArcGIS Online from Esri.

(Figure 4.25). The map opens with a satellite view of the University of Kansas football stadium. Use the measure tool to measure the distance from the 0-yard line to the 100-yard line, or from the line where the blue meets the green, end zone to end zone, first using imperial units (as shown).

Your measured line should be very close to 300 feet, or 100 yards, which is the standard length of an American football field. Take a ruler or pull up Microsoft Word or another software program with a ruler embedded and use any one of these tools to measure the physical size on your device that the field occupies, from north to south. It should be about 3.5 inches or 8.9 cm. Using these two measurements, compute the map scale. Change the units to metric and measure the football field again. The length should be very close to 91.4 m.

Compute the representative fraction (map scale) of this map. If the distance on the map is 100 yards, or 300 feet, and the distance on the computer screen is 3.5″, then converting all units to inches yields 3,600 inches on the map = 3.5″ on the screen, or 1,029 inches on the map = 1 inch on the screen. Therefore, the map scale is 1:1,029. Repeat the process using metric units: 91.4 m on the map corresponds to 8.9 cm on the computer screen, or 9,140 cm = 8.9 cm, or 1,027 cm = 1 cm, yielding a map scale of 1:1,027. Why are the two scales measured in different units (1:1,029 and 1:1,027) not exactly the same? Think about the precision by which you converted inches to centimeters, and the inexact science of measuring with a ruler and with the measure tool on a computer screen. These inaccuracies result in slight and interesting differences.

To the left of the map > Bookmarks > select the Point A to Point B bookmark (Figure 4.26). Measure the east-west distance between Point A and Point B on the map. On the ground, the distance between the two points is 1 mile (1.6 km). Similar to the football stadium activity above, measure the distance across on your computer screen, and compute the representative fraction (map scale). This scale should be smaller than the map of the football stadium.

Lastly, to the left of the map > Bookmarks > select the Lawrence to Kansas City bookmark. Measure the distance between Lawrence Kansas and Kansas City, Missouri (just to the east of the state line that you see as a dashed north-south white line) on the map. Like the two activities above, measure the distance across on your computer screen, and compute the representative fraction (map scale). This scale should be smaller than the map of the football stadium and smaller than the Point A to Point B map.

FIGURE 4.26 Measuring the distance between two points in ArcGIS Online, Esri.

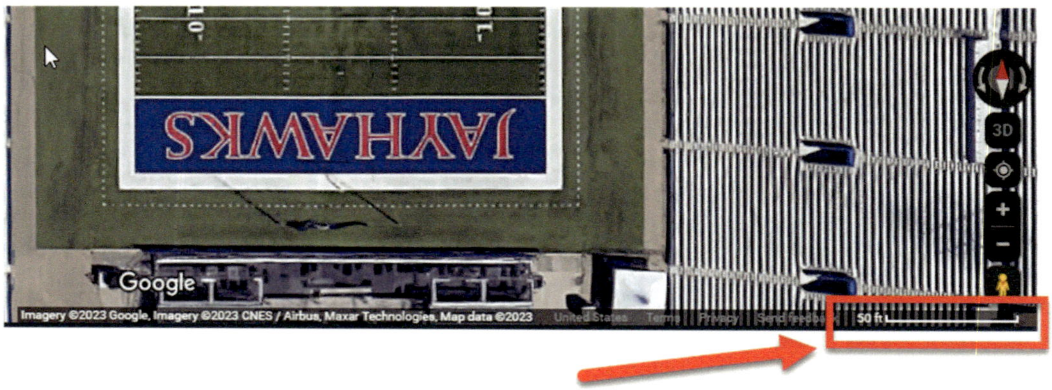

FIGURE 4.27 Reading the map scale on a Google map.

Some maps contain a map scale that is automatically generated as you zoom in and out; for example, Google Maps. Go to https://maps.google.com, change the basemap to imagery, search for the University of Kansas in Lawrence, and zoom to the football stadium until the scale of the map is approximately the same as when you were examining the football stadium in ArcGIS Online. Note the scale (Figure 4.27) in the lower right corner of the map.

Measuring with a ruler, the 50 feet on the map corresponds to 1 inch on the screen. Multiplying 50 by 12 for inches yields a scale of 1:600. Your results may vary depending on the scale of your map, and because the scale in Google maps is meant to serve as approximate.

Activity: Comparing Sizes of Countries

Open the following 3D web mapping application to apply map scale concepts in comparing the sizes of countries: https://arnofiva.github.io/world-sizes/ Click on a country>use the controller bars that appear to move that country to another location on the Earth. In the example here, India is placed

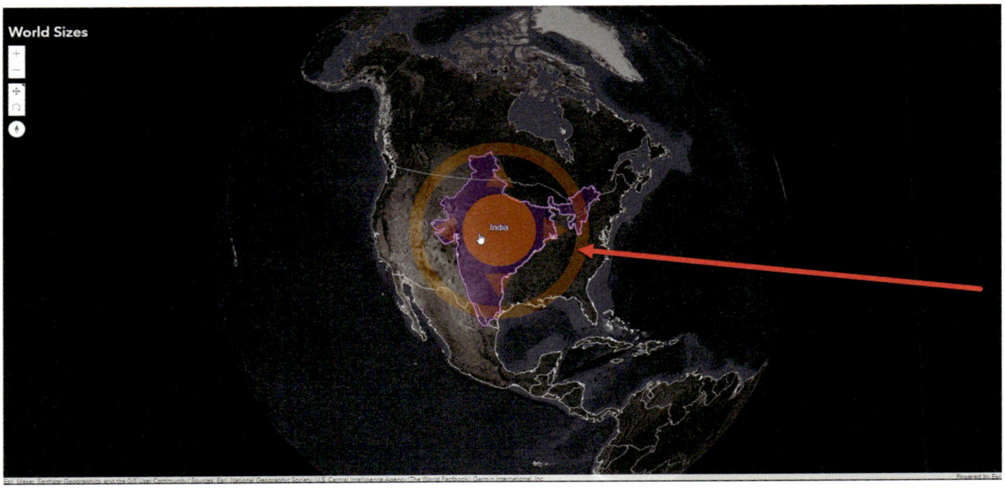

FIGURE 4.28 Comparing India and the USA sizes using an interactive 3D web mapping application. ArcGIS API from Esri.

on top of the continental USA (Figure 4.28): Note that India's size makes it "extend" partway into Canada and Mexico.

If the mapped region is small enough to ignore the curvature of the Earth, then a single value can be used as the scale without causing measurement errors. If a mapped region is large enough for the curvature of the Earth to cause an effect, map scale becomes more problematic to represent. Think back to the discussion of accuracy and precision in Chapter 1; remain alert to fundamental mathematical and mapping issues as you work through real-world activities in this book and elsewhere.

Map Projection

The map projection is critical in understanding how scale varies throughout the map. Generally speaking, a map projection transforms the locations on the curved surface of a globe into locations on a plane. Because the projection of the surface of a sphere to a plane cannot be perfect, there is an infinite number of possible projections and none of them is fully accurate; they all contain distortion of some sort. To imagine this idea, think about trying to flatten an orange peel; there are always bumps, rips, and other distortions. The science of cartography delves deeply into this issue. Read more about the traditional approach to that subject here (link). For an interactive view, consider the web mapping application that you used above (https://arnofiva.github.io/world-sizes/) elsewhere. It provides an easy way to teach about the issue of size and shape distortions for land masses using certain map projections. Use the application to place Greenland atop Brazil (Figure 4.29): Unlike the perception resulting from maps in unprojected spaces or in Web Mercator, the default on most web mapping tools, in reality, Brazil really is much larger than Greenland!

Try comparing other countries using this mapping tool. Which country sizes surprise you when overlaying them on another country?

Tissot Indicatrix

Tissot's indicatrix is often used to visualize the variation of point scale across a map (Buckley, 2011; Laskowski, 1989). Because that scale varies across a map, Tissot took small elliptical snapshots of the stretching of the underlying map coming from that variation. Generally, these ellipses are placed at each intersection of displayed parallels and meridians. In Figure 4.30, all the ellipses are circles,

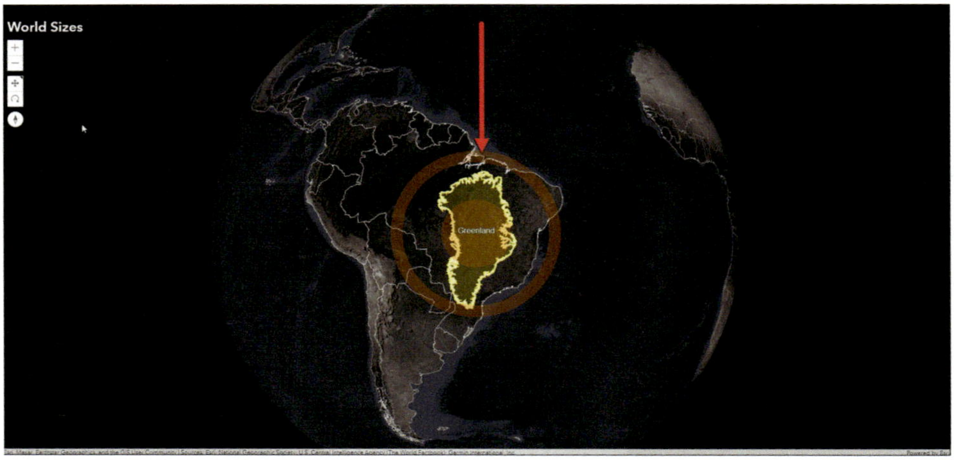

FIGURE 4.29 Comparing the sizes of Greenland and Brazil using an interactive 3D web mapping application. ArcGIS API from Esri.

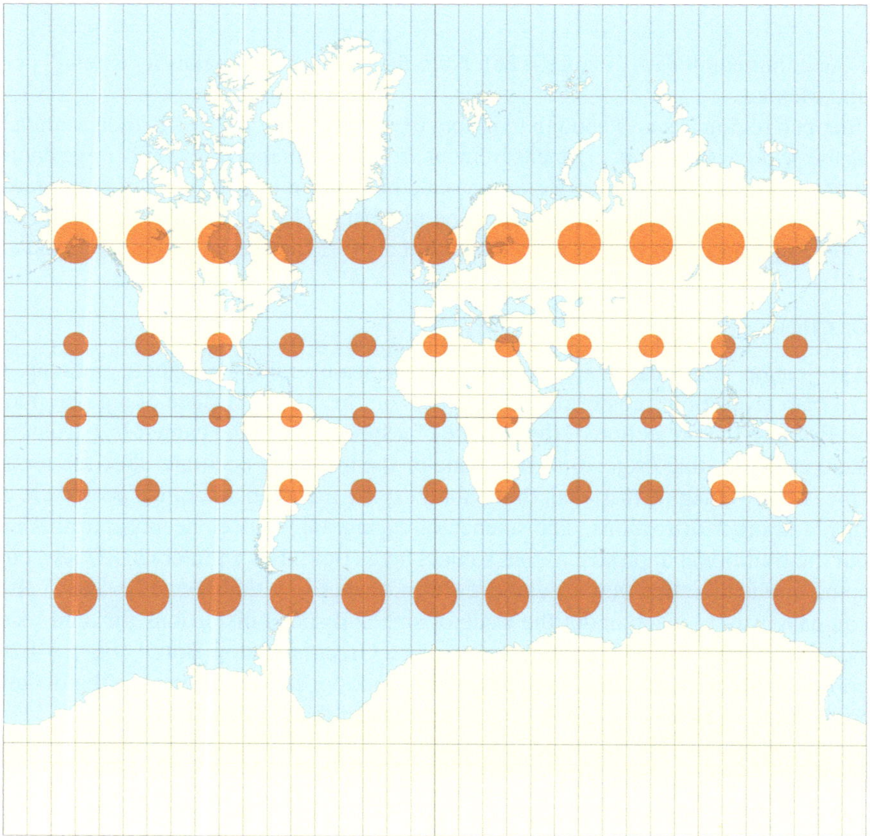

FIGURE 4.30 Map of the world in a Mercator projection (https://en.wikipedia.org/wiki/Mercator_projection) (cropped at 85° of latitude) with Tissot's Indicatrix (https://en.wikipedia.org/wiki/Tissot%27s_Indicatrix) of deformation. Each red circle/ellipse has a radius of 500 km. Scale: 1:5,000,000. By Eric Gaba (Sting - fr:Sting)—Own work Data : U.S. NGDC World Coast Line (public domain), CC BY-SA 4.0, https://commons.wikimedia.org/w/index.php?curid=4677929

with their radii stretched as one moves poleward, demonstrating the exaggeration of land mass size as one moves toward the poles.

Scale and Measurement on Mars

In this activity, you can develop skills in measurement and scale, using a 3D interactive map tool of Mars. Open the following Mars 3D explorer: https://explore-mars.esri.com>Start Now>and you will see the Mars 3D tool (Figure 4.31).

Use the pan and rotate tools to the left of the map to orient yourself with the planet. Then use the basemap tool in the lower right to change the basemap from a satellite image to an elevation map, ranging from blue to green to red to white from low to high elevations. What observations can you make about the elevations around the planet Mars? Change the basemap back to the default imagery.

The Earth's equatorial diameter is 12,756 km. Use the measure tools from one side of Mars to the other, as close to the Equator as you can, to measure the diameter of Mars. Like another activity in this book, the line is curved if you are measuring anywhere but along the Equator, showing the Great Circle route across Mars. Compare your measurement to the Earth's equatorial diameter. How much smaller is Mars compared to Earth's diameter in kilometer? As a percentage?

These same tools allow you to measure height, length, area, and depth. Use Locations > Perseverance Rover>to zoom to the location where the Perseverance Rover touched down in 2021. Use the tools to measure the diameter of the Jetzero Crater, where the rover did its first exploration. The top speed of the rover on flat, hard ground, is 0.152 kilometers per hour. Much slower than even a leisurely stroll for a human! How many meters per hour can the rover cover? How long would it take for the rover to cross the entire Jetzero Crater in hours? In days?

Use the tools to measure the perimeter and the area of the Jetzero Crater (Figure 4.32). The state of Rhode Island in the USA is 3,144 km^2. If the Jetzero Crater sat on top of Rhode Island, what percentage of the state would it occupy?

Use the Locations tool to investigate locations where the Curiosity Rover and the Opportunity Rover landed. Which of the three locations was the flattest? Change the basemap back to elevation. Are there some flatter areas on Mars than the Jetzero Crater? Why do you think the rover did not land in one of the flat lava flows (shown in blue) on the planet?

Mars may be smaller than Earth but does include some **mighty big** features, some of which are the largest in the solar system. Navigate to the Olympus Mons volcano, on the other side of the

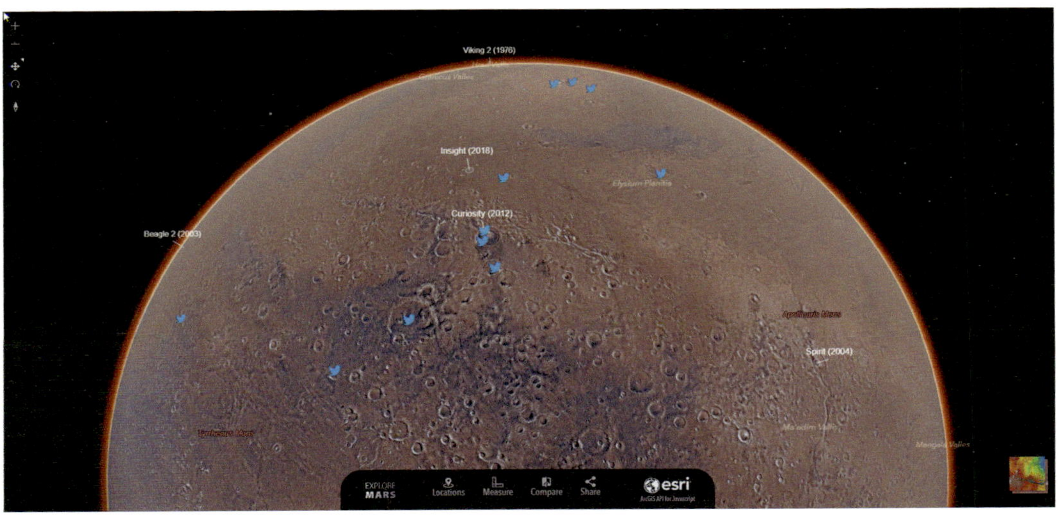

FIGURE 4.31 An interactive 3D Mars web mapping application. Esri.

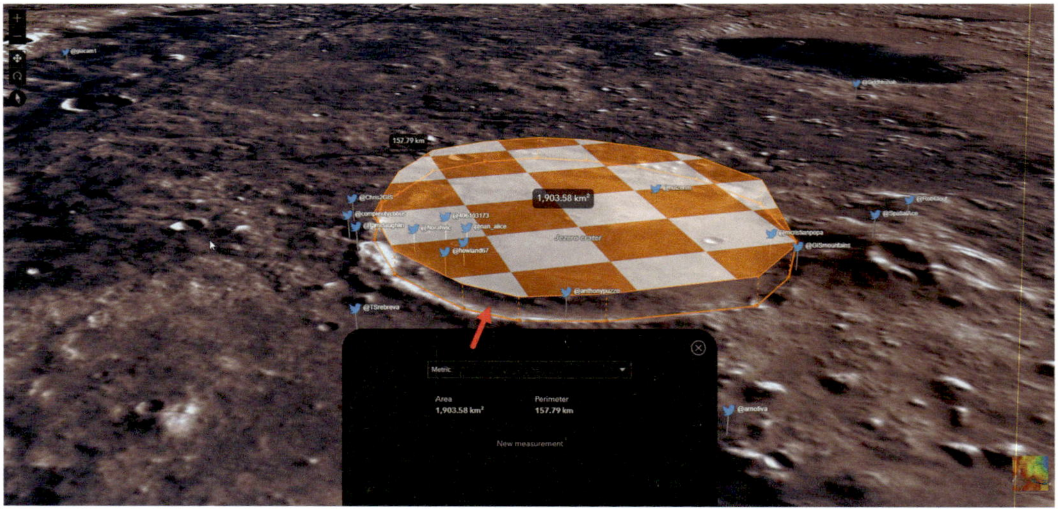

FIGURE 4.32 Measuring Jetzero Crater. Esri.

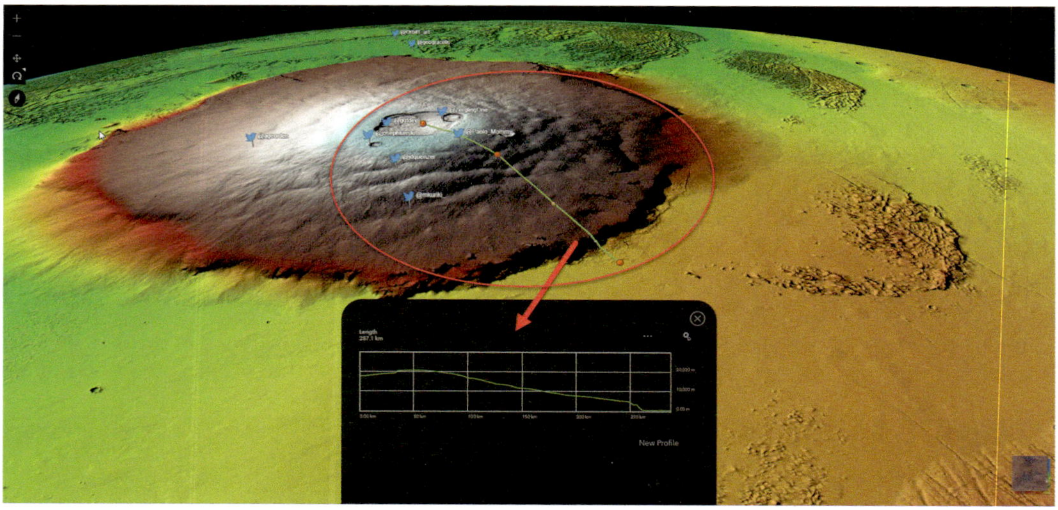

FIGURE 4.33 Creating an elevation profile for Olympus Mons. Esri.

planet from where the Jetzero Crater is (Figure 4.33). Use the elevation tool to generate an elevation profile from the top of Olympus Mons escarpment to the lowlands surrounding it. The elevation descends from over 20,000 m to about 1,000 m over 20 km. By comparison, Earth's Mt Everest is only 8,849 m high, and it lies over 750 km from any land as low lying as 1,000 m. Therefore, as steep as Mt Everest is, if you could stand at the base of the Olympus Mons and look up at the summit, you would have to crane your neck so much that it might hurt.

To get an even better sense of these differences, use Compare > 3D models > select Mt Everest and place it on top of Olympus Mons. What is your estimate of the width of Mt Everest compared to that of Olympus Mons? What is your estimate of the volume of Mt Everest compared to that of Olympus Mons?

Place the Grand Canyon on the deep, wide Valles Marineris. Estimate the depth of the Grand Canyon compared to the Valles Marineris. Use Compare > Regions to compare the size of a country or USA state of your choice to a feature on Mars by placing those political areas on top of Mars. Compare Olympus Mons to France in an estimate percentage and also the raw number difference (in area).

Next, pan to and explore the North Pole and South Pole of Mars. What differences do you observe? Do additional research and explain these differences. What differences do you see in terms of physical features in the polar areas versus elsewhere on Mars? Examine this story map (https://www.arcgis.com/apps/Cascade/index.html?appid=96cc9a6f8df447b4940e3ebca611faba) for further investigation. Just like on Earth, latitudes and longitudes have been created on Mars for exploration and observation purposes. Note the latitude and longitude values on this map of Mars: https://pubs.usgs.gov/imap/i2782/i2782_sh1.pdf

How are these values determined on Mars? How are these values determined on other planetary bodies? The International Cartographic Association Commission on Planetary Cartography and other organizations designed the Mars longitude values to increase from 0 to 360 on Mars. So, unlike the Earth, where longitudes run from 0 to 180 east and 0 to 180 west, there is no eastern and western hemisphere on Mars. The Prime Meridian on Mars (https://www.esa.int/Science_Exploration/Space_Science/Mars_Express/Where_is_zero_degrees_longitude_on_Mars) was fixed at the Airy-0 crater in 1972 (). Use Google Maps to determine the latitude-longitude location of your school or university campus. Determine where this latitude-longitude location would be on Mars (adjusting for the longitude differences as described above). Refer to the USGS Mars map showing the latitude-longitude lines (Figure 4.34). If your campus in Denver Colorado USA was located at 39.7 North and 105 West, in what area would your school in the equivalent Mars location be located? It would be at 39.7 North, the same latitude as Earth, but the longitude would be 180 + (180 − 105) = 255 East. In the case of Denver, the "equivalent" location on Mars is on the Alba Patera plateau, which looks a bit like a kneecap, here (with a blue dot for 39.7 North 255 East). Do the same calculation for your own campus.

To dig deeper (https://www.arcgis.com/apps/mapviewer/index.html?webmap=b164957a19544f55a8d6a001c952a590): Compare and examine craters on Mars using ArcGIS Online. Open and sort the table of craters: Where are the largest three craters and the deepest three craters on Mars? How does the largest Mars crater compare to the size of Meteor Crater, Arizona?

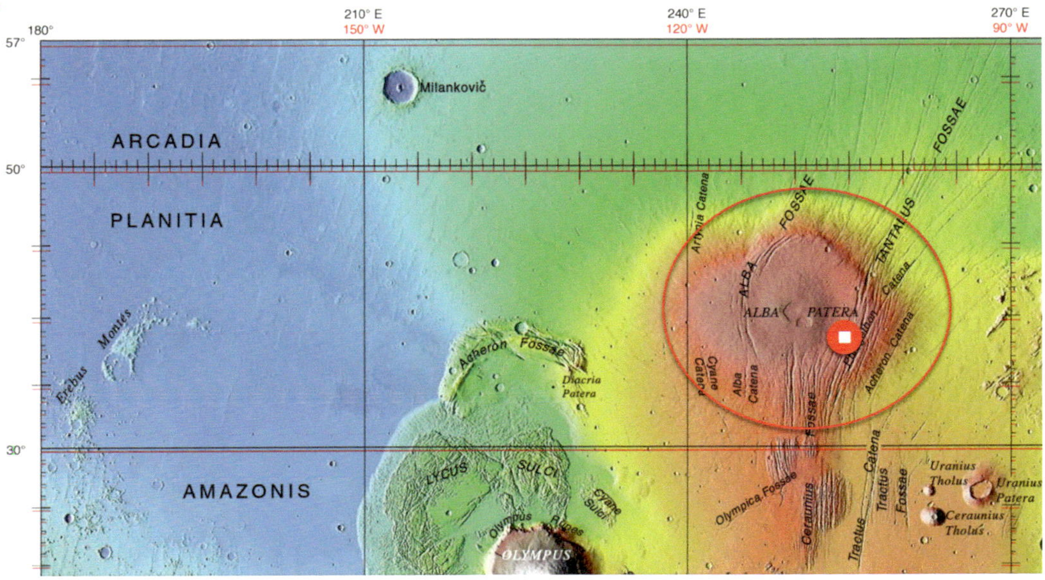

FIGURE 4.34 Analyzing and comparing coordinate systems on Mars to those on Earth. USGS.

QUESTIONS TO THINK ABOUT

Can you think of data sets which, when analyzed mathematically, produce confusing or even incorrect results when mapped? Look around you.

Find examples from your local news media that you think might be suspect in this regard. Do not actually do the analysis; explain why you would consider engaging in such further analysis with justification based on underlying mathematical concepts. Develop your instinct for ferreting out interesting math/geo situations.

5 Data, Variables, and Thematic Maps

The line of images above is a visual abstract of this chapter designed to foster spatial thinking. From the chapter numeral, to the book structure, to the real world, the reader is offered gentle guidance to develop spatial intuition about what might be coming. Those thoughts are then reinforced with a detailed text outline of chapter content below. The images and outline forge as an abstract of chapter content.

CHAPTER OUTLINE

Data
- Data Scale
- Data Types

Attributes
- Nominal Attribute
- Ordinal Attribute
- Interval Attribute
- Ratio Attribute

Generalization
- Selection/Filtering
  - 3D Globe of Extruded Grid Cells Representing Population
  - Understanding Global Population Using Filtering/Selection and a 3D Globe
- Classification
  - Manual Classification
  - Defined Interval
  - Equal Interval
  - Quantile
  - Natural Breaks (Jenks)
  - Geometrical Interval
  - Standard Deviation
- Simplification
  - Grouping
  - Clustering

DOI: 10.1201/9781003305613-5

Hurricane Position: Measurement and Analysis
Symbolization
Symbol Considerations
Visualization Data with a GIS Dashboard
Surveys
Nominal Scale Survey Questions
Ordinal Scale Survey Questions
Interval Scale Survey Questions
Ratio Scale Survey Questions
Thematic Maps
Reference Information
Standardization of Data
Uncertainty
Single Variable Mapping and Data Types: Choropleth Maps
Nominal Data Type
Ordinal Data Type
Interval Data Type
Ratio Data Type
World Earthquakes
Exploring Data Using Bivariate Maps: Symbology and Classification
Heat Map
Relationship Map
Questions to Think About

In Chapter 4, we noted that in the world of maps, two forms of measurement are critical in creating and understanding spatial displays and their analysis. One is the scale at which the map is shown and the other is the spatial data and populations that underlie the content to be mapped. We focused on the former in the previous chapter; in this chapter, we focus on the latter.

DATA

Data Scale

Maps are often based on information gathered from the field, as sets of data about particular variables. A variable is a measurable quantity with a value that changes across an underlying population. An underlying population, for example, might be a set of athletes. Variables associated with the elements of this set might be the name of the sport played, team name, batting average, wins and losses, errors, and others. The values will vary from individual to individual.

When the data is analyzed, it may be sorted into different groupings and those groupings then mapped to detect patterns of interest to help guide decision-making. To analyze sets of data, we need to understand variables and what might be measured using these variables; different levels of data measurement are classified broadly as qualitative or quantitative data. How a variable is measured, and categorized as a type, plays into how it may be analyzed and subsequently mapped.

Data Types

Interactive maps are powered by GIS. As you have seen, the "G" part of GIS is the map—it could be 2D or 3D; it could be at a local scale, a global scale, or any scale in between. Earth features can be represented as vectors (points, lines, and polygons) or raster's (a grid, or mesh).

The "I" part of GIS is the information part, typically stored in a table, or more commonly in a set or tables in a relational database. The fields in these tables contain the attributes, or characteristics, of the mapped features. These fields could describe whether a river is perennial or intermittent, the pH, volume, or dissolved oxygen of the water along specific segments of that river, the name of the river, or other characteristics of the river. Similarly, roads, census tracts, ecoregions, powerlines, or any other mapped thing above, on, or below the surface of the Earth could have a few or hundreds of attributes describing them.

ATTRIBUTES

The fields in the tables underlying a GIS can hold nominal, ordinal, interval, or ratio types of data. Each data type is an incremental level of measurement, meaning, each scale fulfills the function of the previous scale. The order of nominal, ordinal, interval, and ratio is a hierarchy.

Nominal Attribute

A nominal attribute describes the object in terms of its name or some other text—a river name, for example. Other nominal attributes could be city name, tree species, or soil type. Certain nominal attributes are not "more" or "better" than others; you cannot perform mathematical operations on them. Nominal attributes can also include information about images, videos, and sounds. Some nominal attributes can be numbers, such as the Feature ID that is common to all interactive map layers, or a latitude, longitude, or city street address.

Ordinal Attribute

Ordinal attributes imply a ranking or order based on values. These values can be descriptive text or numeric—in the example above, low, medium, or high river flow. Other examples could be low, medium, or high output power rankings for wind turbines, or a 1–10 ranking of community walkability ranking. In a walkability ranking, 10 may be more walkable than 1, but 10 is not necessarily 10 times more walkable than 1. Similarly, in a high-medium-low river flow ranking, high river flow is not three times more than low river flow.

Interval Attribute

Interval attributes imply a rank order and a magnitude or scale. The numbers here have an arbitrary rather than a natural zero. For example, on the Fahrenheit scale, 0°F is not a natural zero point for temperature, but rather is a human-defined one. Thus, 50°F is 10°F more than 40°F, but 50°F is not twice as hot as 25°F. Indeed, some of these rankings could be based on specific numbers, such as the boundary between low and medium river flow, while others could be an arbitrary ranking based on user input, as in the walkability example.

Ratio Attribute

With interval data, addition and subtraction make sense but not multiplication since values are relative from an arbitrary zero. Ratio data, by contrast, implies rank order and magnitude about a natural zero; its numbers allow addition, subtraction, multiplication, and division. For example, for the variable "Volume of the river in cubic feet per second (cuffs)," 200 cfs is 20 times more than 10 cfs.

GENERALIZATION

Maps make understandable the real, complex world. No map can show all the world's complexities and features on a single map, even if that map contains many layers, and even if it is in digital form. Even a single watershed might contain dozens of towns, hundreds of soil sample points, thousands of water wells, tens of thousands of buildings, and hundreds of thousands of trees and plants. Therefore, the real world must be generalized when it is represented on maps. One way to map the world is through the use of symbols. Symbol categories may be grouped into qualitative and quantitative visual variables. These are listed below (though there is some debate in the cartographic community on the "final list"). Think about these in conjunction with the types of data you are mapping—qualitative or quantitative, and if quantitative, whether that data is nominal, ordinal, interval, or ratio data.

Several methods exist to abstract or generalize mapped data. These include:

- Selection/Filtering
- Classification
- Simplification or generalization
- Symbolization.

Selection/Filtering

Selection, or filtering, refers to the process of choosing only a subset of a geospatial data set to show on a map. On a map of resort hotels, for example, you might wish to select only those with four or five stars. On the other hand, if you are only briefly sleeping in the room because you are at a conference or related social events for 18 hours a day, then you might select three star (less expensive) hotels. Consider your needs before selecting map features.

Another way of representing data is not by points, lines, or polygons (vector data), but by a mesh, or grid (raster data). Elevation, slope, temperature, and other continuously changing variables are often represented as grid data on maps. In a GIS, points, lines, and polygons are usually represented as *vector data,* while grids are typically represented as *raster data.* Grids can also be in 3D. Selection or filtering can be performed on raster data just as it can be performed on vector data.

3D Globe of Extruded Grid Cells Representing Population

Examine the following 3D globe, in which population density is represented by a set of raster cells at a specific resolution. Each raster cell has been extruded as a "cylinder" above the surface of the Earth: https://ralucanicola.github.io/JSAPI_demos/world-population/ (Figure 5.1).

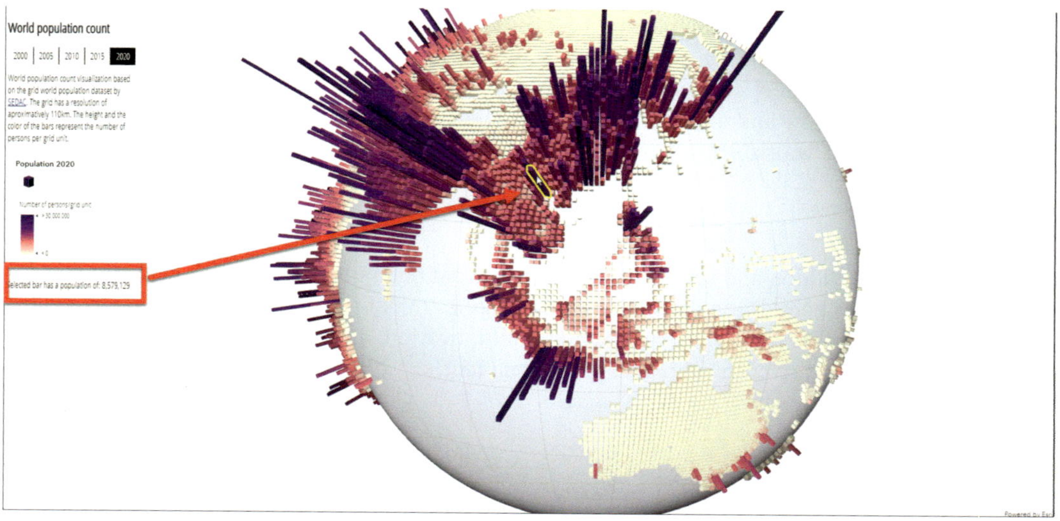

FIGURE 5.1 3D globe of extruded grid cells representing population. World population rendered on 3D globe. Data from SEDAC CIESIN using ArcGIS Maps SDK for JavaScript, created by Raluca Nicola.

This globe shows population density, not raw numbers of people. Each raster cell has a resolution of 110 km on each side, or 12,100 km^2 for each cell. Note the "towers" that represent the very highly densely populated cities of Singapore and Tokyo. Australia also contains some large cities but note how less densely settled those cities are. Maps sometimes challenge our pre-existing world view. How does this extrusion and the fact that it is showing density change your perception of world population? This globe uses data from the NASA Socioeconomic Data and Applications Center (SEDAC) hosted by Columbia University, from 2000 to 2020. Select different years on the globe and consider three changes you observe in Southeast Asia. Next, examine another region of the world, and make an observation. Which regions are experiencing the most rapid population increase? Why? What implications does the tendency of humans to live most densely along coasts have in light of storm surges and potential sea level rise? These maps and globes, and the data behind them, allow you to use mathematics and geography as a tool to engage in smart city planning.

Understanding Global Population Using Filtering/Selection and a 3D Globe

Examine the following globe:

https://ralucanicola.github.io/JSAPI_demos/world-population-2020/

This globe also represents population as grid data, but it allows population to be filtered (Figure 5.2). The grid's resolution is also 110 km; again, 12,100 km^2 is represented by each grid cell. Rotate the globe and observe the patterns. Aside from a few places in northern Canada and central Asia, central Australia, and Antarctica, at least 10 people are living in each grid cell.

Use the filter tool under the globe to filter out the areas of lowest population. Move the filter tool and observe how when lower population areas are filtered out, they become invisible. Pan to different points around the world, observing global and regional patterns of population. What is the value at which only half of Australia is visible? In the graphic shown here, the filter is set to 199 persons per grid cell.

What is the value at which only half of Japan is visible? Think about why the values for Australia and Japan are different. Next, filter out the highest population densities, panning the globe as you do so. What global patterns do you notice where you filter out the *highest* population densities? How does filtering the data help you understand population distribution differently from the globe you examined earlier?

FIGURE 5.2 Population as grid data, with filtering. Data from SEDAC. Original mapped from ArcGIS API for JavaScript.

Classification

Classification is the process of combining observations into bins or classes. Humans can only understand a small finite number of classes on a map. The goal of classification is to reduce the number of unique symbols on a map, as well as to aggregate information to a useful geographic unit, such as a census tract or a country. This goal is generally achieved in one of the following ways:

- Simplify the visualization so that the map-reader will understand what you are trying to convey.
- Group similar observations together.
- Show the differences between the groups.

Suppose you have some data for a set of counties in Ohio of population. Maps usually cannot show every value in a data set, so what sort of simplification and classification strategy can you use? How would you divide up that population data, or classify it, on a map? The main methods of classifying data are enumerated as follows.

Manual Classification

In manual classification, you manually add class breaks and set class ranges that are appropriate for your data (Figure 5.3 shows a comparison of different types of classification).

Defined Interval

In this method, you specify an interval size used to define a series of classes with the same value range. For example, each interval will span 75 units. Your GIS then determines the number of classes based on the interval size and the range of all field values.

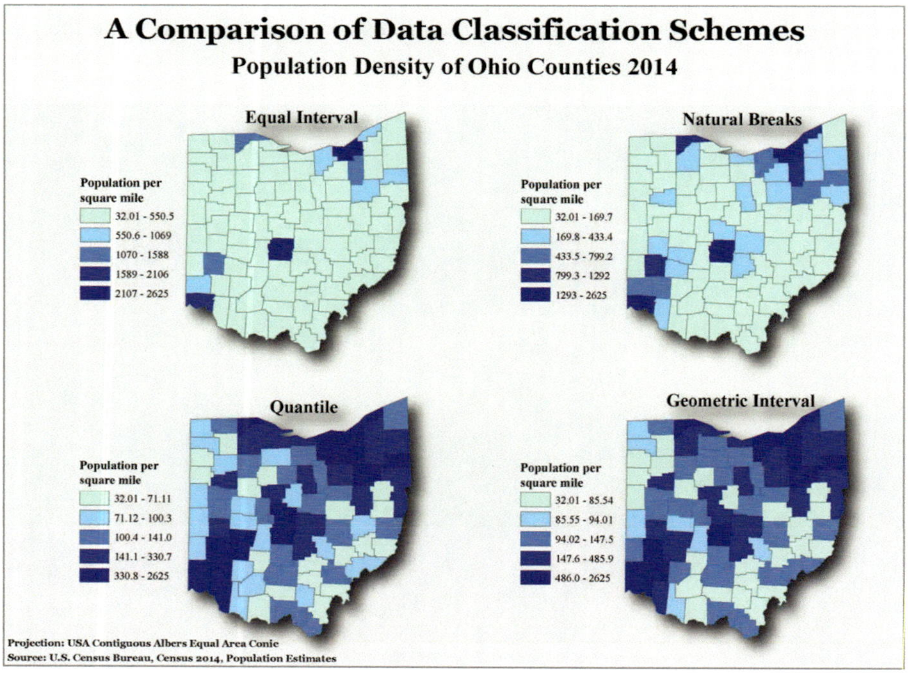

FIGURE 5.3 A comparison of four different classification methods for population density in Ohio by county. US Census Bureau.

Equal Interval

This method divides the range of attribute values into equal-sized subranges. This allows you to specify the number of intervals, and your GIS will automatically determine the class breaks based on the value range. For example, if you specify three classes for a field whose values range from 0 to 300, the application will create three classes with ranges of 0–100, 101–200, and 201–300.

Quantile

In quantile classification, each class contains an equal number (quantity, hence "quantile") of features. A quantile classification is well suited to linearly distributed data. Quantile assigns the same number of data values to each class. For example, if you have 333 weather stations, and map them on temperature by quantile, with three classes, there will be 111 stations mapped in each class. Because features are grouped in equal numbers in each class using quantile, the resulting map can be misleading. On the other hand, quantile can be very useful in many instances. Similar features can be placed in adjacent classes, or features with widely different values can be put in the same class. You can minimize this distortion by increasing the number of classes.

Natural Breaks (Jenks)

Natural breaks classes (devised by cartographer George Jenks of the University of Kansas) are based on natural groupings inherent in the data. Class breaks are identified that optimize the grouping of similar values and that maximize the differences between classes. The features are divided into classes whose boundaries are set where there are relatively big differences in the data values.

Geometrical Interval

This method creates class breaks based on class intervals that have a geometric series. The geometric coefficient in this classifier can change once (to its inverse) to optimize the class ranges. The algorithm creates geometric intervals by minimizing the sum of squares of the number of elements in each class. This ensures that each class range has approximately the same number of values within each class and that the change between intervals is fairly consistent. This was specifically designed to accommodate continuous data. It is a compromise method between equal interval, natural breaks (Jenks), and quantile (Jenks, 1967; McMaster and McMaster, 2002). It creates a balance between highlighting changes in the middle values and the extreme values, thereby producing a result that is visually appealing and cartographically comprehensive.

One example for using the geometrical interval classification could be with a rainfall dataset in which only 15 out of 100 weather stations (less than 50%) have recorded precipitation, and the rest have no recorded precipitation, so their attribute values are zero.

Standard Deviation

This shows how much a feature's attribute value varies from the mean. Your GIS calculates the mean and standard deviation (https://geographyfieldwork.com/StandardDeviation1.htm) of your data, and class breaks are created with equal value ranges that are a proportion of the standard deviation—usually at intervals of one, one-half, one-third, or one-fourth standard deviations using mean values and the standard deviations from the mean.

Again, the classification method you choose for your maps has a direct impact on the way in which the readers of your map perceive and understand (or fail to understand) your data. Consider the graphic below. Which map do you think best communicates population density in Ohio?

Simplification

Simplification refers to visually reducing complexity on your map so that it is understandable. For example, do you really need to show **all** of the islands of the Philippines on your map of Southeast Asia, or **all** of the river meanders in your map of your local watershed? More information is not necessarily better—remember that the goal in mapping and visualization is to make things clearer; to aid understanding. Mapping tools allow you to reduce complexity but still maintain the character of your data, such as by reducing the number of vertices in your line segments or polygon boundaries.

Grouping

Another way of simplifying information for a map is also grounded in mathematics, called grouping.

Open the linked map (https://www.arcgis.com/apps/mapviewer/index.html?webmap=56a9e68b8669430fb7b095b12959f48e) of public libraries in Iowa (Figure 5.4). The data set contains 570 libraries, and you will note that they are currently mapped by floor size. Thus, aggregating the data might enable you to see patterns more clearly. Begin with clustering: On the left, > make the layer of libraries active > on the right, > Aggregation > Clustering. The circles now contain numbers representing how many libraries are within that geographic area, with the size of the clustered libraries indicated by the darkness of the circle and the number of libraries in that geographic space represented by the size of the circle. The number of libraries within that space is given as a number within the circle (Figure 5.4).

Under Clustering > Options > experiment with the cluster radius and the size range. Which radius and size range in your opinion yields the most understandable map? Note that under Cluster fields, you can change the variable that is desired to be mapped and clustered. Binning is another method of simplifying the map. Binning creates polygons such as hexagons or rectangles for data and maps and summarizes variables within those polygons. On the right, still under Aggregation, > Binning > Options > enlarge the bin size a bit (shown in Figure 5.5), and you will see the pattern of the density of public libraries in Iowa. What do you think is the most important factor that influences the presence of public libraries in various parts of Iowa or elsewhere?

Change the symbology to add to your understanding: Under Styles > Style Options > use the "above" theme to easily focus on areas in Iowa where there tend to be more libraries. Note the major cities where these library clusters are (Figure 5.6).

Clustering

The previous activity offered some insight into the process of clustering of data to facilitate ease of map reading and interpretation. The next activity offers more.

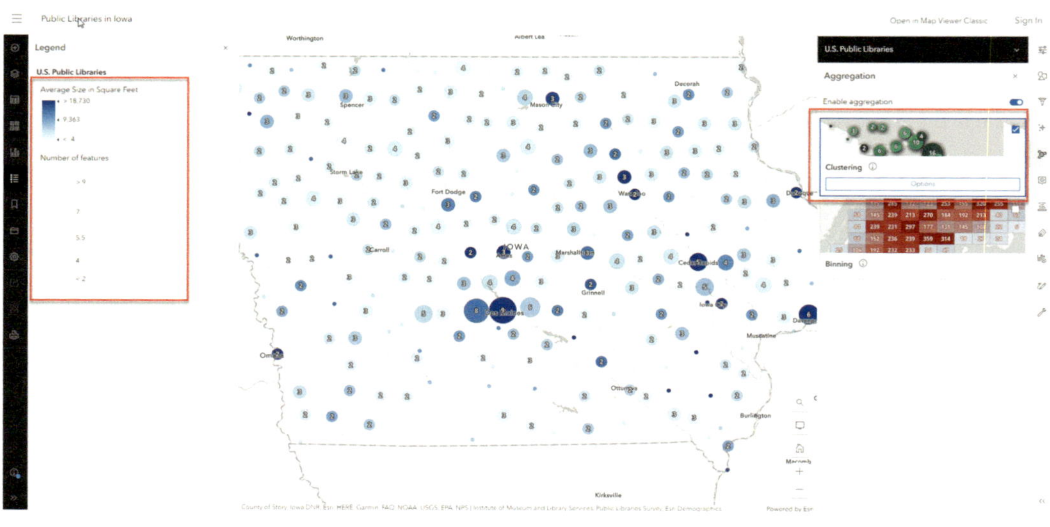

FIGURE 5.4 Public libraries in Iowa shown with a clustering technique. Mapped using ArcGIS from Esri.

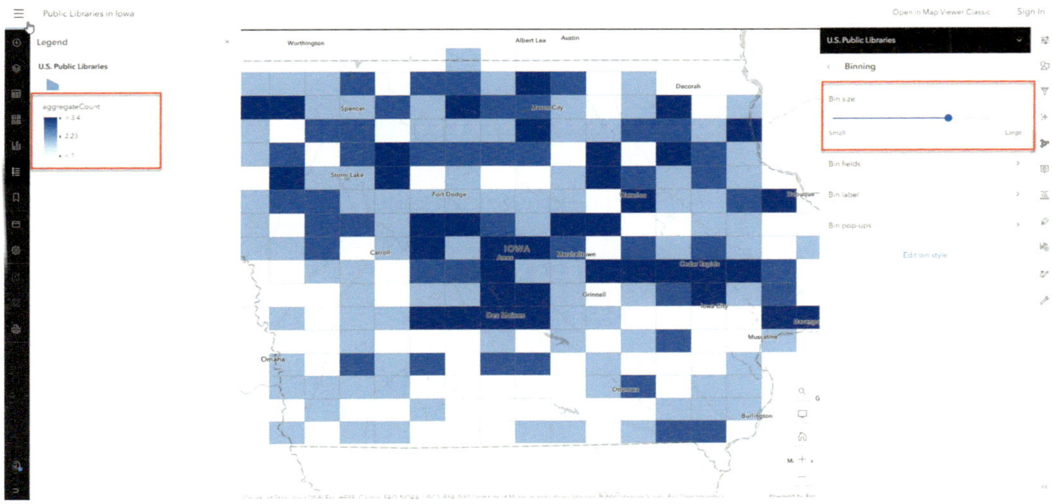

FIGURE 5.5 Representing data on a map using a binning technique. Mapped using ArcGIS from Esri.

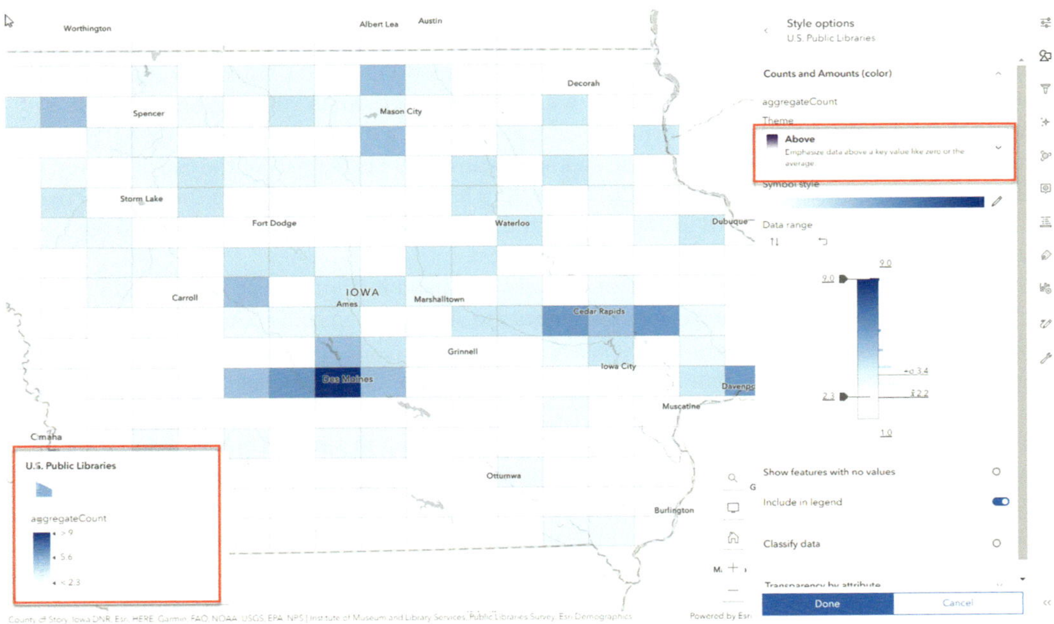

FIGURE 5.6 Libraries in Iowa are shown using binning technique and the above symbology. Mapped using ArcGIS from Esri.

Hurricane Position: Measurement and Analysis

Access the National Geographic MapMaker at https://mapmaker.nationalgeographic.org/ and then > Create A Map (Figure 5.7).

You will see a list of layers on the left side of this interactive mapping tool. Search for the layer "Longitudes and Latitudes." Add it to the map. Note that you need to zoom in to a larger scale to see the layer. Note that you cannot see every line of latitude and longitude, but rather, they have been filtered to show every 5th line of latitude and 5th line of longitude. That is, the lines of latitude and

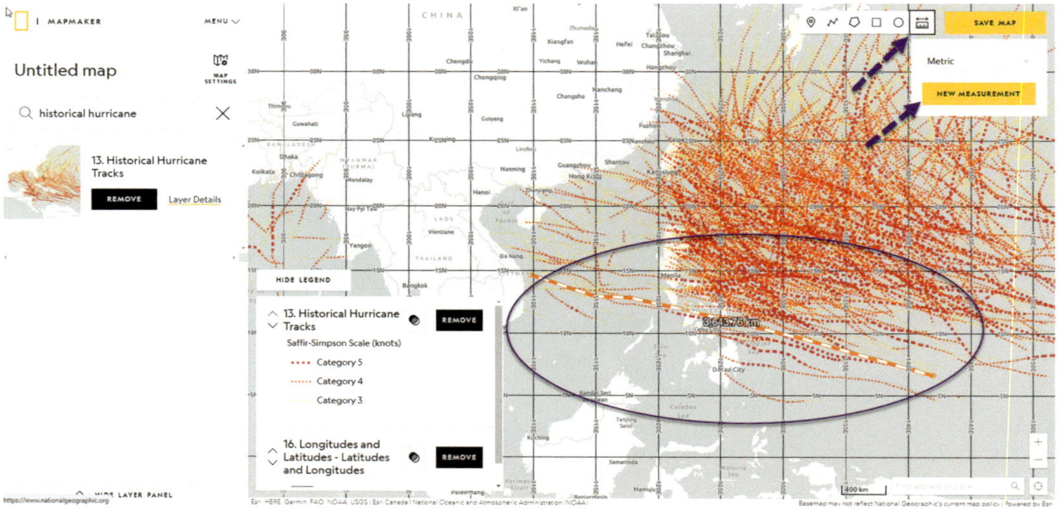

FIGURE 5.7 Historical hurricane tracks in the Pacific Ocean. National Geographic Society MapMaker and Esri.

longitude are spaced apart in 5° increments. The Equator is the line between 5 N and 5 S, having the latitude of 0°. The International Date Line is the line between 175 W and 175 E, having a longitude of 180° East and 180° West simultaneously, with the Prime Meridian running North-South at 0° East and West simultaneously. The northern hemisphere is the part of the Earth between 0° latitude proceeding northward to 90 North, and the southern hemisphere is the part of the Earth between 0° latitude proceeding southward to 90 South. The eastern hemisphere is the part of the Earth between 0° longitude proceeding eastward to 180 East, and the western hemisphere is the part of the Earth between 0° longitude proceeding westward to 180 West.

Search for the layer "Historical Hurricane Tracks" and add it (Figure 5.7). Notice in the legend that **only** the Category 4 and Category 5 hurricanes are shown. Even so, according to the metadata for this map layer, during the long time span of this data set (1842–2020), there have been plenty of hurricanes. They are more commonly known as cyclones in the Pacific and Indian Oceans, and hurricanes in the Atlantic Ocean.

Maps often help us to see clusters of data that may be hidden in a data table. When certain phenomena are clustered on maps, they help us take notice and encourage action to be taken. Clusters revealed by mapping data such as hurricanes are of concern to many people, from ship captains to those living near coasts. These clusters help us predict the location of future storms.

Observe the patterns of hurricanes on the map. Which two locations on Earth would you say are the locations of the two most significant clusters of hurricanes? Certainly, the western Atlantic Ocean and the western Pacific Ocean are prime candidates for the most clusters. Zoom in so that you can see the latitude-longitude lines and the hurricane tracks. Note that the hurricanes are clustered in two bands, one to the north of the Equator and one to the south of the Equator. Between which lines of latitude would you say 50% of the hurricanes occur in the western Pacific Ocean in the northern hemisphere? Between which lines of latitude would you say that 90% of the hurricanes occur in the western Pacific Ocean in the northern hemisphere?

Due to the position of land masses, ocean temperatures, and other factors, note that the bands of hurricanes in the southern hemisphere contain far fewer hurricanes than in the northern hemisphere. In fact, you could probably count all of the southern hemisphere hurricanes on this map. Often, sampling gives us some keen mathematical insights. Count the number of hurricanes that crossed into the part of the Earth between 10 South and 15 South, and between 135 East and 140 East. Then, count the number of hurricanes that crossed into the part of the Earth due north of this "square," between 10 North and 15 North, and between 135 East and 140 East. You should zoom to a larger scale when counting the northern hemisphere hurricanes for a more accurate count. Add your southern hemisphere total to your northern hemisphere total. What percentage of the total number of hurricanes occurred in the southern hemisphere?

Knowledge of an additional mathematical fact can assist in the understanding of a phenomenon. In the northern hemisphere, hurricanes move from east to west. In the southern hemisphere, they generally move from west to east.

Consider the southeast coast of the USA. Given the discussion above, where do you think is the origin for most of the hurricanes in the North Atlantic Ocean? Note that many of the hurricanes change direction once they strike land, and curve to the north and to the east. This is due to the Coriolis Effect, because the Earth is rotating from west to east on its axis. The Coriolis Effect also affects the rotation of the hurricanes themselves, as we will shortly see. This effect increases the potential wind and flood hazards from hurricanes if it lengthens the hurricane's duration.

Consider the southeast coast of Asia. Where do you think is the origin for most of the hurricanes in the North Pacific?

Westward-flowing air masses pick up warmth and energy as they blow across the Atlantic and Pacific Ocean, respectively. They are dangerous in part because they spend most of their lifespans over water, gaining energy from warm waters, and then striking the southeast coast of North America and Asia, respectively. These often pose grave danger for those sailing those waters or living on islands and continents. Indeed, due to the storm surges and winds coupled with the heavily populated coastlines along the USA and along Southeast Asia (China, Taiwan, Japan, Philippines, Vietnam), they pose persistent threats to the people living there and to the mariners sailing those offshore waters.

Next, use the measure tool (Figure 5.7) to measure a hurricane of your choice. For example, the hurricane measured in the example was 2,567 km long. The average forward speed of a hurricane is about 16 kph. If this hurricane traveled at the "average forward speed," how many hours did the hurricane exist as a category 4 or 5 storm? How many days did the hurricane exist? Perform these calculations for a hurricane you find and measure and compare to the hurricane shown. Can you find a hurricane that was over 4,000 km long?

However, one of the most dangerous components of hurricanes and other natural hazards is that they often do not follow an "average" in anything, whether it is lifespan, wind speeds, forward motion, or even a "standard path." For example, Hurricane Dorian did not move forward for 52 straight hours as it stalled out over the Bahamas and lashed the islands as a Category 4 or 5.

Furthermore, the lines you are examining on the map represent the central path of each storm: They have been generalized for clarity on the map. In reality, hurricanes are not thin lines at all, but are composed of a swirling mass of clouds and high winds. This cloud bank and associated severe weather is much wider than these narrow lines on the map indicate. Hurricane Kate of 2003, for example was 300 km wide—the eye alone was 32 km wide (Figure 5.8). How does knowing an additional mathematical fact—in this case, the width of these hurricanes, change your assessment of the amount of ocean and coastline that is in danger during these storms? Further compounding the issue is that hurricanes rotate or spin as they move, which helps them gain and sustain energy, increasing winds, precipitation, and storm surges. As they move forward, they also spin counterclockwise in the northern hemisphere and clockwise in the southern hemisphere. Finally, a central theme of this book is to always think about the data, and how that data is gathered and represented on maps. Knowing that this data set spans nearly 180 years, and knowing that for most of those years, the locations of the hurricanes were not derived from satellites and GPS but rather, from observations from ship captains and damage reports on land. Thus, the data set represents a mixture of plotted data in high precision and interpolated lines with a low degree of precision. Again, maps are representations of reality—very useful representations, but need to be viewed critically and thoughtfully. All maps and resulting analyses are dependent on the quality of the data that are being mapped.

FIGURE 5.8 A satellite image of Hurricane Kate in the mid-Atlantic on 4 October 2003. NASA, Public Domain - https://en.wikipedia.org/wiki/Tropical_cyclone#/media/File:Hurricane_Kate_(2003)-_Good_pic.jpg.

Think about: how did the combination of mathematics and geography with some hands-on work with mapping tools help you deepen your understanding about the origin, clustering, and direction of movement of hurricanes? How did they help you to consider once again "what is where, why is it there, and why should we care?"

Symbolization

The fourth method of abstraction and generalization is symbolization—this is the process of assigning such visual variables as shape, color, size, and orientation to a feature (Figure 5.9). Real-world features, such as mountains, powerline transmission towers, or entire cities, are far too complex to be represented in high detail on maps. Each must be represented as symbols, and these symbols can greatly aid in map interpretation. For example, choosing a school symbol for each school that you are mapping will help your reader to quickly take note wherever they see one on your map, and thus focus on your main message, instead of needing to pause and learn about a new symbol every time they encounter another school.

Visualizing Data with a GIS Dashboard

Earlier, you read and reflected on an article entitled *Geography in Everyday Life*, https://www.directionsmag.com/article/11897. You also participated in a survey where you were asked to think about how you might think spatially throughout a typical day and examined an infographic.

Now, visualize (https://www.arcgis.com/apps/dashboards/cf80c2d91102482782fc1c1bd81052c4) the results of your survey and the results of others who have filled out the survey in a dashboard (Figure 5.10 shows a screen shot of a dashboard).

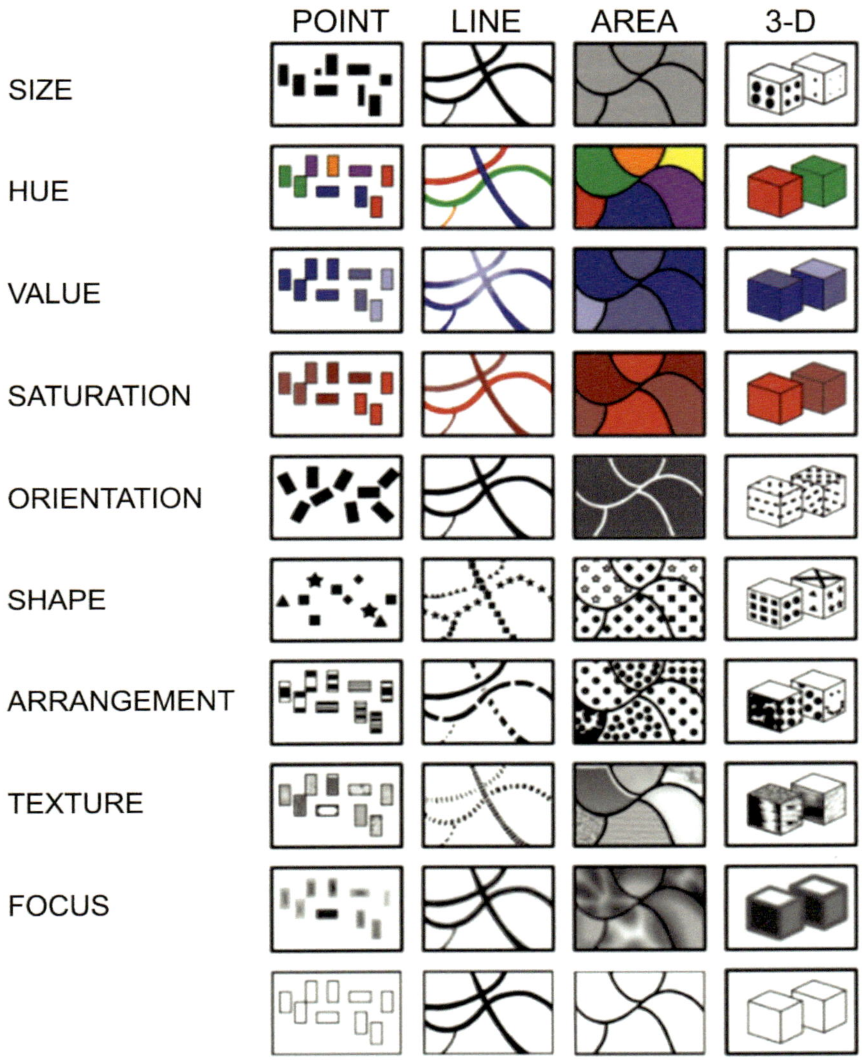

FIGURE 5.9 Symbol considerations. From Affordable Learning Georgia: https://alg.manifoldapp.org/read/introduction-to-cartography/section/21773ec3-7eb3-4197-b46d-8a7c2ebaa25f Introduction to Cartography open course. Open licensing: https://www.affordablelearninggeorgia.org/resources/create/open-licensing-and-copyright/

Like the dashboard in your car, a GIS dashboard provides a great deal of information in a small amount of space. Also, as with a car dashboard, a web GIS dashboard provides information in real time. This dashboard provides the *current results* of the survey that is linked to the article about geography in everyday life.

As people fill out the survey, the dashboard will continue to update and reflect current results, until the creator of the survey decides to close further input. Like many web GIS dashboards, this ArcGIS dashboard contains several familiar mathematical objects—a map, a serial chart, pie charts, and an indicator—a number indicating the total number of surveys filled out. Which elements in this dashboard do you find most informative, and why? Which elements do you find least informative, and why?

Next, compare the two pie charts in the middle of the dashboard. The top chart contains labels for the responses, with numbers provided representing the percentage of the total survey respondents

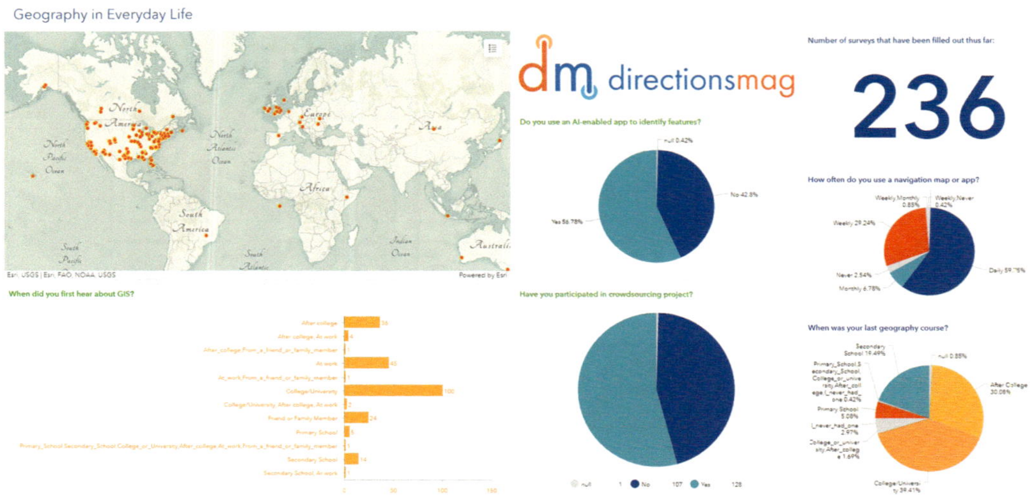

FIGURE 5.10 A dashboard showing a real time representation of information gathered in a survey. Esri.

who answered Yes and No to that question. The bottom chart also represents the responses to a Yes/No question, but it is formatted with a legend underneath the pie chart, rather than labels, and the raw numbers of respondents rather than the percentage. Which of these two pie charts do you find most informative, and why?

Are there any pie charts that you feel would be more clearly understood as a serial chart, or vice versa? Note that the colors on the dashboard, such as the blues, teals, and oranges, use the same color palette as those in the *Directions Magazine* (the publishers of the article) logo at the top of the dashboard. How do these color choices affect the readability of the dashboard?

Note that the dashboard includes a map indicating each respondent's location. What symbol on the map for each respondent's location might be better understood than an orange point? The points could be symbolized according to some attribute, such as "how often do you use a navigation map or app." However, given the number of mapped points and its small size in the dashboard, mapping by attribute instead of the same symbol used for all points might make the map more difficult to interpret.

Dashboards, infographics, and other visualization tools have proliferated over the past decade. Dashboards are powerful and easy to create, to embed in other media, and to share. They are increasingly able to be tied to mapped data and are increasingly populated from real-time data feeds. The most popular dashboard of the past decade, and indeed of all time, is the Johns Hopkins University (JHU) Coronavirus Resource Center's COVID dashboard: https://coronavirus.jhu.edu/map.html (Figure 5.11). This dashboard has been viewed in the trillions by almost everyone on the planet with a phone, tablet, or computer, and rightly so, for it was and remains a useful tool for individual decision-making and community policymaking.

Spend a few moments re-familiarizing yourself with the JHU COVID dashboard. What is one similarity and one difference that you notice between the geography in everyday life dashboard and the JHU COVID dashboard? Note that even though the JHU COVID dashboard is still presented as a single screen of data, the dashboard's tabs allow the user to access related visualizations and maps at different scales and covering different regions. Note also that the Everyday Life Geography dashboard receives data from a single data feed—that of the survey that the reader of the article fills out, while the JHU COVID dashboard receives feeds from many sources, including national and regional health agencies.

Dashboards are easily misinterpreted. Dashboards are also so easy to create that proper thought and care can be neglected. As you create your own dashboards, infographics, and other visualization

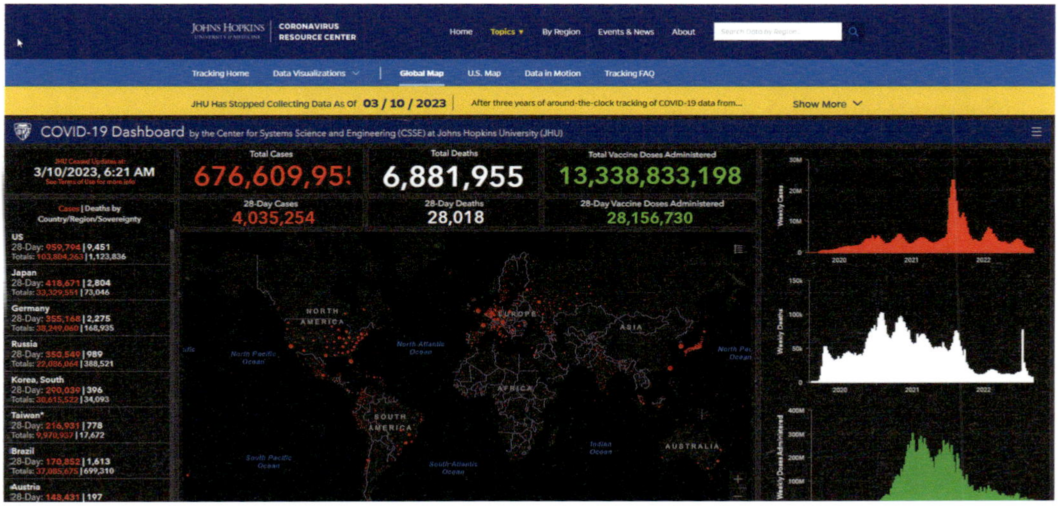

FIGURE 5.11 The Johns Hopkins University COVID-19 dashboard. Powered by Esri.

tools, you will have many choices regarding the maps, images, charts, and other elements that you could include. These choices involve the number of elements, the type of each element, the size of each, the colors in each, the categories in each, and many other choices. More of anything (elements, charts, and so on) is not necessarily better. The objective of any of these visualization tools is to communicate clearly and effectively. Make these choices wisely: seek feedback on what you create. Regularly examine and evaluate others that are regularly produced, across many themes and scales, deciding what you like and do not like about each.

SURVEYS

Nominal, Ordinal, Interval, and Ratio data types are measurement scales that can be used to capture data in the form of surveys and questionnaires using multiple-choice questions.

Nominal Scale Survey Questions

Nominal Scale is used for labeling variables and placing them into distinct classifications; it does not involve a quantitative value or order. An example of such a question is:

Where do you live? 1. City; 2. Suburb; 3. Rural area.

Another use for nominal scale might be for a question where only the variable labels matter.

For example, a customer might be asked "Which brand of toothpaste is your first choice? 1. Brand X; 2. Brand Y; 3. Brand Z.

In this survey question, only the names of the brands are significant; order is irrelevant. There is no need for any specific order for these brands. In analyzing the results, the most frequently chosen brand will emerge: the mode of the distribution of answers.

Nominal data is often collected in multiple-choice questions such as the ones above. Open-ended questions might also be used.

Ordinal Scale Survey Questions

Recall the hierarchical nature of the four Data Types. Thus, Ordinal Scale may be thought of as Nominal Scale with Order added.

Market Survey Questionnaires often employ Ordinal scales, with questions such as:

Were you satisfied with our product? 1. Dissatisfied; 2. Somewhat dissatisfied; 3. Neutral; 4. Somewhat satisfied; 5. Satisfied.

Ordinal scale is a step above nominal scale: the order, as well as the naming, is relevant in the results.

Interval Scale Survey Questions

Recall the hierarchical nature of the four Data Types. Thus, Interval Scale may be thought of as Ordinal Scale where the difference (interval) between variables is known.

For example, consider a Celsius/Fahrenheit temperature scale: 80° is always higher than 50° and the difference between these two temperatures is the same as the difference between 70° and 40°. The value of 0 is arbitrary, if present.

The Celsius/Fahrenheit temperature scale is a classic example of an interval scale. This scale permits the use of statistical analysis of provided data. Mean, median, or mode can be used to calculate the central tendency in this scale. The Interval scale quantifies the difference between two variables whereas nominal and ordinal scales are solely capable of associating qualitative values with variables. All the techniques applicable to nominal and ordinal data analysis are applicable to Interval Data. In addition to those techniques, descriptive statistics apply as well.

Ratio Scale Survey Questions

Again consider the hierarchical nature of the four Data Types. Thus, Ratio Scale may be thought of as Interval Scale with a True Zero (origin or starting point). With the option of true zero, varied inferential, and descriptive analysis techniques can be employed.

Some examples of ratio scales involve weight and height. In market research, a ratio scale might be used to calculate market share, annual sales, or the number of consumers.

Ratio scale provides the most detailed information because researchers and statisticians can calculate the central tendency using statistical measures such as mean, median, mode, as well as methods involving geometric mean, the coefficient of variation, or harmonic mean. To decide when to use a ratio scale, the researcher must observe whether the variables have all the characteristics of an interval scale along with the presence of the absolute zero value.

THEMATIC MAPS

A thematic map portrays the spatial pattern of a specific theme, or subject matter, within geographical polygons. It is distinct from a reference map which focuses on the location and names of places or features. Thematic maps often involve the use of map symbols or shading to visualize attributes of geographic features that are not visible, such as temperature, language, or population. There are many different types of thematic maps; a course on mapping would likely group them all together into a single chapter. Our focus in this book is on the use of these maps to teach mathematics (and reinforce areas of mathematics that typically cause problems to students) using maps. Thus, we introduce many types of thematic maps in various chapters forward from this one as they support our primary focus: on mathematics.

By focusing on a single subject, the thematic map can be used for a narrower range of tasks than a reference map. These tasks tend to fall into three categories:

1. Provide specific information about specific locations.
2. Provide general information about general locations.
3. Compare spatial patterns on two or more maps.

Reference Information

While thematic information is the primary feature of a thematic map, other geographical features may also be included as reference information. The purpose of reference information is to establish the location of the thematic information in a context understood by the map-readers (i.e., to answer questions such as "where is this red region in the real world?"). Common reference layers include government administrative boundaries, roads, cities, a latitude/longitude graticule, or even terrain. These layers play a secondary role in the use of the map, so they are usually included sparingly, but not so faded that they cannot be used.

Standardization of Data

An important consideration in general thematic mapping is whether to present data as raw counts or totals (e.g., the population of each county) or to standardize (normalize) data to create rates or ratios (e.g., number of people in each county *divided* by the area of that county=number of people per square mile/km). In this population example, the raw counts tell you how many people exist in each county, and the ratio tells you how tightly packed-in those people are (population density). One of the primary reasons to standardize (or normalize) data is to allow readers to compare places that are very different. Compare a large place (like Canada) with a smaller place (like Switzerland). Although Canada has more people than Switzerland, it has far lower population density; without standardization, that fact might not be obvious. When making a map showing data aggregated over predefined geographical regions by coloring or shading these regions, it is generally advisable to use only standardized data if the maps are to be used for comparative purposes. If map-readers are to understand *magnitudes*, use map totals/counts. If map-readers are to understand *relative differences*, use standardized data. Be sure to check to see if the data sets are already standardized when you receive them!

Uncertainty

Uncertainty is a complex concept which has been defined differently by various authors. For example, Longley (2005) define uncertainty as "the difference between a real geographic phenomenon and the user's understanding of the geographic phenomenon." Many variations of uncertainty may emerge during mapmaking—during data collection, data classification, visualization, map-reader interpretation, and more (Kinkeldey and Senaratne, 2018).

Uncertainty is a rich topic; consider the following examples:

- Read the linked discussion (https://spatialreserves.wordpress.com/2015/11/22/understanding-your-data-it-is-critical/) and discuss mapping earthquakes offshore off the coast of Peru: Are the earthquakes on or offshore? It depends on the scale of the map you use for the shoreline. A large-scale detailed shoreline might show a particular earthquake as on shore, while a small-scale shoreline might show the same earthquake as occurring offshore. Why does this matter? Consider the different impact from an offshore earthquake producing a tsunami versus an on-shore earthquake producing ground shaking.
- Read the linked discussion (https://spatialreserves.wordpress.com/2012/11/05/maps-as-representations-of-reality-the-deciduous-coniferous-tree-line/) about the "boundary" between the deciduous versus the coniferous tree "zone" in Wisconsin, Michigan, and Minnesota. Rather than a sharp boundary on the land between two areas or polygons, this "boundary" is really an "ecotone"; a zone, where vegetation gradually changes from predominantly deciduous tree cover to the south to predominantly coniferous to the north. How to best represent this and other lines that are in reality "transition zones" accurately on maps is the subject of much discussion and research.

- Open the linked web map (https://www.arcgis.com/apps/mapviewer/index.html?webmap=771b15f99bb24e96b5bc4d7617cc2f51) of a GPS collected set of tracks from a hike: Use Bookmarks>Water Tank, and observe one of the tracks up and down a steep slope with one of the legs of the track that is mapping 10 m off of the trail, even though the data collector was hiking on the trail for both the uphill and downhill portions of the hike. Why is the collected track not "on" the trail as depicted in the satellite image? It could be due to GPS satellites in the western part of the sky being hidden by the steep ridge behind the person collecting the data, or hidden based on where the hiker was holding the GPS unit, or other factors. In addition, the satellite image base map itself contains uncertainties because it has been warped and modified through photogrammetric methods to fit the oblate sphere that is the Earth. And ultimately, all coordinate systems and projections that maps are based on are based on a calculated value of the Earth's shape. This value's accuracy increases over time as mathematical models improve and as Earth sensors become more numerous and more accurate, but lacking a tape measure to drape around the whole Earth, our estimate of the true shape and size of the Earth is just that—an estimate. Therefore, all maps, which are based on that shape, are imperfect, and all positions on those maps uncertain.

It is therefore appropriate to assume that all geographic data contain some level of uncertainty, and it is equally appropriate to consider the trust and ethics of one's sources.

What examples of uncertainty do *you* see in the world around you?

SINGLE VARIABLE MAPPING AND DATA TYPES: CHOROPLETH MAPS

A choropleth map shows data aggregated over predefined geographical regions, such as counties or states, by coloring or shading these regions. For example, counties with higher rates of coronavirus disease might appear darker on a choropleth map. (Please note the spelling of the word "choropleth"; the first syllable is derived from the Greek word, choros, meaning "region." It has nothing to do with a similar, but not identical, spelling of a green plant pigment).

The success of many thematic maps, including choropleth maps, depends on matching the right data type with the right map symbols. Not all geographic data types are the same, nor can they all be mapped in the same way.

Nominal Data Type

Nominal data types have no numbers attached to them. They simply represent different types of things. They cannot be ranked or analyzed numerically. Compare and contrast Reference Maps with Nominal Data Types. What types of geographic entities might be represented by nominal data types? How might they be mapped? What might be the role of statistical measures of central tendency?

Ordinal Data Types

Mapping ordinal data (and nominal data that, in addition, can be ordered) might be represented using a choropleth map in which the number of classes representing shading of the variable is equal to the number of data categories. It is often useful to have the color scheme be sequential to match the data pattern: lighter blues might represent lower population densities and darker blues might represent higher population densities, with the change in color shifting gradually through the associated mid-ranges in data values.

Interval Data Type

Any attribute that can be counted will make a fine thematic map. With interval data, there may be no zero value, so this mapping type might include positive and negative values.

Ratio Data Type

Again, anything that can be counted will make a fine choropleth map. With ratio data, there is a true zero, or starting point.

World Earthquakes

You will now have the opportunity to examine data on a map—of the world's most intense and deadliest earthquakes. Open the linked map (https://www.arcgis.com/apps/mapviewer/index.html?webmap=167402e476f84e2e95b52e2b19cd10f7) of these earthquakes (Figure 5.12).

Open the table for the most intense earthquakes. Which fields represent nominal, ordinal, interval, and ratio data? For interval or ratio data, you should be able to change the style, or symbology, to indicate a larger or different color symbol for something that is "more" than something else, such as magnitude or fatalities. Indeed, if you open the tables for each layer, notice that both the most intense and deadliest earthquake layers are in fact just pointing to the same layer—each layer simply focuses on one of the attributes and maps it. Try changing the color and symbols for earthquakes. Sort the table in ascending and then in descending order for selected fields such as magnitude and deaths. Where was the most intense earthquake according to this data set (magnitude) and what was its magnitude? Where was the deadliest earthquake according to this data set and how many were killed? Note the latitude and longitude fields. These are numbers, and are necessary to map each earthquake, and can be sorted. However, is latitude and longitude really an interval or ratio data type? What data type do you believe that it is?

EXPLORING DATA USING BIVARIATE MAPS, SYMBOLOGY AND CLASSIFICATION

A thematic map or choropleth map usually focuses on visualizing the distribution of values of a single type of feature (a *univariate* map), occasionally including two (bivariate) or more (multivariate) feature types that are hypothesized to be statistically correlated or otherwise closely related. These multivariate maps may include reference map layers as well as multiple data layers.

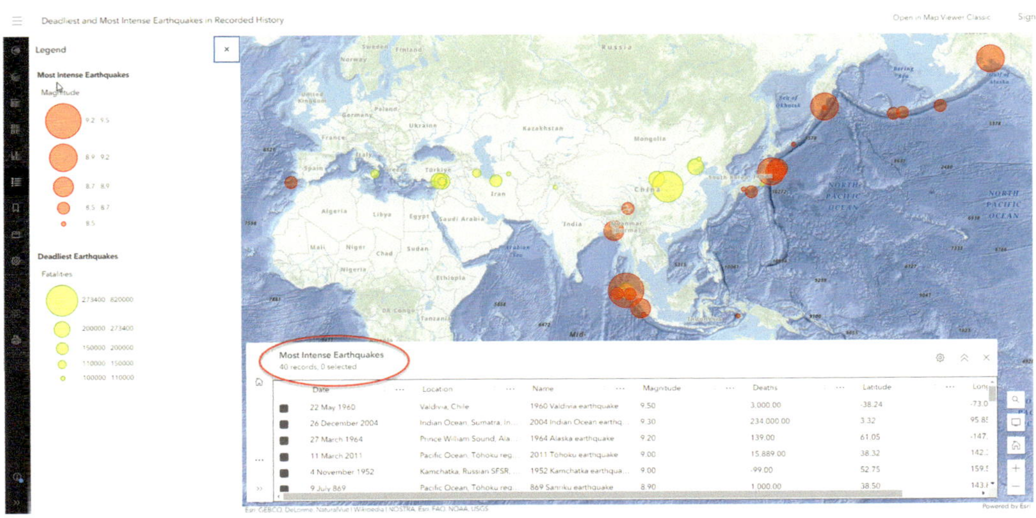

FIGURE 5.12 An interactive map showing the deadliest and most intense world earthquakes. Source: Joseph Kerski, using ArcGIS Online from Esri.

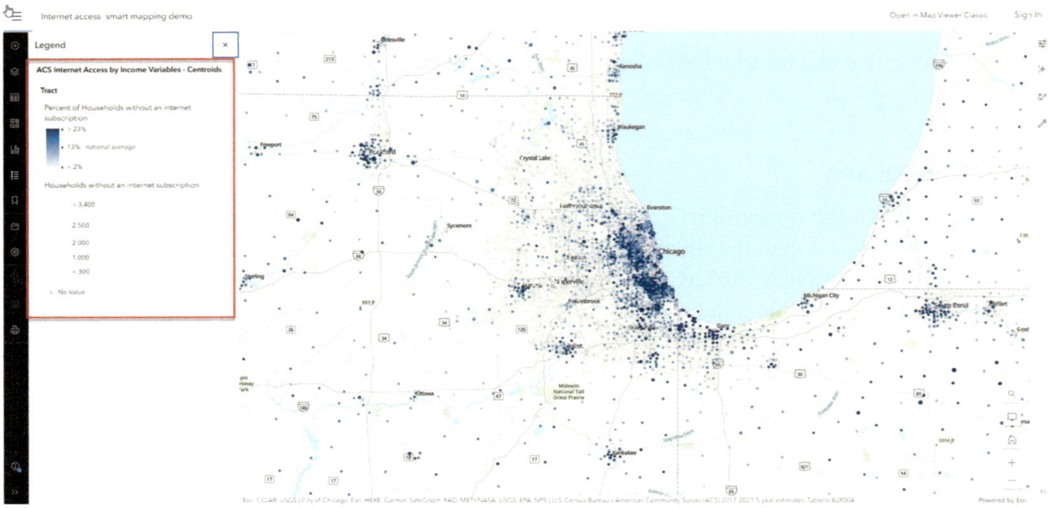

FIGURE 5.13 Internet access by income. Mapped using ArcGIS from Esri.

In the following activity, you will explore and understand data through classification, creating heat maps, and creating relationship maps.

Access to the internet has rapidly become an important component of modern life, tied to education, health care, job opportunities, and much more. Uneven access to the internet, evident at multiple scales, is an ongoing challenge to societies around the world. Open the linked ArcGIS Online map (https://www.arcgis.com/apps/mapviewer/index.html?webmap=485dd7cbf0a4405094f7b1c53ef4a14b) showing internet access by income variables, here (Figure 5.13):

The map shows data from the US Census Bureau's American Community Survey (ACS) with each point representing the centroid of a census tract. Since the data comes from the ACS, the map extent only covers the USA. The data are shown as a bivariate map—showing two variables—the *percent* of households without an internet connection, by color, and the *number* of households without an internet connection, as graduated symbols (in this case, circles). A census tract with a large dark blue circle therefore contains a high percentage *and* a high number of households without an internet connection, as measured by the ACS.

Use the left panel > Layers > Expand the ACS Internet Access by Income Variables group. Note the state, county, and census tract scales are each represented by different layers, so make sure you make each of these visible if you wish to see them. Note the patterns at different scales by zooming in and out, making sure that you zoom out to county and to the state level. How does your perception of internet access change from the census tract to the state level? Zoom to the census tract level for an area you are interested in.

The choice of the symbols used in mapping has a great impact on the map-reader's perception of the data and the engagement in an issue. Change the symbology of this map at the tract level: Use > Left panel: Layers > Tract > Right panel > Styles > Choose attributes > delete the Households without an Internet subscription field so you are just mapping the percent field as a single variable map: Now, you are no longer mapping as bi-variate. Select > Counts and Amounts (color) > Style Options > Theme > Change from High to Low to Above and Below. Above and below style shows data above and below a value such as 0 or the average. The default is dark blue for high and red for low, with white near the mean. Scroll down to access the histogram for the data (Figure 5.14). Note the mean value of 13.9. Of all the census tracts in the USA, the mean according to this ACS data for the year the data were collected was 13.9% of households in a census tract.

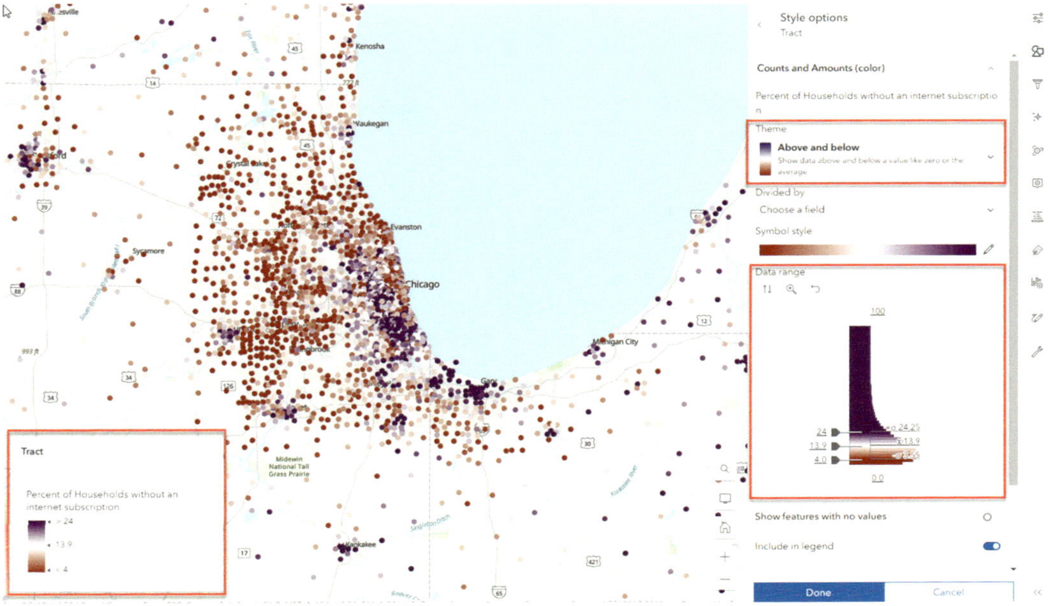

FIGURE 5.14 Using Above and Below symbology for percent of households without an internet subscription and examining the data's histogram. Mapped using ArcGIS from Esri.

Note the resulting patterns on the map. The above and below style highlights census tracts where there are internet access challenges, and those challenges seem to exist not only in some urban areas, but in rural areas too.

The symbol size also affects perception. Click on the pencil to the right of the color ramp and adjust the symbol size, noting how the map changes, before clicking Done.

The classification method also affects how the data and issue is perceived. Use > Styles > return to Style Options > scroll down > Classify Data. Keeping the classification method on Natural breaks, change the number of classes to 2. Note the break point (17%) and the map's appearance. Then, change the number of classes to 6. What in your judgment is the ideal number of classes for this data?

Change the classification method to standard deviation. Note the map's appearance with 1 standard deviation classification on the map. Change it to 0.5 standard deviation. Which standard deviation do you think results in the most understandable map?

As you have learned, quantile is a classification method that places the same number of observations in each category. Change the classification method to quantile with five classes. Here, the number of census tracts in the USA (over 84,000) is divided into five classes. How many census tracts therefore exist in each category? With the map and the histogram visible, note how you can drag any of the break points to help you understand any of the categories. If you drag the break point in your current histogram toward 100, you will be able to see census tracts with a very high percentage with no internet access, which might help you work with local authorities and internet providers to come up with a solution for those areas. Realize though that when you change the break points, you are no longer adhering to quantile or whatever method you began with; you are using the manual classification method, instead.

Change the style to > Color and Size. Now, the census tracts with a low percentage of households are not only dark blue, but they are larger in size. Change the size range to 3–33. Does color and size add value to your map over just color alone?

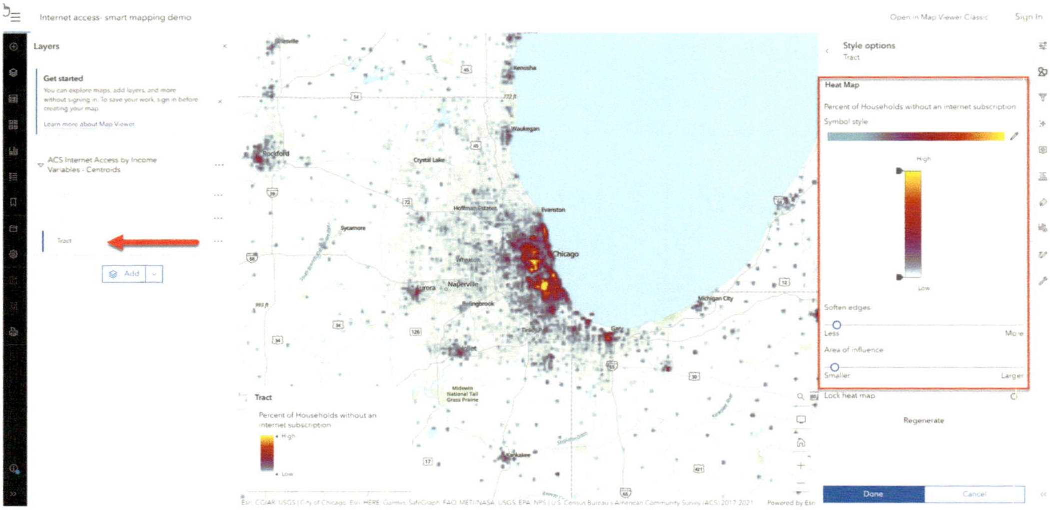

FIGURE 5.15 Heat map. Mapped using ArcGIS from Esri.

Heat Map

Next, change your map to a heat map (Figure 5.15). A heat map uses "hot" colors such as reds and oranges to highlight certain variables and help you pick out patterns: Use styles > heat map. What patterns do you now notice (as shown)? Do you notice anything now that you did not notice earlier by using color maps and color and size maps? Use > Styles > Style options > Adjust the area of influence to Smaller to help you to pick out more details.

It is important to note that a heat map is not a hot spot map showing areas of significant difference. Moreover, a heat map does not show temperature. Rather, a heat map shows areas where certain variables of a certain value are clustered.

A relationship map, as the name implies, is another way to see patterns across different variables.

Relationship Map

Earlier, you created a bivariate map. Now you will create a relationship map (Figure 5.16). Right panel: Styles > Add Field: Households whose income is \$10,000–\$19,999 > Now that you are mapping two variables, > Style > Relationship. Make the legend visible and pay particular attention to the legend for this very interesting style of map. The relationship map's 3×3 grid shows nine possibilities of combinations of the percent of households without an internet subscription to households whose income is low; that is, \$10,000–\$19,999 (shown here). What relationship do you notice between these two variables? What spatial patterns are evident in the area you are examining? In the area north of Evanston, Illinois, along the Lake Michigan shoreline, for example, there are very few households with a low income and a low percent without an internet subscription; conversely, there are households north of this area, in Waukegan, and south of this area, in Gary Indiana, where both variables are high. Of particular interest, and possible with the relationship map style, may be the blue cells—those where the percent of households without internet is high but the income is also "not low" (we cannot, with this classification, say the income is average or high without further investigation).

Next, change the symbology of the relationship map: Use Styles > Options > Change grid size from 3×3 to 2×2. What is one advantage and one disadvantage of simplifying the relationship map symbology from a 3×3 grid to a 2×2 grid? Make the state level layer to the left of the map

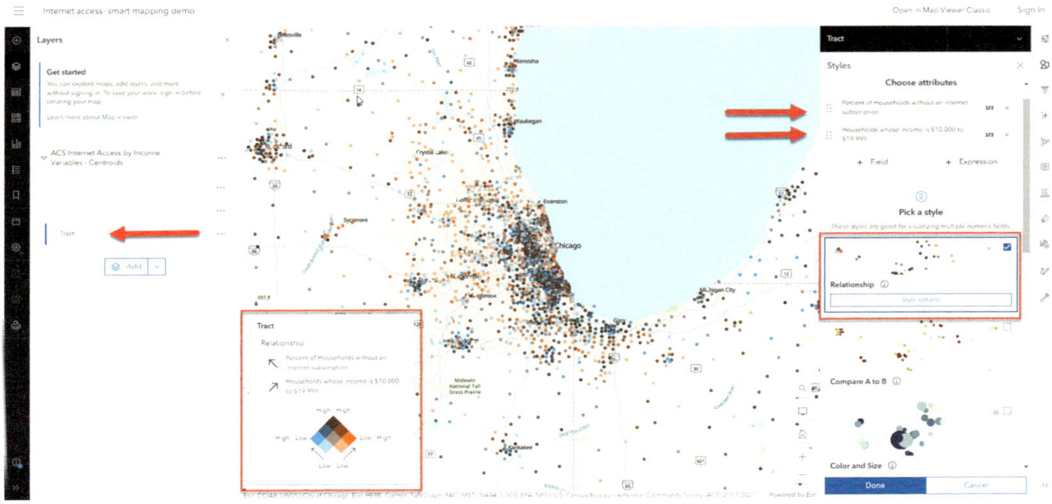

FIGURE 5.16 Examining a relationship map of percent of households without an internet subscription, and households whose income is $10,000–$19,999. Mapped using ArcGIS from Esri.

active, and zoom out to a smaller scale. Add the same two variables that you were analyzing above to analyze these variables, this time at the state level: Create a relationship map with a 3 × 3 grid using the same method you used earlier. After doing so (Figure 5.17), answer the following questions: (1) What patterns are interesting or surprising to you at the state level? (2) What is one caution about analyzing this type of data at such a macro (small scale) level?

This link (https://pro.arcgis.com/en/pro-app/latest/help/mapping/layer-properties/bivariate-colors.htm) explains how to create the complex two-dimensional legend display. Think of other variables that might benefit from this sort of display; create your own relationship map from other variables in this data set that you have at your fingertips.

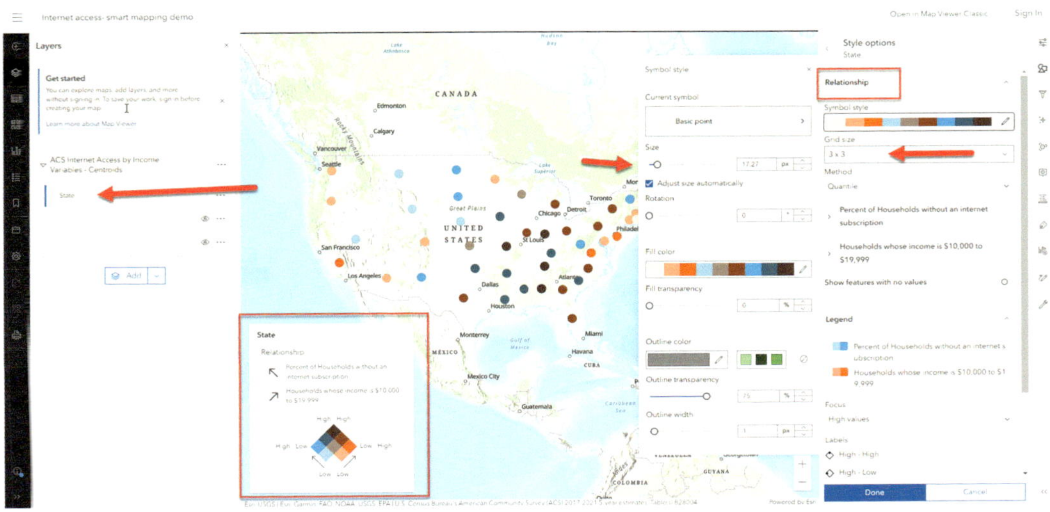

FIGURE 5.17 Examining a relationship map of internet access and income, this time at the state scale. Mapped using ArcGIS from Esri.

Underpinning the single-variable maps, the heat maps, the bivariate maps, and the relationship maps that you have been examining is mathematics—these maps would not exist without mathematics—the raw numbers, the percentages, and other variables in these relational databases. Indeed, these maps would not exist at all without the science of geodesy—understanding the size and shape of the Earth, which again, rests upon mathematics.

TO THINK ABOUT

The bivariate map summarizes many of the concepts in this chapter, and elsewhere. That is precisely why part of the legend, the diamond shape, is used to represent the structure of this entire book. Notice that the concept of scale is present in the chapter heading, from the local symbol representing a single chapter, to the bivariate symbol representing the whole book, to the dashboard representing the complexity of the real world: a shift from local to global scale. Look for scale transformations, and their associated implications, in the world around you.

6 Mathematics' Foundations
Set Theory and Algebra

The line of images above is a visual abstract of this chapter designed to foster spatial thinking. From the chapter numeral, to the book structure, to the real world, the reader is offered gentle guidance to develop spatial intuition about what might be coming. Those thoughts are then reinforced with a detailed text outline of chapter content below. The images and outline forge as an abstract of chapter content.

CHAPTER OUTLINE

- Existence and Foundational Mathematics
  - Law of Excluded Middle
  - The Axiom of Infinity
  - The Axiom of Choice
- Set Theory
  - Set Fundamentals
  - Subsets
  - The Empty Set and the Universal Set
  - Basic Set Operations
  - Venn Diagrams
  - Set Theory in Hazards Analysis
- Algebra
  - Fundamental Proofs
    - Field Axioms
    - Order Axioms
    - Sample Proofs
  - Simple Linear Equations and Graphs; Inequalities
    - The General Linear Form
      - One Variable
      - Two Variables
    - Inequalities
  - A Unifying Map
  - Simple Quadratic Equations and Graphs; Inequalities

DOI: 10.1201/9781003305613-6

The General Quadratic Form
One Variable: The Parabola
Two Variables: Conic Sections
Two Variables: Rapid Multiplication
Inequalities
Questions to Think About

EXISTENCE AND FOUNDATIONAL MATHEMATICS

In Chapter 5, we discussed data and offered the reader focused activities that involved investigating a variety of data types and scales. The activity on dashboards touched on the notion of "metadata": data about data! In all that we have been doing so far, we have emphasized the role that mathematics plays as the foundation for mapping, data, and spatial analysis. But what are the foundations of mathematics, itself: as "metamathematics"?

LAW OF EXCLUDED MIDDLE

In Chapter 1, we made brief mention of the Law of Excluded Middle; that Law is an Existence statement. It does not offer any algorithm or means of construction of answers. Yet it underlies much of what we do. Consider the donut/ring map made to track the pattern of litter in a region of downtown San Diego (Chapter 3). There were a number of types of litter; indeed, for some purposes, too many. We can simplify the data set by choosing to view the litter as either made from plastic, or not made from plastic: two values, yes or no, with no choice for a middle value. Binary switches, themselves, the foundation of the digital world are two-valued: on or off. There is no middle ground. In the non-digital world, decisions might be made in situations where the middle is excluded, but perhaps should not be: as in, here is my agenda; if you fail to support it, then it means you are opposed to all of it. Perhaps. However, perhaps there is middle ground that can be discussed or negotiated by people with sincere intentions at resolution. The Law of Excluded Middle arises in numerous instances; some are more transparent than others. It is wise to keep it in the back of our minds.

Think about The Law of Excluded Middle the next time your opinion differs with that of another. Is someone trying to use this Law when it does not apply? Can you create a situation to move the discussion from one extreme, into the Middle, where perhaps it belongs—as a transition from one extreme to another?

THE AXIOM OF INFINITY

Axiom of Infinity: There exists at least one infinite set; specifically a set containing the natural numbers.

With regard to mapping, we need this to know that we always have "enough" numbers, whether we are building datasets, or subdividing them.

We will never run out of numbers!

THE AXIOM OF CHOICE

Axiom of Choice: Given an arbitrary collection of sets, finite or infinite, each of which contains at least one element, it is always possible to construct a new set by arbitrarily choosing one element from each set.

This axiom tells us that this process can be done. We do not necessarily invoke it deliberately, but we know that such a choice can always be made. In a mapping context, we assume it any time we make a selection, such as selecting the smallest value from a set of values to be mapped. It is at work behind the scenes and is often invoked deliberately only in cases involving infinite numbers of values.

SET THEORY

To this point, we have been using the word "set" in an intuitive or conversational way; for most mapping purposes, that is sufficient. For example, "this set of apples is Honey Crisp," or "this set of points in this region of the map indicates a low-pressure system." The concept of "set" is intuitively understood and has been throughout time. Euclid employed ideas about sets as did Eratosthenes in his calculation of the circumference of the Earth. The foundational axioms discussed above, along with others, underlie formal set theory. Set theory, as a formal discipline, emerged in the latter part of the 19th century (Cantor, 1874, 1915) and appreciation of it, in that timeframe, came only from within pure mathematics. In today's world of mapping, there are operations with sets that underlie mapping operations; thus, it is prudent to be aware of what they are and to master how to work with them.

Set Fundamentals

The concept of a set is a well-defined collection or grouping of elements. Typically, all members of a given set share a common characteristic. It is straightforward to identify elements that are, or are not, members of a given set.

As arithmetic is based on binary operations on numbers, set theory is based on binary operations on sets. Set theory presumes a binary relation between an object o and a set A. If o is an element of A, the notation $o \in A$ is used to denote membership of o in A.

When displaying sets by enumerating members, it is conventional to enclose the members in braces: {1, 2, 3} or {2, 4, 6, …} with ellipsis used to denote continuation in pattern. A more abstract, but compact, notation is to express a set in "set-builder notation" as $\{a \mid a \in X\}$ or $\{a : a \in X\}$ to be read as "the set of all a such that a is an element/member of X".

Subsets

The membership relation can relate sets to each other. If all the members of set A are also members of set B, then A is a subset of B, denoted $A \subseteq B$. For example, {0,1} is a subset of {–1, 0, 1}, while {1, 2} is not. Or, the set of even positive integers is a subset of the set of positive integers. The full, given, set is a subset of itself (hence the equal sign on the inclusion symbol). If the full set is to be excluded from the set of subsets, then we use the term, proper subset, for cases where only subsets smaller than the full set are considered (and notationally, the equals part of the symbol is removed).

The Empty Set and the Universal Set

There are two special sets.

- The importance of zero in arithmetic, also has a parallel in set theory with the empty (or null) set (a set with no elements), denoted $\varnothing$. Note the difference between {0} and the null set. The first set has one element, the element 0. The null set has no elements.
- Another special set is the "universe of discourse" or the "universal set" or the "universe." The universal set might be denoted as U, or some other symbol where context makes the designation obvious. The universal set serves as a context within which analysis takes place.

Basic Set Operations

Sets can be combined in a variety of ways; it is likely that the reader is familiar with some of these but perhaps not with others. We summarize them, diagrammatically, below. Figure 6.1 shows basic operations of union, intersection, and power set.

Figure 6.2 shows a simple view of the ideas of union and intersection, using Venn Diagrams on two sets (Venn, 1890).

Real-world examples of the union of two sets are abundant. Union might be seen as a list of all guests at a party for which you and your partner each had separate lists. There might be some overlap, but it is the full set of both lists/sets, of the total number of different invited individuals, that will determine how much food and beverage you need to buy to plan for the party. The intersection of the two lists shows which friends you have in common. Social media platforms often show shared "friends"—think about how the concept of intersection fits in there.

The Hasse diagram (Figure 6.3), based on ordering the relation of containment, shows a hierarchy of all members of the power set of the abstract set {1, 2, 3, 4} (classical reference samples; Vogt, 1895; Birkhoff, 1948). There are four elements in the original set so there are $2^4 = 16$ elements in its power set.

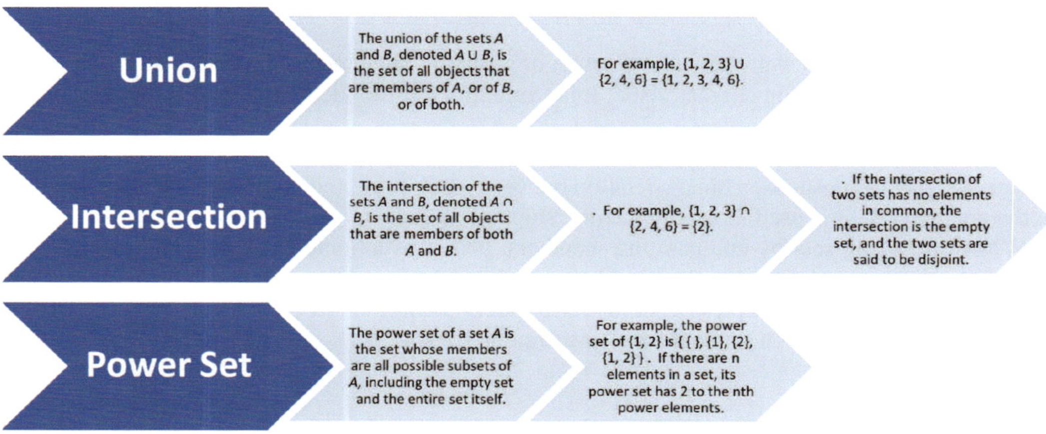

FIGURE 6.1 Basic set operations.

FIGURE 6.2 Venn diagrams illustrate union and intersection of two arbitrary sets. Source: Public Domain, https://commons.wikimedia.org/w/index.php?curid=3437020.

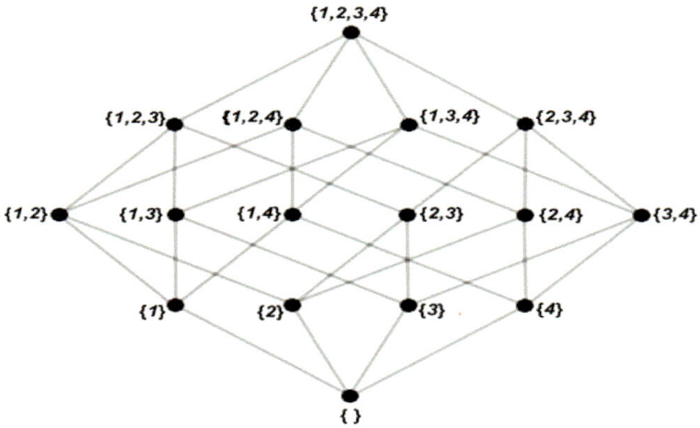

FIGURE 6.3 Hierarchy showing all members of the power set of {1, 2, 3, 4}.

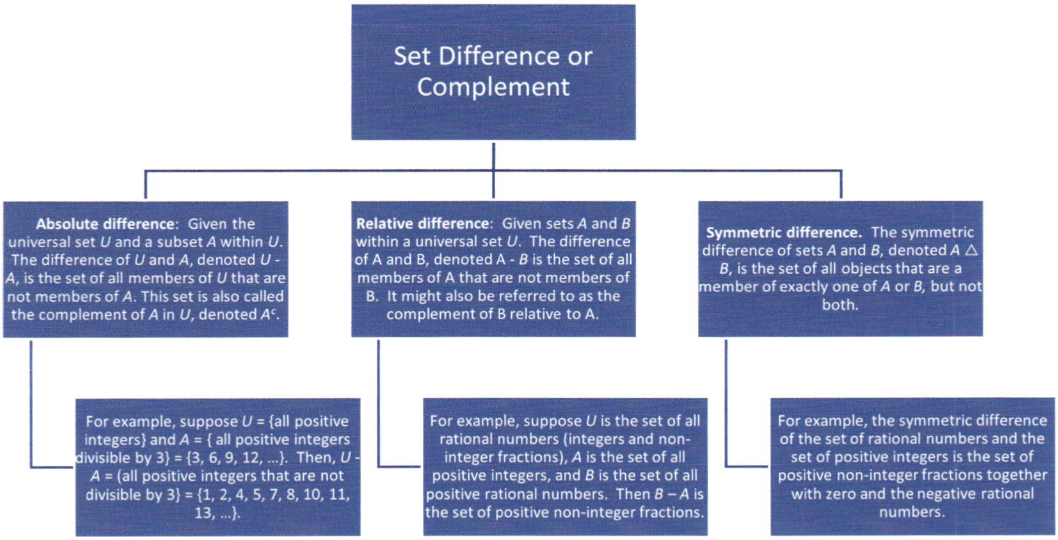

FIGURE 6.4 Set operations based on the idea of difference.

Hasse diagrams get very complicated very quickly. In a real-world example of power set, consider that you have downloaded 100 songs to your smartphone. The number of members in the set of all playlists that can be formed, including the empty set (playlist) of "peace and quiet," is

$$2^{100} = 1{,}267{,}650{,}600{,}228{,}229{,}401{,}496{,}703{,}205{,}376.$$

More basic set operations, all based on a concept of difference, are shown in Figure 6.4.

One visualization of these set theoretic differences is offered below (Figure 6.5). To see other variations in combining sets, consider viewing the linked page (https://commons.wikimedia.org/wiki/File:Venn0001.svg) which draws together the similarities between Set Theory and Symbolic Logic.

The Cartesian product (like the Cartesian Coordinate system, again named for French Philosopher and Mathematician, René Descartes) of A and B, denoted $A \times B$, is the set whose members are all possible ordered pairs (a, b) where a is a member of A and b is a member of B (Figure 6.6).

FIGURE 6.5 Venn diagrams illustrate the complement of one set relative to another and the symmetric difference of two sets. Source: Public Domain, https://commons.wikimedia.org/w/index.php?curid=3437435 Public Domain, https://commons.wikimedia.org/w/index.php?curid=3437442.

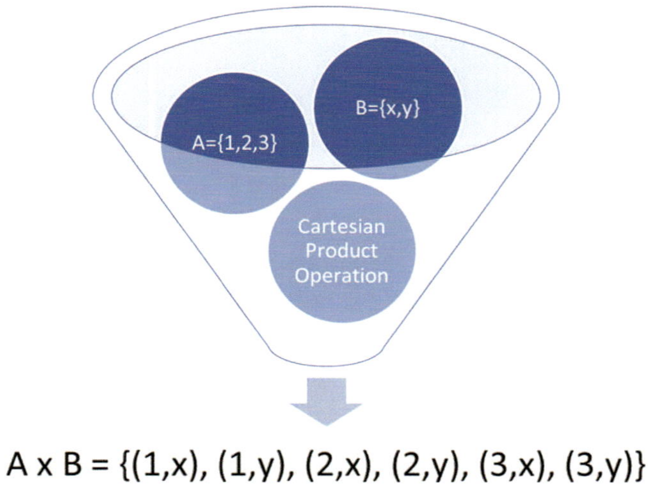

FIGURE 6.6 The Cartesian product of two abstract sets.

The Cartesian Coordinate system, used elsewhere in this book in an intuitive manner, is simply the Cartesian product of the real numbers, R, with itself: $R \times R$. One real-world example of the Cartesian Product is a deck of standard playing cards: 13 alphanumeric symbols crossed with four card pips (♣, ♦, ♥, ♠) producing a $13 \times 4 = 52$ card deck (Figure 6.7).

Look around you. What real-world examples of set theory do you see? Consider the following evidence of set theory in the world of mapping!

Venn Diagrams

A useful visual display (as we have seen above) for looking at relationships among sets and their elements is a Venn diagram (Figure 6.8). In it, typically, circles representing sets within a universe are intersected to look for various patterns of set intersection. In the diagram below, the intersection of all three sets, A, B, and C, is the small dark part in the center, $A \cap B \cap C$. Practice labelling various parts of the diagram using the various set operations; point to $(A \cap B) - (A \cap B \cap C)$. What do you think is the order of operations with set operations? Where is the Universe of Discourse?

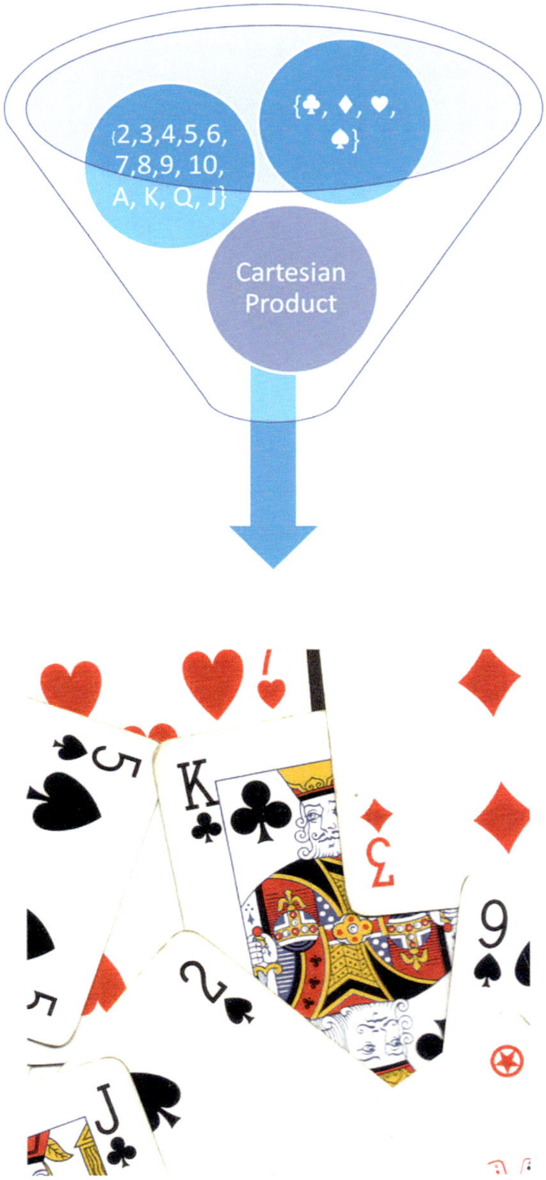

FIGURE 6.7 A real-world example of the Cartesian product. Source for playing cards image: https://commons.wikimedia.org/wiki/File:Playing_cards_collage.jpg WhiteDr4ag0n, CC BY-SA 4.0 <https://creativecommons.org/licenses/by-sa/4.0>, via Wikimedia Commons.

In commonly used language, often the idea of "union" is viewed as equivalent to the word "or" while the word "intersection" is viewed as equivalent to the word "and." While this observation is often helpful, one needs to take care with the word "or": there are two forms of the word. The inclusive form of the word "or" means "either *a* or *b*, or both" while the exclusive form of the word "or" means "either *a* or *b*, but not both." In set theory, and in mapping, the word "or" always is the inclusive form. That is clear from looking at the Venn diagram for "union" (Figure 6.2).

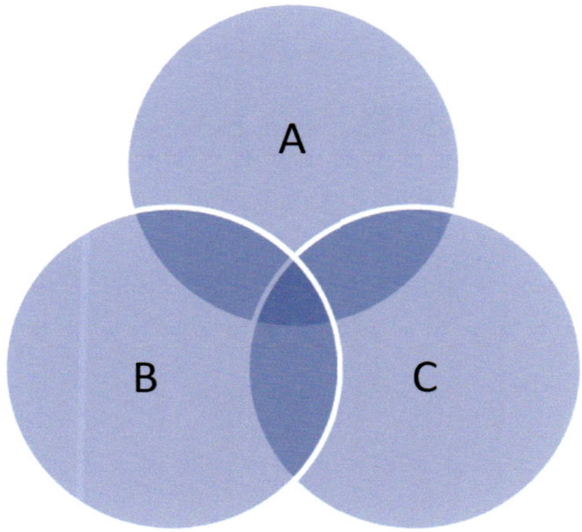

FIGURE 6.8 Venn diagram of three intersecting sets, A, B, and C.

Set Theory in Hazards Analysis

In this activity, you will have the opportunity to build your understanding of set theory, Venn Diagrams, and your skills in spatial analysis. Open the linked map (https://www.arcgis.com/apps/mapviewer/index.html?webmap=28aa62c3d07b45b2a1b5f5b4ac36f5c3) in ArcGIS Online entitled, Boulder County Colorado Hazards Analysis for Set Theory (Figure 6.9). The map contains seven layers and two bookmarks: When the map opens one of the layers is visible, the geologic hazards of Boulder County, Colorado. This county, like many areas of the world, is prone to natural hazards, and set theory is critical to identifying areas most at risk to specific hazards—in this case, from geology (landslides) and water (flooding). Suppose that Boulder County, hearing of your excellent spatial analysis skills, hires you to assess which areas are threatened by **both of** these two types of hazards—landslides and flooding. When you open the map, use the bookmarks>click on Boulder County to see the entire county at once. Next,>use the Layers tool at the left side of the map to see the names of the layers that you will examine one at a time. The visible layer when you open the map is located at the bottom of the stack of layers, showing all types of geologic hazards (Figure 6.9).

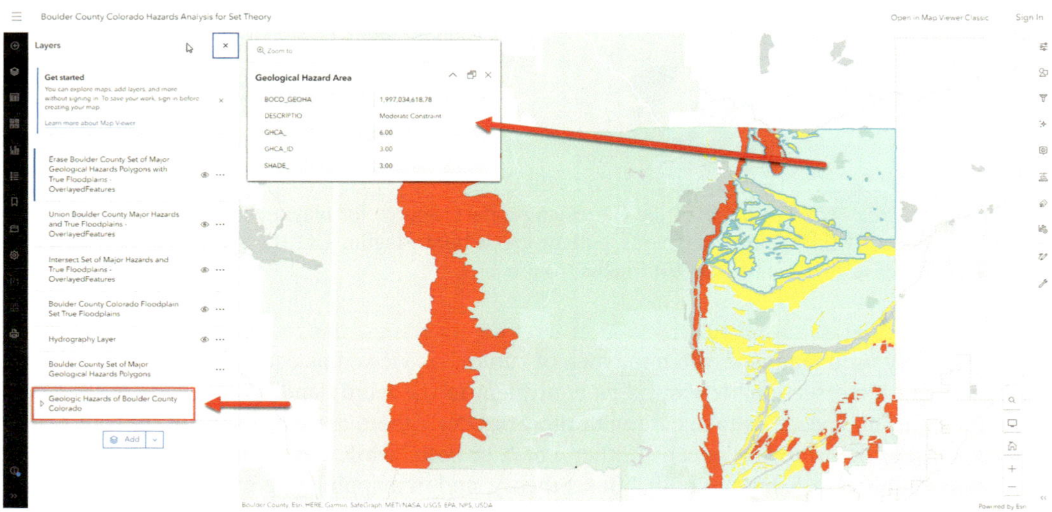

FIGURE 6.9 Considering geologic hazards for risk analysis: selecting an area of moderate geologic hazard. Mapped using ArcGIS from Esri.

Click on different areas on the map to access from the popup information which areas of the county are ranked major, moderate, and minor in terms of its geologic hazards.

You will see that the major hazards are shown in a reddish color, and accessing the map legend will confirm this. Turn the legend off and access the map layers. Make the layer "Geologic Hazards of Boulder County" invisible and make the layer "Boulder County Set of Major Geological Hazards Polygons" visible. The major geologic hazards represent a **set** within all the geologic hazards (Figure 6.10)—the hazards have been filtered to only show the major hazards. Open the table for all geologic hazards and open the table for the major geologic hazards. Which layer contains fewer features—all geologic hazards or major geologic hazards? Why? In a similar way, toggle on and off the next two layers up the stack—the hydrography layer and the Boulder County Colorado Floodplain Set of true floodplains. The true floodplains are a set within the hydrography layer. The hydrography layer, which includes water-related areas that are not floodplains, has been filtered and saved in the "true floodplains" layer, representing just what the name indicates. You will recognize the shape of the floodplains as closely following rivers in the county, which flow largely from the western mountains toward the east, onto the Great Plains. Open the table for each layer. Which layer contains fewer features—hydrography or true floodplains? Why?

Recall that your assignment is to find areas threatened by both major geologic hazards and flooding. Turn off all layers except for the major geologic hazards layer and the true floodplains layer. By looking at the map, you can see the areas where these two layers *intersect*. But having mapping tools driven by a Geographic Information System (GIS) allows you to do far more than just "eyeball" the map—you can do spatial analysis and determine exactly which areas are affected by both hazards. Make the "Intersect" layer visible>turn on the legend>use the bookmarks>access Southeast Boulder County. Zoom and pan in this portion of the county, and you will notice that the intersect layer indeed represents another set—the intersection of the major hazards and floodplains (shown here)—the intersect layer is mapped in a blue color, the floodplains are in a light blue/gray, and the major hazards are in red).

To create this intersect layer, a specific type of map overlay operation was created—one that would retain, from the two input layers (major geologic hazards and floodplains)—areas common to both. Results from overlay operations on maps are visual representations of set theory. You can perform intersects on points, lines, or polygons. Again, intersect is analogous to "AND."

Two other types of overlay operations also result in sets—union and erase. Each type is useful, depending on your objectives. Start with union. Turn all layers off and turn these three layers on:

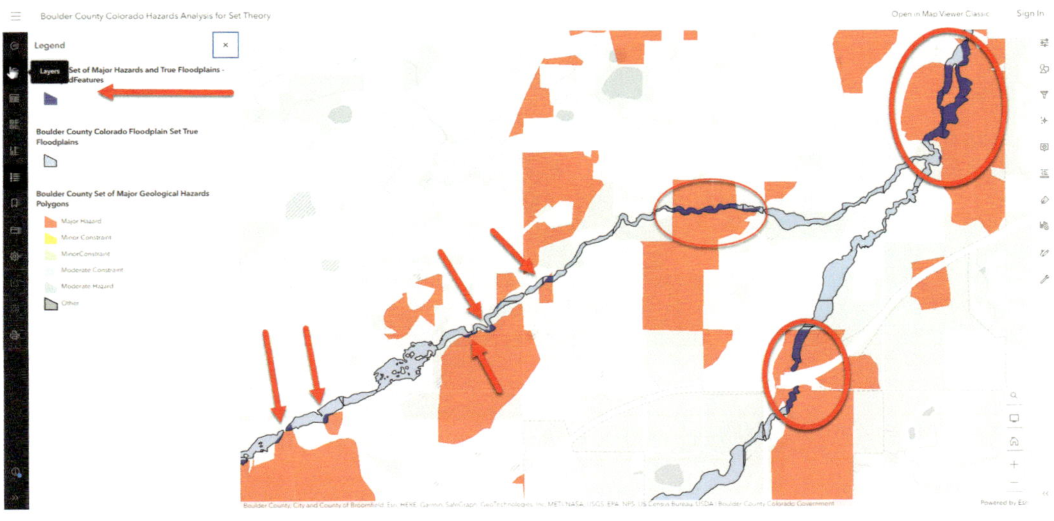

FIGURE 6.10 Assessing hazards risk: note how the intersection of major geologic hazards and true floodplains appears in dark blue. Mapped using ArcGIS from Esri.

Major Geologic Hazards, True Floodplains, and Union (Figure 6.11). Compare the union layer to the other two layers. The result of the union operation contains features from **both** the input layer and the overlay layer. It is analogous to "OR," the inclusive form.

Finally, examine the map layer that resulted from an Erase operation. Turn all layers off and turn these three layers on: Major Geologic Hazards, True Floodplains, and Erase (Figure 6.12). Compare the union layer to the other two layers. It may be most helpful to turn the hazards and floodplains off—then it is clear that erase contains areas **not** in a major hazard or in a floodplain. The result of the erase operation contains features in neither the input layer nor the overlay layer.

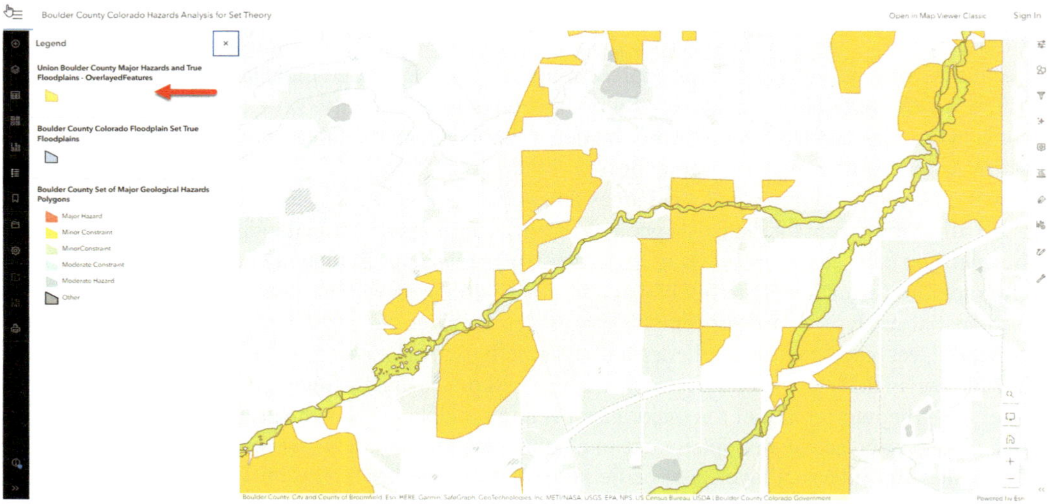

FIGURE 6.11 Assessing hazards risk: note how the union of major geologic hazards and true floodplains appears in orange. Mapped using ArcGIS from Esri.

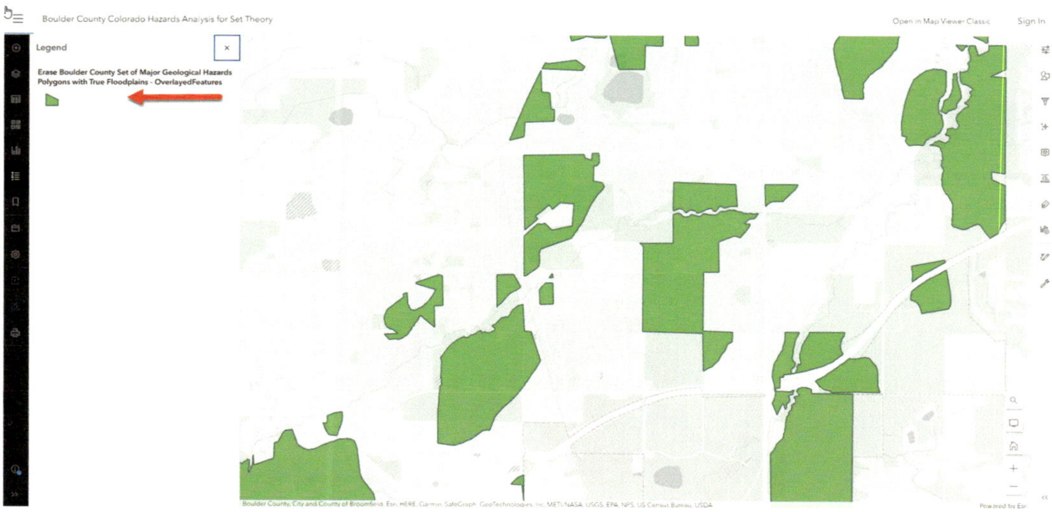

FIGURE 6.12 Assessing hazards risk: note how the erase results of major geologic hazards and true floodplains appear in green—those areas in a major geologic hazard zone but not in a floodplain. Mapped using ArcGIS from Esri.

Overlay and other spatial analysis operations are used frequently with interactive maps and GIS to result in sets that are important inputs in making decisions on a daily basis for individual cities to entire regions. The above analysis focused on natural hazards, but set theory is integral for making decisions about water, energy, public safety, housing, habitat, and many other applications in our world.

ALGEBRA

Fundamental Proofs

In Chapter 1, we proved that $(-x) * (-y)=xy$. Here we offer more proofs of other fundamental algebraic facts. Such a proof is a list of reasons why we can move from one step to the next, leading to the eventually desired outcome in a clear and accurate manner. Creating an algebraic proof is an art; the best ones often take the fewest, and most clever, steps, and is deemed, "elegant."

One view of algebra is simply that it is a rearranged form of arithmetic: $2+2=x$ is arithmetic while $2+x=4$ is algebra. Elementary arithmetic/algebra, follow from the following sets of axioms.

Field Axioms

There exist concepts of addition and multiplication, along with those for additive and multiplicative identities and inverses, such that the following set of laws holds.

- Associative law for addition: $a+(b+c)=(a+b)+c$
- Additive Identity: There exists 0 such that $a+0=0+a=a$
- Additive Inverse: $a+(-a)=(-a)+a=0$
- Commutative law for addition: $a+b=b+a$
- Associative law for multiplication: $a * (b * c)=(a * b) * c$
- Multiplicative Identity: There exists 1 such that $a * 1=1 * a=a$ where a is not 0
- Multiplicative Inverse: $a * a^{-1}=a^{-1} * a=1$ where a is not 0
- Commutative law for multiplication: $a * b=b * a$
- Distributive law: $a * (b+c)=a * b+a * c$.

Order Axioms

There exists a subset, P, of positive integers such that the following laws hold.

- Trichotomy: Either
  - a is an element of P, or
  - a is not an element of P, or
  - $a=0$

    where the word "or" is used in an exclusive manner; that is, exactly one of the three conditions above is true.
- Closure under addition: if a and b are elements of P, then $a+b$ is an element of P
- Closure under multiplication: if a and b are elements of P, then $a * b$ is an element of P.

Sample Proofs

The point of showing these proofs is to expose the reader to the rigor of the foundations upon which mathematics is built. Learning to appreciate, and understand, such rigor will lead to a stronger foundation for one's own mind, and for acquiring clarity in thought of the myriad complicated processes one will encounter in the real world.

- Prove that: if $a+b=a+c$, then $b=c$.
  - $a+b=a+c$ [hypothesis]
  - $(-a)+(a+b)=(-a)+(a+c)$ [existence of additive inverse]

- $((-a)+a)+b=((-a)+a)+c$ [associative law for addition]
- $0+b=0+c$ [additive inverse]
- $b=c$ [additive identity]
- QED (Quod erat demonstrandum, thus it is shown).

In this proof, the "art" enters the process in the first step after the hypothesis: it is probably natural to recognize that the hypothesis is longer than the desired outcome, so introducing an additive inverse might also be natural, and if so, it would be obvious which one to pick. However, what seems to matter most, in terms of "art," is where to insert the term $(-a)$. If it had been inserted on the right-hand side on either side of the equal sign, rather than on the left-hand side (which is what was done), then the proof would have been one step longer because we would have needed to use the commutative law to move the $(-a)$. As it is, we did not need the commutative law. The proof with the commutative law would not have been as "elegant" as the one not using the commutative law. Thus, we learn to begin to appreciate the art in mathematics and its associated clarity/purity of thought: hence, so-called "pure" mathematics.

- Prove that: if $a * b=0$ then $a=0$ or $b=0$

  Suppose that, without loss of generality, a is not equal to zero. Then, it remains to prove that $b=0$.

  If $a * b=0$, then $a^{-1} * (a * b)=a^{-1} * 0$ [Existence of multiplicative inverse when a is not 0]
  But, $a^{-1} * (a * b)=(a^{-1} * a) * b$ [Associative law of multiplication]
  $= 1 * b$ [Multiplicative inverse]
  $= b$ [Multiplicative identity]
  And $a^{-1} * 0=0$ [Multiplicative identity].
  Thus, $b=0$.
  QED.

  In the remaining sections, we move forward into solving algebraic equations of various kinds, keeping in mind the root principles, and the care that needs to be taken in justifying, not guessing, one's answer—as was exhibited in this section.

Simple Linear Equations and Graphs; Inequalities

Linear equations are formed from constants and variables raised to the 0 or 1 power.

The General Linear Form

The general form of a linear equation in:

- one variable, x, is $ax+b=0$ where x is a variable, a and b are arbitrary constants, and a is not zero
- two variables, x and y, is $ax+by+c=0$, where x and y are variables, a, b, and c are arbitrary constants, and both a and b are not zero
- with n variables is $a_1x_1+a_2x_2+\ldots+a_nx_n+C=0$, where $x_1, x_2, \ldots, x_n$ are variables, $a_1, a_2, \ldots, a_n$ are arbitrary constants, and none of the latter is zero; C is a constant.

In this book, we confine our interest to simple linear equations of one and two variables. Examples are given below; try using the graphing calculator on them. Figure 6.13 illustrates the graph of $y=3-2x$. What difficulties do you encounter? In the two variable case, isolate y on one side of the equation and then graph the other side.

Try using this graphing tool to graph equations and inequalities, as they occur, and practice until graphing becomes second nature. Use it throughout multiple sections to follow and practice.

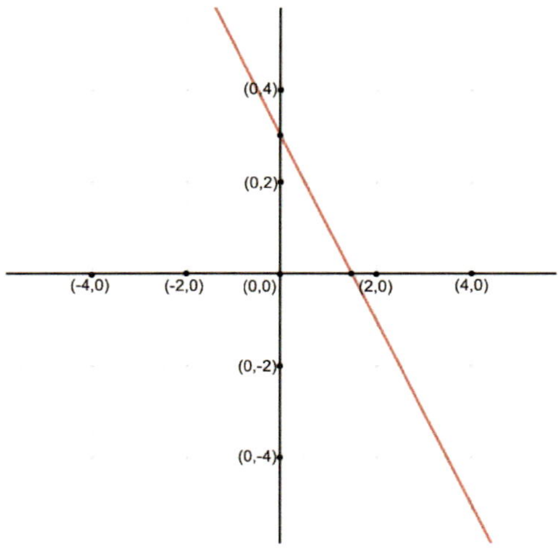

FIGURE 6.13 $y=3-2x$. Powered by desmos.com.

One Variable	Two Variables
$2x+17=0$	$5x+2y-3=0$
$(5/4)x+3=2$	$(1.7)x+47y=7$
$15x=33$	$y=3-2x$

One Variable

The one variable equation, $ax+b=0$ (a not zero), has a graph in Cartesian Coordinates that is a straight line that is vertical: parallel to the y-axis. That is easy to see by solving the general equation, $ax+b=0$ so that $x=-b/a$ which is a constant. When $b=0$, it follows that $x=0$ since a is not zero.

Solving linear equations in one variable is straightforward, as indicated above. The goal in solving simple linear equations is to isolate a variable on one side of the equal sign. Thus, in the equation $3x=5$, isolate the variable, x, by dividing both sides by 3. The equation becomes $3x/3=5/3$ and since $3/3=1$, $x=5/3$. Notice how the axioms from earlier in the chapter become involved in the background—they make it all happen consistently from one example to the next. As another example, solve $5x+17=0$. To isolate x, first move 17 to the other side of the equal sign from where the x is. The equation becomes, adding −17 to both sides: $5x+17-17=0-17$ or $5x=-17$ so that $x=-17/5$. Again, notice which foundational ideas lie behind the rearrangement?

Two Variables

The two variable linear equation, $ax+by+c=0$ (a and b not zero) has a graph that is a straight line. If we allow b to possibly be zero (thereby including the one variable case), then all simple linear equations of the latter form have graphs in Cartesian Coordinates that are straight lines, and, all straight lines in the xy-plane have equations that are linear in this form. There is a one-to-one correspondence between the set of all lines in the plane and the set of all linear equations of this form. Equations in the general linear form may be rearranged into other forms that might be more useful for specific purposes.

Generally, straight lines in the Cartesian plane may be thought of intuitively as having a certain amount of "slope" with respect to the xy-axes and as intersecting one or both axes. The slope is conventionally denoted with the letter m. The slope of a straight line is often defined as $m=(y_2-y_1)/(x_2-x_1)$ where (x_1, y_1) and (x_2, y_2) are points in the xy-plane. In words, the slope is the difference of the rise in the line divided by the distance over which it rises: a rise over a short distance is steeper than that same rise over a longer distance. The location where a straight line intersects the y-axis is called the y-intercept and that point has typical coordinates (0, *b*), where *b* is some constant. The location where a straight line intersects the x-axis is called the x-intercept and that point has typical coordinates (*a*, 0).

Lines and angles are all around us on the landscape. Open the web map (https://www.arcgis.com/apps/mapviewer/index.html?webmap=c97b1c3300ef4ad4be868c37d04eaa2d). On the left, access the bookmark>Powerline in Kansas, to view a powerline running northwest-southeast at an angle relative to roads running in the cardinal directions north, east, south, and west, on the map (Figure 6.14).

To the right of the map, >Map Tools>Location>Change format to UTM>touch points A, B, and C>note the UTM coordinates for each. The UTM coordinates are given in meters. The powerline's slope looks similar to a line with the formula $y=3-2x$. You will be able to predict the location where the powerline intersects the road, empowered with some geometric formulas in upcoming sections of this book.

The slope-intercept form of the equation of a straight line is $y=mx+b$ where m is the slope and b is derived from the y-intercept. The equation $y=3x+7$ has slope 3 and y-intercept (0, 7). Notice: the intercept is not itself a single value—it is an ordered pair. What does the equation look like when the line is parallel to the x-axis or to the y-axis and why? When the line is parallel to the x-axis, all values of y are the same; the line is the same height above or below the x-axis. So a typical example might be $y=5$; x ranges across all values but y is always 5 units above the x-axis. Similarly, a line parallel to the y-axis is always displaced the same amount from the y-axis to the left or right. A typical example might be $x=-3$. The equation of the x-axis itself is $y=0$; that of the y-axis is $x=0$. Use the graphing calculator to graph some examples.

Two points determine the position of a straight line in the plane so another useful form is created by using the definition of slope given above in terms of two points and generalizing one of them to an arbitrary point with coordinates (x, y). The two-point form is $y-y_1=m(x-x_1)$. Use the graphing calculator to create some suitable graphs.

Suppose we have an equation such as $y=3x+5$. There is an infinite number of solutions to this equation: $x=1$ and $y=8$; $x=-1$ and $y=2$; name some more. To find a single solution, introduce another equation and then isolate one variable and reduce the set of two equations to the previous

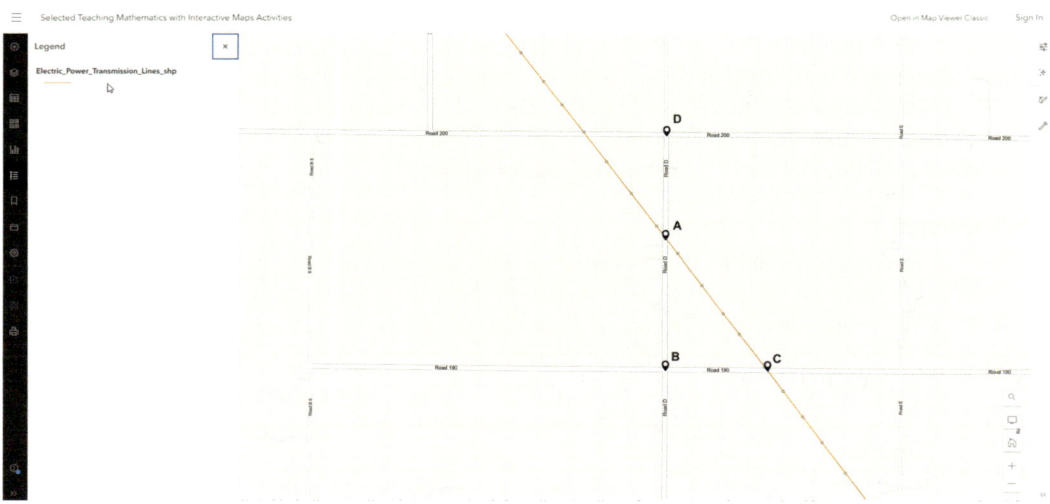

FIGURE 6.14 A powerline running northwest-to-southeast and intersecting roads that run in the cardinal directions north, east, south, and west. Mapped using ArcGIS from Esri.

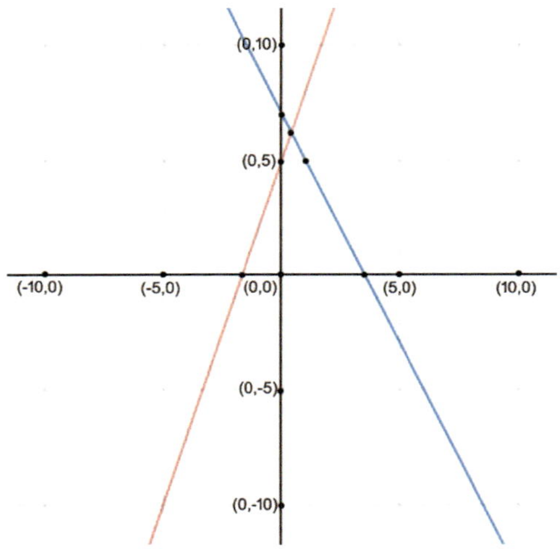

FIGURE 6.15 $y=3x+5$ (red) and $y=-2x+7$ (blue). Powered by desmos.com.

case of one equation in one variable. So, suppose we also have that $y=-2x+7$. In this case, y is isolated in both equations, so we have $3x+5=-2x+7$, an equation in the single variable, x. Solve: $5x=2$, $x=2/5=0.4$. Now substitute $x=2/5$ back into one of the original equations to solve for y: $y=-4/5+7$ or $y=6/5+5$. In either case, $y=6.2$. If the values had not matched, there would have been a mistake, probably in arithmetic. Then, one would go back and check. Use the graphing calculator to graph the results (Figure 6.15). Your graph should show two intersecting straight lines. What are the coordinates of the point of intersection of the two lines? From the above analysis, solving a system of equations simultaneously, the intersection point has coordinates (0.4, 6.2) as shown in Figure 6.15.

The general strategy, suggested by this easy example, works for all systems of two linear equations in two variables: isolate a variable and reduce the problem to solving a linear equation in one variable. Then, make appropriate substitutions to determine the value of the eliminated variable and check your work. The strategy is simple; the procedure can become more complex. However, as long as you keep the overall strategy in mind, the solution will not be difficult. Do not get distracted by the detail; keep the big picture in mind. Remember that the word "linear" contains the word "line."

Inequalities

In some applications, we want to be able to consider not only points along a line but also all points on one side of that line: the line is the edge of a half-plane. To visualize, for example, the inequality $y \leq 3x+5$, first graph $y=3x+5$. Then, select coordinates of any point that does not lie on that line; test it in the inequality. Here, choose (3, 3) for example. The inequality is true for this value: $3 \leq 3 * 3+5$. So, the point (3, 3) lies in the half plane for which the inequality holds and it determines that set completely. One point, not on a line, and a line, determine a half-plane. The same strategy works for strict inequality, as well.

The figure below shows a map in which inequality is critical to portraying a particular theme. It shows, for countries of the world, the ratio of (female life expectancy at birth/male life expectancy at birth) * 100. The darkest color is where female life expectancy most greatly exceeds male life expectancy (Syria and Georgia display prominently in Figure 6.16). Open this web map (https://www.arcgis.com/apps/mapviewer/index.html?webmap=5a91ed9e44e84fbaaf0e84e4e61ddaff/) to interact with the map and data. What observations do you note? Change the classification method under style>style options>from natural breaks to another classification. In which additional countries do you detect another gender disparity?

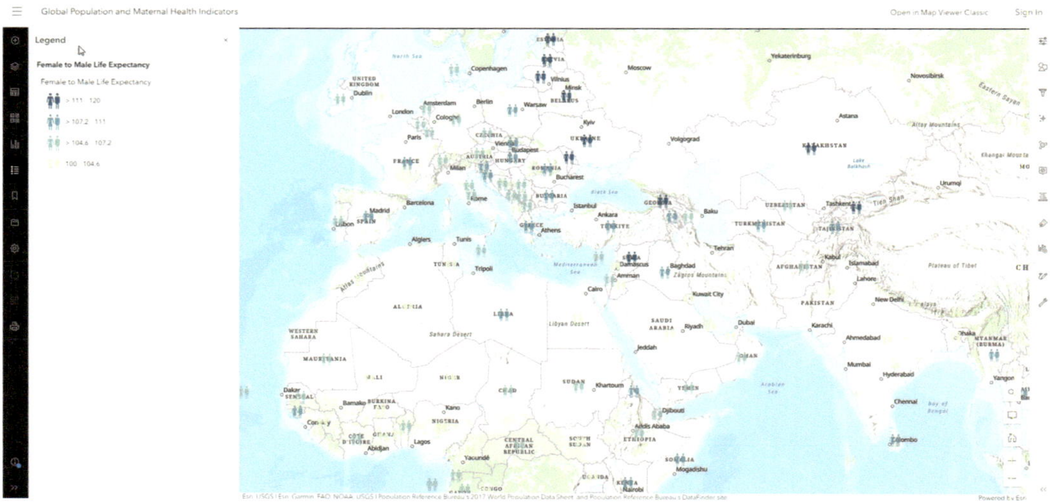

FIGURE 6.16 Inequality as mapped and measured by maternal health. Mapped using ArcGIS from Esri.

An analysis such as this one, based on straightforward mathematics, can offer insight to guide practical application: an international agency delivering medical supplies to outposts in developing nations might wish to know, for example, where to allocate varying percentages of medical supplies for females of child-bearing years. In this case, they might decide to allocate a larger than normal fraction to Syria. Inequality can be as revealing as equality and maps can take the lead in offering guidance based on that inequality.

A Unifying Map

Single mapping activities can illuminate roadblocks in mathematics and lead to mastery of them. Other mathematical concepts and procedures are not themselves roadblocks, but an assembly of them may be. In such instances, often a simple single map may demonstrate organization of material not readily visible to students, otherwise. For example, students may often think of linear equations and linear inequalities as totally separate topics. That is an understandable position given that texts and teachers often do separate them as a way to learn process. It might be worthwhile, therefore, to take a step back and look at the whole system in a single map.

The map in Figure 6.17 shows an image of the road network in Detroit, Michigan. The map is centered on the downtown area and one street, Woodward Avenue, is highlighted for attention. Notice that the road network in the map is composed of a grid pattern overlain with a radial pattern. Everything to the east of Woodward Avenue, Detroiters refer to the "East Side" without offering a line of reference (and similarly for West Side). Compare and contrast the map in Figure 6.17 with other maps in this book:

- Scale bar, orientation arrow and implied cardinal compass points are present throughout.
- Cartesian Coordinate (Chapter 4) system is present in the grid street pattern here.
- Polar Coordinate (Chapter 4) system is present in the radial street pattern here.
- Slope of straight lines, shown Figures 6.13 and 6.14 and singled out here as Woodward Avenue, draw on properties of general linear equations
- Inequalities. Woodward Avenue separates Detroit into two sides: one side East of Woodward Avenue and one side West of Woodward Avenue. In the same way, any straight line separates the plane into two half-planes. The half-plane on one side of the line represents an inequality associated with the linear equation of the line.

FIGURE 6.17 Detroit Michigan. Notice Woodward Avenue. The area of Detroit to the west of Woodward is known to locals as the "West Side." Mapped in Google Earth Pro, 2023.

Look for analogues in the world around you! Tie the concepts you learn to places you know. Then, consider broader questions:

- Are there other cities that have similar road patterns? Look at Washington D.C. and Paris, France. Notice streets running at acute angles from selected prominent monuments and street intersections in the city. Read about the designs implemented from Pierre L'Enfant. Consider a project that integrates the history of urban planning, road networks, and coordinate systems.
- Notice that Canada is actually south of the USA at Detroit rather than north of the city. Consider the towns of Angle Inlet. Minnesota, and Point Roberts, Washington, both reachable by Americans by car only by first entering Canada. What are other geographic anomalies of other cities and places you might map?

This single map unifies six roadblock concepts into a whole that should be easy to grasp and it suggests directions for future geographic research. A lot of power can be embedded in a single map!

Simple Quadratic Equations and Graphs; Inequalities

Quadratic equations are formed from constants and variables raised to the power of 2.

The General Quadratic Form

The general form of a quadratic equation in:

- one variable, x, is $ax^2+bx+c=0$ where x is a variable, a, b, c are arbitrary constants, and a is not zero
- two variables, x and y, is $ax^2+bxy+cy^2+dx+ey+f=0$, where x and y are variables, a, b, c, d, e, f are arbitrary constants, and both a and c are not zero
- n variables, the generalized form contains multiple terms with products of multiple powers of the variables.

In this book, we confine our interest primarily to quadratic equations of one variable with examples of quadratic equations in two variables. Examples are given below, try using the graphing calculator on the one variable cases. Figure 6.18 illustrates the graph of $y=x^2+5x+6$ and Figure 6.19 contains the graph of $y=5x^2+3x+1$. What difficulties do you encounter?

Try using this graphing tool to graph equations and inequalities, as they occur, and practice until graphing becomes second nature. Use it throughout multiple sections to follow.

One Variable	Two Variables
$x^2+4x+4=0$	$x^2+2xy+y^2=0$
$x^2+5x+6=0$	$x^2-y^2=0$
$5x^2+3x+1=0$	$x^2+3xy+2y^2+4x-5y=7$

The graphs of quadratic equations in one variable are parabolas. Experiment with graphing some different equations. When is the parabola concave up (would hold water)? When is it concave down?

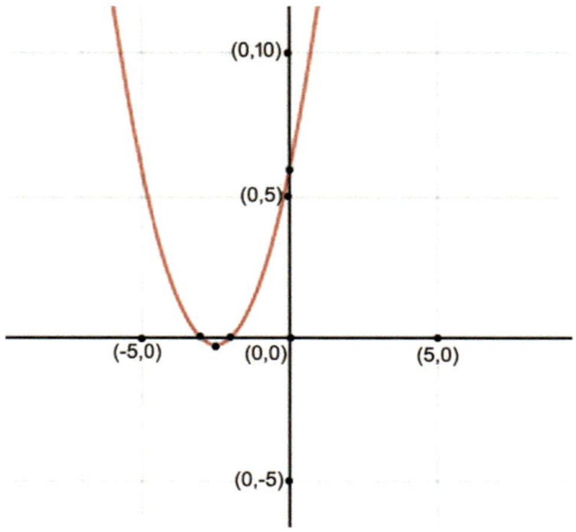

FIGURE 6.18 $y=x^2+5x+6$. Powered by desmos.com.

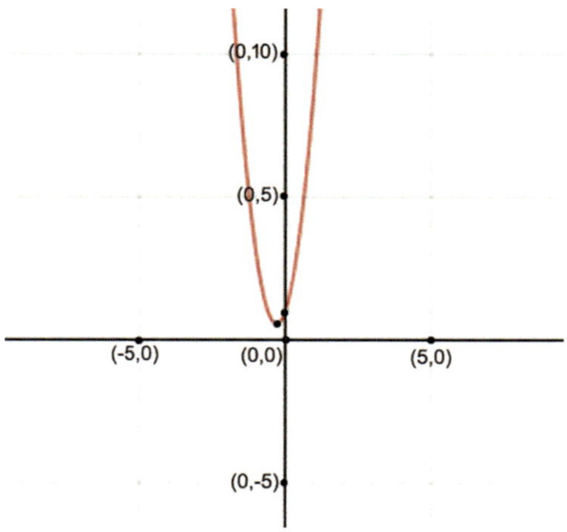

FIGURE 6.19 $y=5x^2+3x+1$.

Look at positive and negative signs of the coefficients. How many times might it intersect the x-axis? Under what circumstance might it be tangent to the x-axis? Notice where the minimum and maximum values are.

One Variable: The Parabola

To solve the equation $x^2+5x+6=0$, we must find values of x that lie on the parabola in Figure 6.18 and on the line $y=0$ (the x-axis). These values are precisely the locations where the parabola intersects the x-axis: in this case, at $x=-2$ and $x=-3$. Find the solution(s) of the next equation in the list using the graphing calculator. Then, move on to the last equation in the list of one variable quadratic equations (given above). Notice that the parabola representing that equation (Figure 6.19) does not cross the x-axis. What does that mean? It means there is no value, in the set of real numbers, that simultaneously lies on the parabola and on the x-axis: there is no solution within the set of real numbers to the given equation!

Other than by graphing, one might notice that the expression x^2+5x+6 equals $(x+2)(x+3)$, using the distributive law to write the three terms linked with plus signs into the product of two factors. This conversion process, using the distributive law, is called "factoring":

$(x+2)(x+3)$

$=x(x+3)+2(x+3)$, first use of the distributive law

$=x^2+3x+2x+6$, second use of the distributive law and commutative law for multiplication

$=x^2+5x+6.$

Remember: the distributive law reads forward and backward!

Skill at factoring comes with practice: the distributive law is at the heart of all factoring procedures. No matter how skillful one might become at factoring, there will still be quadratic equations which defy simple factoring by observation. Thus, the next material develops a systematic algebraic process which always works: it is called the quadratic formula. Use it when needed; but look first to see if a given equation factors in a straightforward manner: factoring, when obvious, is easier than the quadratic formula.

In the list of example equations above, the equation $x^2+4x+4=0$ factors as $(x+2)^2=0$; it factors as a single squared term, called a "perfect" square. Generally, $(x+A)^2=x^2+2Ax+A^2$. Perfect squares are easy to deal with: thus, in seeking a universal quadratic formula, we return to the general quadratic form and see if we might be able to rearrange it into a perfect square. Consider the general quadratic:

$ax^2+bx+c=0$. Move c to the right side and divide through by a, which we know is not zero: $x^2+(b/a)x=-c/a$. Now from observing the form of $(x+A)^2=x^2+2Ax+A^2$, we see that the value of (b/a) is playing the role of $2A$. So to complete the left side of the rearranged general quadratic to a perfect square, we need to divide (b/a) in half, to become A, and then square the result and add the square of that result also to the right side, so the equation balances. Thus, $x^2+(b/a)x=-c/a$ becomes $x^2+(b/a)x+((1/2)(b/a))^2=-c/a+((1/2)(b/a))^2$. The left side now factors to the perfect square $(z+b/2a)^2$. Simplifying, $(x+b/2a)^2=-c/a+b^2/4a^2=(b^2-4ac)/(2a)^2$. Take the square root of both sides so that $x+b/2a=\pm(\sqrt{(b^2-4ac)})/2a$. Finally, isolating x, $x=(-b\pm(\sqrt{(b^2-4ac)})/2a$, the quadratic formula. Note the importance of assuming a is not zero. The form b^2-4ac appears elsewhere in mathematics and is called the "discriminant" of a quadratic form.

If you felt lost in the arithmetic, try to figure out where you first had a problem. Then review arithmetic associated with areas that caused discomfort until they no longer cause a problem. The only way to fix the problem is to confront it. And remember even if you received marks of 90% in math, that means there is 10% you did not master—and that 10% will keep coming back to haunt you because mathematics is linear in its structure: everything depends on a sound foundation.

Two Variables: Conic Sections

In the case of one variable, the graph of the general quadratic was a parabola. The general two variable quadratic forms the class of conic sections, when graphed. Just as we look at a plane slicing through the Earth sphere to generate small circles and great circles as sections of a sphere, so we can think of the parabola and the ellipse (two figures we have seen earlier) and other shapes as being formed when intersecting a plane with a cone. Here we consider how those patterns align with algebra.

A cone is formed by rotating a line in three dimensions about a fixed point, the apex, through a fixed angle. This line is called the generator of the cone, which is generated in two pieces, each referred to as a frustrum. When the fixed angle is 90°, the cone is called a right circular cone. There are many interesting classic books concerning conic sections as well as modern works, both print and digital. We suggest a few in the References section at the end of the book.

Given a cone and a plane (Figure 6.20): the following sections can be produced by intersecting the cone with the plane.

- If the intersecting plane is parallel to one side of the cone (the generator) then the section is a parabola; it lies completely within one frustrum of the generated cone.
- If the intersecting plane is parallel to the base of cone, then the section is a circle; it lies completely within one frustrum of the cone.
- If the intersecting plane is not parallel to the base, and has slope less than that of the generator, then the section is an ellipse; it lies completely within one frustrum of the cone.
- If the intersecting plane has slope greater than the slope of the generator (thereby intersecting the base of the cone), then the section is a hyperbola; it lies within both frustrums of the cone and each piece is referred to as a "branch."

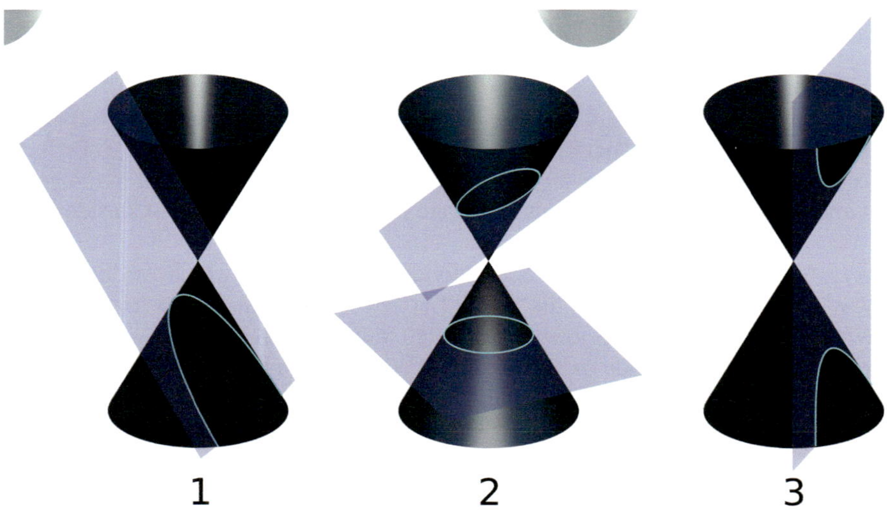

FIGURE 6.20 Conic sections. 1: Parabola, the slicing plane is parallel to the generating line for the cone. 2: Ellipse (including circle), the slicing plane does not cut through the cone; when it is parallel to the base, the ellipse is a circle. 3. Hyperbola, the slicing plane cuts the base of the cone. Source: Pbroks13, CC BY 3.0 <https://creativecommons.org/licenses/by/3.0>, via Wikimedia Commons.

Further, there are "degenerate" cases:

- If the intersecting plane has slope the same as the generator, and passes through the apex of the cone, then the section is a straight line with slope the same as the generator.
- If the intersecting plane passes through the apex but does not intersect either frustrum (is parallel to the base of the cone), then the section is a single point.

Algebraically, the general equation for any conic section is the general quadratic equation in two variables, x and y, $ax^2+bxy+cy^2+dx+ey+f=0$, where x and y are variables, a, b, c, d, e, f are arbitrary constants, and both a and c are not zero.

- If the discriminant is less than zero, then the conic associated with a given equation is an ellipse.
- If the discriminant is equal to zero, then the conic associated with a given equation is a parabola.
- If the discriminant is greater than zero, then the conic associated with a given equation is a hyperbola.

The use of the discriminant, when easy to observe, can offer quick insight into graphical form.

Two particular quadratic equations in two variables are helpful in engaging in rapid mental calculation (do all work in your head and only in your head) and both are direct uses of the distributive law:

$$x^2-y^2=(x-y)(x+y)$$
$$(x+y)^2=x^2+2xy+y^2$$

Two Variables: Rapid Multiplication

- Difference of Two Squares

  Find the product: $(x-y) * (x+y)$. Use the Distributive Law (DL), $a(b+c)=a * b+a * c$. Suppose a in the DL is $(x-y)$; $(b+c)$ is $(x+y)$. Then, using DL:

$(x-y) * (x+y)=(x-y) * x+(x-y) * y$ One use of DL

$=x * x - y * x+x * y - y * y$ Another use of DL

$=x^2 +0-y^2$ Notice that $y * x$ is the same as $x * y$ and that adding zero causes no change

$=x^2 - y^2$

- Squaring a Number Ending in 5

  Generally:

$$(10x+5)(10x+5)=(10x+5) * 10x+(10x+5) * 5=10x * 10x+5 * 10x+10x * 5+5 * 5$$
$$=10x(10x+5+5)+25=10x(10x+10)+25=10 * 10x(x+1)+25=x(x+1) * 100+25$$

So, a quick way to express it: 45 * 45 ends in 25 and the front part is 4 * 5, so 2,025. Or, 65 * 65 ends in 25 and the front part is 6 * 7 so 4,225.

Use the quick way; but make sure you understand WHY it works, that is, PROVE it... the answer for all of these is the Distributive Law. When you know WHY things work, you remember them and can be creative with them.

Putting things together...
What is 24 * 26?

$$(25-1)(25+1)=25 * 25-1 * 1=625-1=624$$

What is 32 * 38?

$$(35-3)(35+3)=35 * 35-3 * 3=1{,}225-9=1{,}216$$

Keep looking for the distributive law...see if you can make up other interesting uses of it... the distributive law will come up in arithmetic, algebra, geometry, trigonometry, calculus, and many other places...it is truly a root concept.

- Book Pricing Example
  One book costs $7.99. How much do eight books cost? Use DL:

$$8 * (800-1)=8 * 800-8 * 1=6{,}400-8=6{,}392 \text{ or } \$63.92$$

Here, using the DL enables solving the problem quickly in your head; it might be hard to solve 8 * 799 in your head otherwise. Practice all of these rapid calculations as you go about your daily life; put them together in different ways.

Inequalities

To graph quadratic inequalities proceed as in the case for linear inequalities. Graph the associated curve. Pick a point not on that curve. If the coordinates satisfy the given inequality, then all other points on the same side of that curve do so, as well. If there are multiple inequalities, look for set intersections of regions. The process is simple; however, it draws on multiple foundational concepts.

REAL-WORLD SHAPES AND CHALLENGES

Ellipses are all around us on the Earth and beyond it. Consider the latitude-longitude foundations laid in this book, and in particular, imagine yourself looking down at the Earth from the international space station at a position far above the North Pole. The North Pole, directly below you, is a point on the globe at 90° North Latitude. The line of 89° North Latitude would, if you could draw it on the Earth, make the shape of something close to a circle. Outside that, slightly closer to the Equator, would be 88° North Latitude, and so on, for all lines of latitude down to 0° North, at the Equator (Figure 6.21). Why are these lines circles as viewed from the North Pole? The Earth is an oblate spheroid, slightly wider around the Equator than around the Poles, but the lines are still circles. Open the linked 3D web scene (https://www.arcgis.com/home/webscene/viewer.html?webscene=5c3a2807b34841f2afa4c383fb57b98a) and investigate these lines yourself. Then use the 3D scene viewer to rotate the globe so that you are looking straight down on the Equator and you are seeing the North and the South Poles at the top and bottom of your screen, respectively. Because of the oblateness of the Earth, the lines of longitude that you see are curved, but they are not perfect circles—they are non-circular ellipses.

Similarly, ellipses are commonplace outside the Earth as well: The orbits of the planets around the sun are ellipses. In fact, the orbit of Pluto was so much more an ellipse than the orbit of the other planets (that is, its eccentricity is pronounced) that this is one of the factors that eliminated Pluto from the official list of planets! The orbit of the solar system around the center of the Milky Way is also an ellipse as are many other examples in our universe.

Throughout this chapter, we have drawn analogies back and forth between algebraic concepts and other elements of our world: to sections of the Earth sphere and associated surface measurements being similar to sections of a cone and associated equations, to algebraic equations employed

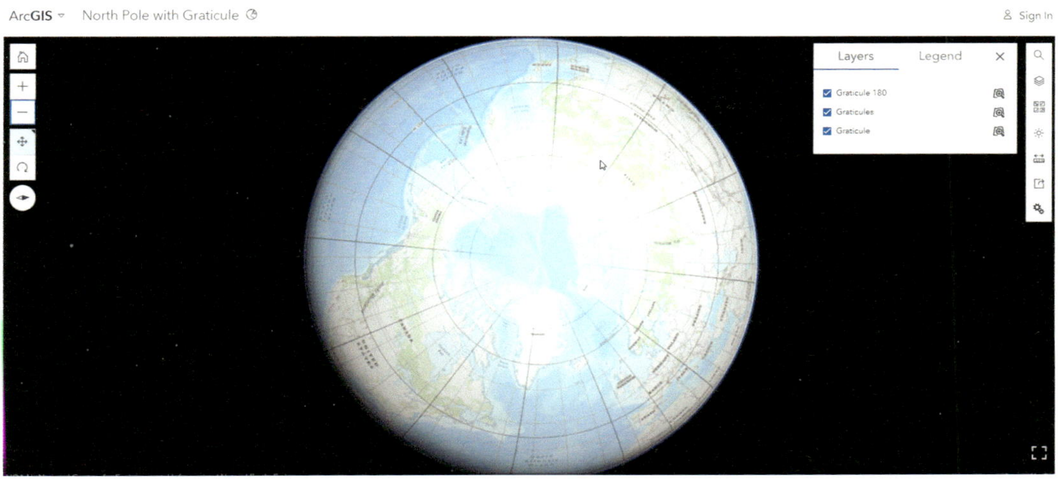

FIGURE 6.21 Lines of latitude as seen from a position above the North Pole. Mapped using ArcGIS from Esri.

to check the cashier in a grocery store while shopping. This chapter offers a particularly stiff challenge in that regard; the algebraic foundations presented are quite abstract. See how you might consider realizing them in your own world; they are there and they are critical—the challenge is to recognize them as they are often quite subtle.

7 Mathematics' Foundations
Dimension and Geometry

The line of images above is a visual abstract of this chapter designed to foster spatial thinking. From the chapter numeral, to the book structure, to the real world, the reader is offered gentle guidance to develop spatial intuition about what might be coming. Those thoughts are then reinforced with a detailed text outline of chapter content below. The images and outline forge as an abstract of chapter content.

CHAPTER OUTLINE

Synthetic Geometry
- Euclid: Selections
- Axioms for Euclidean Geometry in the Plane
- Selected Concepts and Theorems
- Solid Geometry
- Whole-Earth Measurements
  - Determining the Mass of the Earth
  - Determining the Volume of the Earth

Transforming Dimensions: Approximations
- Area Under a Curve: Approximation Techniques
  - Graph Paper Approximation to Area
  - Interpolation on a Map: Simpson's rule
- Representing Numbers as Isolines
  - Symbolizing Contours
  - Investigating the Depths of the Great Lakes

Dimension
- Overview
- Beyond the Usual—Fractional Dimension
- Testing the Coastline Paradox: Fractals

Non-Euclidean Geometry

DOI: 10.1201/9781003305613-7

Error Issues
Digitizing
Jordan Curve Theorem
Geometric Models of the Real World?

In Chapter 6, we concluded by showing how algebraic forms derive from cross sections of a cone. In previous chapters, especially in Chapter 4, we saw how coordinate systems (Descartes, Smith, Latham, 1954) worked in relation to a globe, a spherical representation of the Earth in order to create a scheme for measuring that representation. In order to motivate the reader to integrate the Foundations of Mathematics approach of why things work, based on axioms, postulates, and theorems (from Chapter 6) with the real-world interpretation of geometric forms, we expand our understanding of Earth measurement by unifying a few forms from the past that will lead to those in the rest of this final set of chapters (Heath, 1956; Coxeter, 1961).

We have noted throughout the book that the Earth is not a sphere. A globe is a sphere. In fact, the shape of the Earth is an oblate spheroid; it bulges at the Equator. The science of Geodesy is centered on this fact; we just take a quick peek at it to see how important implications from one basic assumption might yield varying, possibly parallel, results. Building on the end of Chapter 6, where a 3D scene of the Earth was examined, Figure 7.1 shows what the globe graticule (the latitude and longitude lines) looks like when transformed to fit an oblate spheroid.

Note that the small "circles" have become non-circular ellipses. Why is that? Does the bulge at the equator push those small circles out more to become flatter? As we approach the pole, the ellipses become slightly more circular. The pole serves as a limit to such flattening. Thus, we see integration of concepts from the mapping world with a concept from the great mathematical feat of Leibniz and Newton (Palomo, 2021)—that of the "limit," as the basis of the subject of calculus which is at the root of many Engineering applications. Algebra and geometry intertwined at the end of the last chapter;

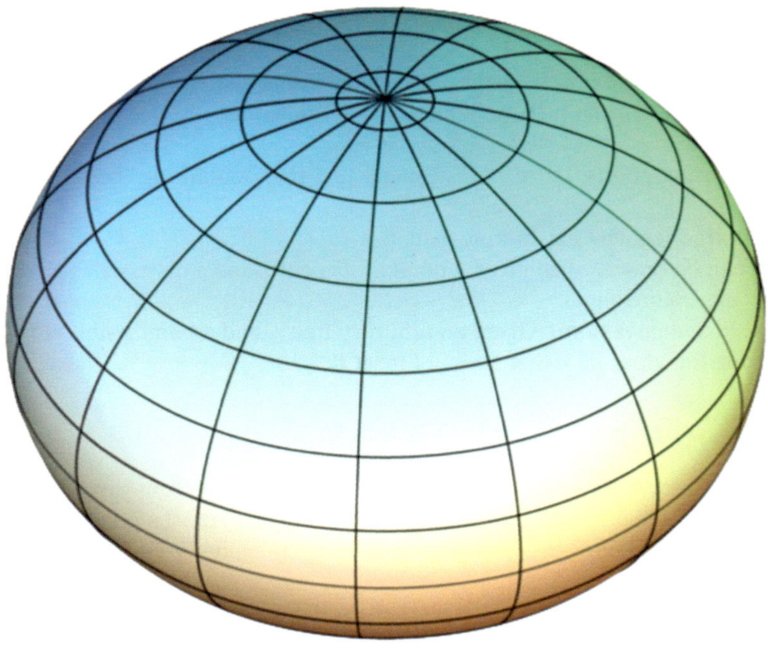

FIGURE 7.1 Graticule on an oblate spheroid. Source: Wikimedia Commons, https://commons.wikimedia.org/wiki/File:OblateSpheroid.PNG. This image was made by en:User:AugPi using *Mathematica* (https://commons.wikimedia.org/w/index.php?title=Mathematica&action=edit&redlink=1).

that characteristic carries over into this chapter to suggest that we will see the foundational approach yield familiar geometry which will then transform to calculus form, much of it illustrated through mapping as in the simple example of Figure 7.1.

SYNTHETIC GEOMETRY

For much of history, the word "geometry" has been taken to refer to the geometry of Euclid, a geometry free from coordinates (Steiner, 1867). Yet, much contemporary use of geometry does employ Cartesian coordinates. Typically, we measure the distance, D, between two points (x_1, y_1) and (x_2, y_2) as $D = \sqrt{[(y_2 - y_1)^2 + (x_2 - x_1)^2]}$, plotted on graph paper or in the Cartesian coordinate system and we expect that software is using this sort of calculation to make measurements behind the scenes to produce requested results. We study the geometry of Euclid in secondary school. But that study is merely an introduction. What are some of the other elements of the geometrical foundations of mathematics that perhaps we do not see, or do not appreciate, when first exposed to the rigor of axiomatic geometry? That rigor is designed, among other things, to foster and develop skills with clever and concise approaches to mathematics: far more than previous training may have. It displays the difference between "education" (learning to think in creative ways that might find application in various directions) and "training" (learning focused skills designed to lead to a small, predictable set of results).

EUCLID: SELECTIONS

Euclid's *Elements* begin, as we did in considering foundations more generally, with some intuitive terms that make sense in terms of the way we see the world around us. He defines terms such as "point" and "line"; he creates axioms from which to build his consistent development of geometry.

AXIOMS FOR EUCLIDEAN GEOMETRY IN THE PLANE

Over the centuries, these axioms have been restated using the vernacular of the times, but their meaning has remained constant. His axioms are often stated as:

- For two distinct points, there exists exactly one line containing them: two points determine a line.
- A straight line segment has infinite extent.
- A circle can be drawn when given a point as its center and a distance as its radius.
- All right angles are equal.
- (Parallel Postulate/Axiom). Given an arbitrary line L and a point p not on L. There exists exactly one line through p that does not meet L.

SELECTED CONCEPTS AND THEOREMS

The following is a very brief summary of concepts and theorems from Euclidean geometry: some as they have evolved and others as originally (for the most part) presented. The reader interested in more is referred to references at the end of this work and to numerous sites online.

- Rigid motion: a transformation of the plane to itself that preserves distance. Relative distances between the points and relative position of the points remain the same, before and after transformation. Shapes are not changed although their location in the plane quite likely does change. Examples of rigid motions are given below.
  - Translation: shapes in the plane are moved by the same amount and in the same direction. A translation has both direction and distance.

- Rotation is a transformation of the plane to itself that fixes one point (the center) and one angle: everything rotates by the same amount around the center.
- Reflection is a transformation of the plane to itself that fixes a line in the plane, as a mirror in which to reflect points. It exchanges points from one side of the mirror with points on the other side of the mirror, at the same distance from the mirror.
- Glide reflection is a rigid motion composed of two rigid motions: a reflection and a translation.

Composing transformations often leads to new ones that are also useful. Suppose, for example, that I set an analog watch when I change time and then do the same for someone else; I synchronize the two watches side-by-side and then hand the second watch to someone else. That action might be viewed as a rotation (of watch hands) followed by a translation (sliding the watch across the table to someone else). Look around you; synthetic geometry is everywhere.

- Congruence: two triangles are congruent if one of them can be moved, through a rigid motion (no stretching), to superimpose perfectly on the other. Theorems involving conditions of the shape and size of the triangles, involving side-angle-side, angle-side-angle, and side-side-side relations create a body of work.
- Symmetry. "Pons Asinorum." Two sides of a triangle are equal if and only if the angles opposite those sides are equal. Such a triangle is called an isosceles triangle. Could other shapes have isosceles characteristics? Consider states of the USA that historically were defined along latitude and longitude lines: What are their true shapes? Watch this video (https://youtu.be/50sLWA8EPVU) where Joseph Kerski discusses the shape of Colorado.

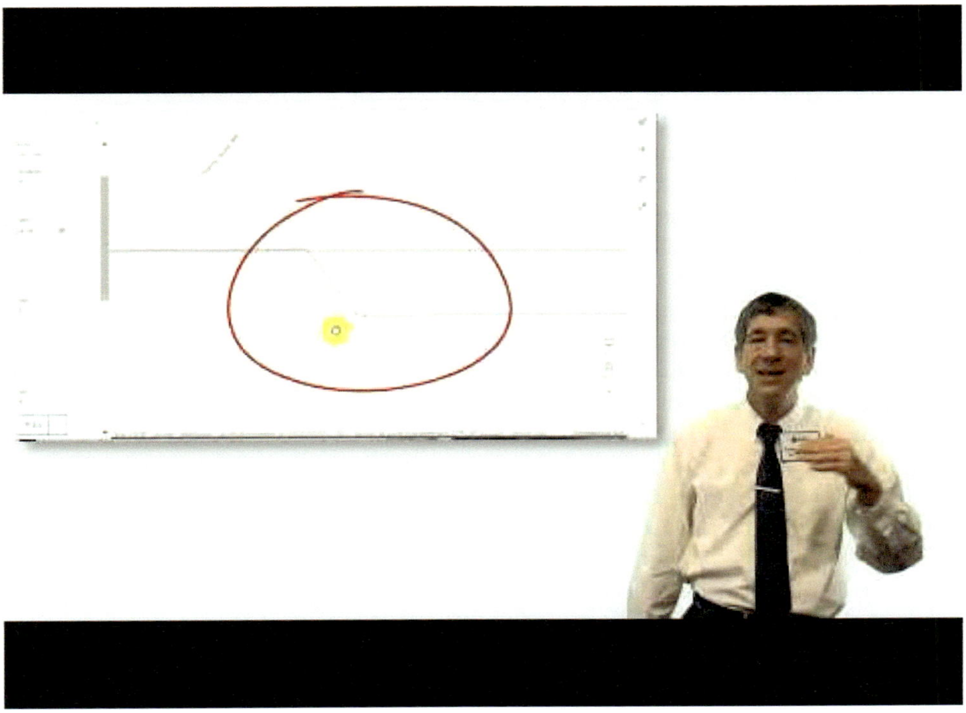

On a topographic map, the Quadrangles are actually isosceles trapezoids. Why? The left and right sides of a topographic map are usually lines of longitude. Remember that these lines of longitude (Chapter 4) converge at the Poles. Thus, they are not parallel along the sides of a topographic map, even if they look that way.

When photographing a tall building be aware that what looks like a rectangle in shape is, if well-centered, an isosceles trapezoid. Straighten out the photograph in Adobe Photoshop and make it a true rectangle when seeking to apply the photograph to an extruded parallelepiped representing the underlying skyscraper structure.

- Measurement. The sum of the degrees in the three angles of a triangle is 180°. This concept is based on the Parallel Postulate.
- Similarity. A line segment, $|L|$ partitions two sides of a triangle into proportional segments if and only if $|L|$ is parallel to the third side of that triangle. Similar triangles are related by a scaling factor.
- Areas of geometric figures (repeated here for convenience):
  - The perimeter of a circle is $\pi \times$ diameter $= 2\pi \times$ radius.
  - The area of a circle is π times the radius squared ($A = \pi r^2$).
  - The area of a triangle is ½ its base times its height: $A = (1/2)bh$.
  - The area of a polygon: determine it by dissecting it into triangles and rectangles.
- The Pythagorean Theorem.
  - For a right triangle, $\triangle ABC$:
    - if $\angle ACB$ is a right angle, then $a^2 + b^2 = c^2$;
    - if $a^2 + b^2 = c^2$, then $\angle ACB$ is a right angle.
  - For a general triangle, $a^2 + b^2 = c^2 - 2ab$ (cos ($\angle ACB$)) (the Law of Cosines). When $\angle ACB$ is 90°, the Law of Cosines reduces to the Pythagorean theorem for a right triangle because cos (90°) = 0.
- Watch Saunders Mac Lane's 1989 Lecture (https://youtu.be/qyBBD32pT98) as he addresses a general audience and offers a proof of the Pythagorean Theorem.

  The Pythagorean Theorem is important because it links geometry and algebra which in turn permits systematic analysis of distances on Earth.
- Invariance of angles subtended by a chord in a circle.
- In a circle, equal chords (line segment from one point on the circle to another point on the circle) determine equal angles and the other way around. Thus, if AB is a chord in a circle, the vertex angle of any triangle with third point on the same arc is the same for all selections of third point made from that same arc.

  Open this web map (https://www.arcgis.com/apps/mapviewer/index.html?webmap=c97b1c3300ef4ad4be868c37d04eaa2d). A powerline runs from northwest to southeast in this map (Figure 7.2), through the points A and C, forming a right triangle $\angle ABC$. Use Map Tools>measurement to measure the distance in meters from point A to point B. Use the measure tool>measure the distance in meters from point B to point C. Using the Pythagorean Theorem as described above, calculate the distance from point A to point C. Now, use the measure tool to measure the distance from point A to point C. How close was your calculated value to the measured value? Name one reason why your calculated value might not be exactly the same as the measured value. Now, using the measured value of the distance from A to B and B to C, calculate the area of the triangle. Use the measure tool>change to area>measure the area of the triangle. Again, name one reason why your calculated value might not be exactly the same as the measured value. Determine the percentage that your calculated value is different from your measured value for the perimeter and for the area. The value of the interior angle at point B is 90°. See if you can determine the value of the interior angles at points A and C. Then, determine the value of the *exterior* angles at points A and C. Lastly, use the measure tool to measure the perimeter of the triangle. What is the perimeter of the triangle? Let us say you were able to walk on the perimeter of the triangle, but at a slow speed of 3 km/hour, allowing for the fact that you would be walking on gravel roads and then through a field. Based on your speed and the perimeter of the triangle, how long would it take you to walk around it in minutes? In hours?

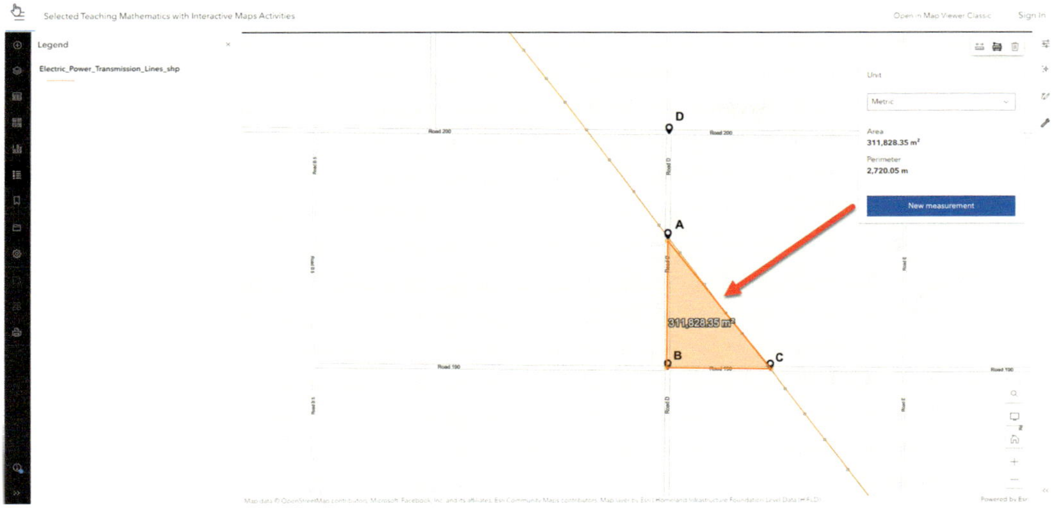

FIGURE 7.2 Calculating and measuring angles and sides of a triangle on a map. Mapped using ArcGIS from Esri.

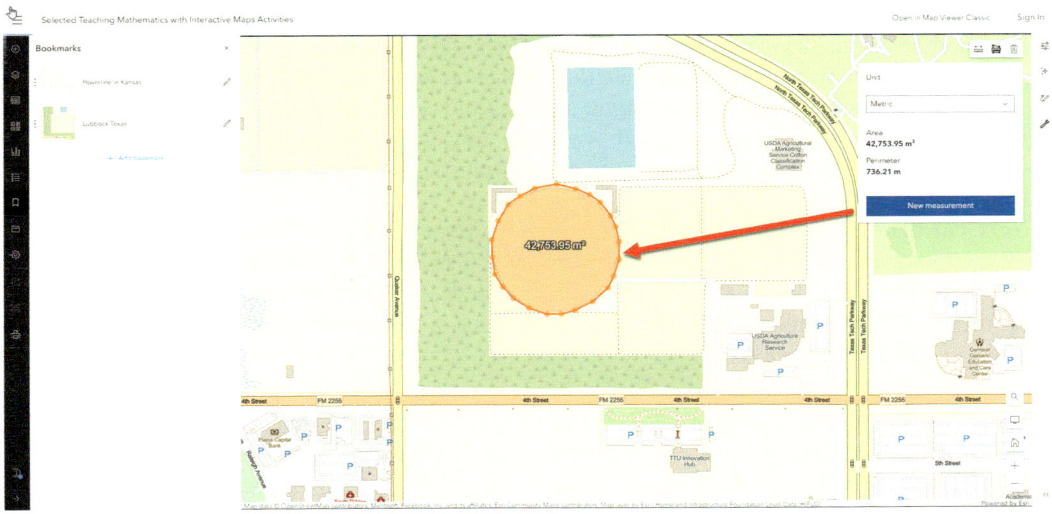

FIGURE 7.3 Calculating and measuring a circle on a map. Mapped using ArcGIS from Esri.

Using this same map, use Bookmarks>Lubbock Texas and observe the circle that is outlined by a road at Texas Tech University (Figure 7.3). Using the measure tool, measure the diameter of the circle. Using this figure, calculate the perimeter and the area of the circle. Then, use the measure tool to measure the perimeter and then the area of the same circle. Name one reason why your calculated value is likely to be different from the measured value. Determine the percentage that your calculated value is different from your measured value for the perimeter and for the area.

Solid Geometry

Intuition in solid geometry can be much more difficult to deal with than it is in plane geometry. We live in three-dimensional space and cannot get outside it to take a look at it and guess what might be true about it. We can do that with the plane.

Some concepts remain the same as we move from the two dimensions of the plane to the third dimension of space. Others do not. Two that are clearly different involve the use of volume (in three dimensions) for area (in two dimensions) and three-dimensional shapes such as sphere or tetrahedron for two-dimensional shapes such as circle or triangle.

The geometry that underlies mapping is a combination of two-dimensional and three-dimensional geometry. In return, the mapping technology helps to reveal the workings of the underlying mathematical benchmarks.

Let us do some measurements and calculations with 3D objects on the Earth's surface using a 3D map. Open a 3D web map (https://www.arcgis.com/home/webscene/viewer.html?webscene=b6be327daad64537ab682fe16c2c5daa) of North Table Mountain in Golden Colorado (Figure 7.4). North Table Mountain was formed from lava that slowly oozed from a fissure on the Earth to spread over a section of land; the surrounding land then experienced erosion, leaving North Table Mountain. As its name implies, and as you can see in the 3D scene, the top is very flat indeed. Use the measure tools to measure the area of the top of the mountain, clicking as you follow the top as best you can and noting the contour hints that follow you as you go, as shown below. Double click to end the polygon and you will receive an area and a perimeter returned from the GIS software. Record these numbers and calculate the area in square meters. It should be around 4 km^2 or 4,000,000 m^2. Take the square root of this number, so that you essentially have a length and a width of the mountain if it were a perfect square.

Next, use the Measure>Elevation profile to measure North Table Mountain from its top to the surrounding terrain (Figure 7.5).

Subtract the lower from the upper elevation. It should be around 1,910–1,740 m, or 230 m in height. Record this number. Next, calculate the volume of North Table Mountain: Use your number for the square root of the area for the length, and then the square root of the area for its width, and then the height, and since volume = length × width × height, this yields: 2,000 × 230 × 2,000 = 920,000,000 m^3, or 920 million m^3. These numbers are approximate: Use your own measured numbers instead of these numbers.

Next, calculate the weight of North Table Mountain. Except for a thin layer of crushed lava on the top of the mountain, most of North Table is solid rock. Solid rock is estimated at 2.5 to 3.0 tons/m^3. Knowing your value of the number of cubic meters for North Table Mountain, what is the weight of the mountain? If a standard car weighs 1,300 kg, what is the weight of North Table Mountain in numbers of cars? That's heavy!

FIGURE 7.4 Measuring 3D landforms on the Earth's surface. Mapped using ArcGIS from Esri.

FIGURE 7.5 Measuring elevation and creating a 3D profile of terrain on the Earth. Mapped using ArcGIS from Esri.

WHOLE-EARTH MEASUREMENTS

Determining the Mass of the Earth

In a previous activity, you determined the polar circumference of the Earth. In this activity, you will have the opportunity to determine the Earth's mass and volume (Figure 7.6).

FIGURE 7.6 Earth as seen from space. Image Source: NASA, from Apollo 17, Public Domain.

1. Jot down the length of the polar circumference from your work in the previous activity.
2. Determine the Earth's radius based on the circumference.
3. Compute:

   Mass = acceleration * radius^2/G

   where acceleration due to gravity = 9.8 m^2, and G, the constant of proportionality, which was computed by Henry Cavendish in 1798, is 6.67×10^{-11}/kg seconds^2.

   For example: Radius = 6.4×10^6 m

   $M = 9.8 * (6.4 * 10^6)/6.7 \times 10^{-11}$ kg

   $M = 6.0 \times 10^{24}$ kg
4. How close is your computed figure of mass to the accepted figure of the mass of the Earth?
5. Name one reason your figure might not match the accepted figure.

Determining the Volume of the Earth

1. Jot down the length of the polar circumference from your work in the previous activity.
2. Determine the Earth's radius based on the circumference.
3. Compute the Earth's volume from the formula below:

   Volume = 4/3 * π * r^3
4. How close is your figure to the accepted volume of the Earth?
5. Name one reason why your figure might not match the accepted figure.

TRANSFORMING DIMENSIONS: APPROXIMATIONS

Areas that are triangular or rectangular or another shape that we know how to measure are easy to deal with. But what about one that is curved in an irregular pattern? We might wish to know how much area there is under that curve. The mathematics developed by Leibniz and Newton, the Calculus, made such measurement possible. We consider a few techniques for making measurements that might be useful in real-world applications.

Area under a Curve: Approximation Techniques

Graph Paper Approximation to Area

The simplest way to approximate area under a curve is to use graph paper, or a gridded image of some sort, to count number of small cells within the curve and count overfit and underfit of these cells at the boundary and try to balance the amount of over/under area. The finer the mesh of underlying cells, the better the approximation. One generalization of this idea is the so-called Trapezoidal Rule studied in calculus, using a set of trapezoids to measure the area between a curve in the xy-plane and the x-axis, with the underfit and overfit of the trapezoids minimized using the calculus concept of a limit to infinite process that converges to a single answer.

Interpolation Using Simpson's Rule

Simpson's Rule, which employs pieces of parabolas to approximate curves, allows us to estimate areas under curves even if a formula for the curve is not known, but the values of the function at an even number of equally spaced points along the axis are known. Only those values and the length Δx of the spacing are needed. Notationally, these ideas are captured in Simpson's 1/3 Rule (there are a number of different Simpson Rules) often taught in Calculus courses but which really requires no direct knowledge of calculus in order to use it:

$$\int_a^b f(x)\,dx = S_n = \Delta x/3\Big[\big(f(x_0)+4f(x_1)+2f(x_2)+\cdots+2f(x_{n-2})+4f(x_{n-1})+f(x_n)\big)\Big]$$

where n is even (otherwise a value gets left out) and $\Delta x = (b - a)/n$, the length of one of the small intervals. Basically, use strips of varying length/height, evenly spaced, to fill the area under consideration. Multiply them with weights of two or four, alternating, to fill in space between strips. The more strips the better the approximation; so the larger the value of n, the better the approximation.

What we show here, in detail, is an activity created and developed and classroom-tested, by co-author W. Arlinghaus, in Calculus II Workshops held in support of Calculus II courses at Lawrence Technological University over a decade or so from 2000 to 2009 (for a full retrospective, see Arlinghaus, 2021).

In this activity, we use Simpson's Rule to measure the area of the State of Iowa; use the map of Iowa shown in Figure 7.7.

On this map, evenly spaced points have been selected along the Eastern and Western boundaries along with two extra points. The following facts are needed to help compute the area and find the numerical approximation solution:

- The circumference of the earth is approximately 24,800 miles.
- The circumference of the small (non-equatorial) circle at latitude α is $24{,}800 \cos \alpha$.
  - This circle is divided into 360° of longitude.
  - So, the length, in miles, of 1° of longitude at any given latitude, is $(24{,}800 \cos \alpha)/360 = 68.88 * \cos \alpha$.
- We assume the length of a degree of latitude is constant at $24{,}800/360 = 68.88$ miles (that the Earth is a perfect sphere).
- The distance between points of equal longitude is measured along great circles, which have the same circumference as the equator.

In the case of Iowa, we look to calculate its area in two separate pieces:

- One large piece bounded by two arcs, one as a northern boundary and one as a southern boundary; use the ten selected points on each of the eastern and western boundaries as the Simpson's Rule guide points.

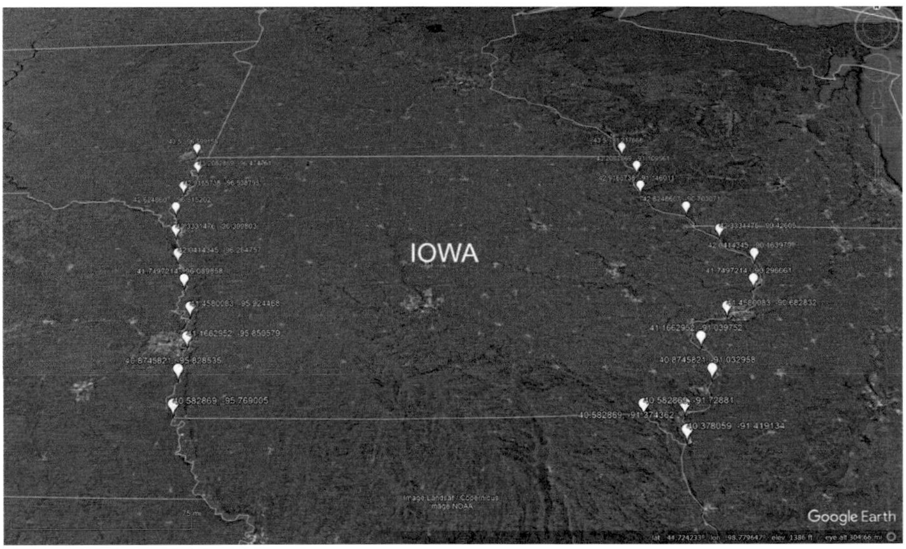

FIGURE 7.7 Simpson's rule using interpolation to approximate the area of Iowa. Implementation of William C. Arlinghaus: first mapped in 1997, subsequent version, shown here, in Google Earth, 2015.

- One small triangular piece that lies to the south of the long southern border, at the southeast corner of Iowa.

1: Latitude, α.	2: Long. West Boundary	3: Long. East Boundary	4: Difference between E. and W. Long.	5. Length in Miles of 1° Long. (24,800 cos α)/360 = 68.88 * cos α	6. Distance, in Miles, across State: Column 4 Times Column 5	7. Weight for Simpson's Rule Times Column 6
43.5	−96.5992	−91.2177	5.3815	49.72	267.57	1: 267.57
43.21	−96.4748	−91.1096	5.3652	50.2	269.33	4: 1077.32
42.92	−96.5388	−91.146	5.3928	50.44	272.01	2: 544.02
42.63	−96.5152	−90.7031	5.8121	50.68	294.56	4: 1178.24
42.33	−96.3998	−90.4261	5.9737	50.92	304.18	2: 608.36
42.04	−96.2648	−90.164	6.1008	51.16	312.17	4: 1248.68
41.75	−96.0899	−90.2967	5.7932	51.39	297.71	2: 595.42
41.46	−95.9245	−90.6828	5.2417	51.62	270.58	4: 1082.32
41.17	−95.8506	−91.0398	4.8108	51.85	248.65	2: 497.3
40.88	−95.8285	−91.033	4.7955	52.38	251.19	4: 1004.76
40.60	−95.769	−91.3744	4.3946	52.30	229.94	1: 229.94
Total						8,333.93

The chart shows latitude/longitude values and associated derivative information for each of the ten points on the eastern and western boundaries.

Here, $\Delta x = (43.5 - 43.20829)24{,}800/360 = 20.0955777778$. We might wish, for the sake of ease of calculation, to use the value of 20. Add the values in Column 7 to obtain a total of 8,333.93 as a total length of the weighted E–W strips across Iowa and multiply by 20/3, as Simpson's strip width giving a value of 55,559.5333 which we round to 55,560 square miles. This latter area is the approximate area of Iowa without the small triangle to the southeast.

Next, calculate the area of the small triangle. We add two points to the south boundary to do so: (40.58287, −91.7288), (40.37806, −91.4191). Use these to create a right triangle to approximate the area as one-half the base times the width of that triangle. The base lies between −91.72881° and −91.419139° west longitude, a difference of 0.309671° of longitude at north latitude of 40.582869°. The length of 1° of longitude at this latitude is (24,800 cos 40.582869)/360 = 24,800 * 0.75946584/360 = 52.3187579 miles. So, the length of the base is 52.3187579 * 0.309671 = 16.2016021 miles. The height lies between 40.378059° and 40.582869° north latitude, a difference of 0.204810°. So, the height of the triangle is 0.204810 * 68.88 = 14.1073128 miles. The area of the triangle is 16.2016021 * 14.1073128/2 = 114.280534 square miles which we will round off to 114 square miles. Lee County, containing the triangular piece, is about 500 square miles; this area appears to be, very roughly, about a quarter of Lee County, or about 125 square miles. The estimate accounts for about 91.2%, (114/125) * 100 = 91.2, of the area.

Thus, the entire area of Iowa, approximated using Simpson's 1/3 Rule, is

$$55{,}560 + 114 = 55{,}674 \text{ square miles.}$$

The published area of Iowa is 56,363.3 square miles. So, our estimate accounts for (55,674/56,363.3) * 100 = 98.78% of the area. Our estimate underfits the actual area. We might think that an approximation that is almost a 99% fit is good enough; perhaps for many purposes it is. But if it is not, how can we improve it? Why is our estimate an underfit? There are many possibilities here:

- The estimate is an underfit in the large area because:
  - We used only ten sampling points.

    - We rounded off consistently to be less than the actual calculated value.
    - Even if we used all the decimal values that we could see, that were returned by the calculator, we still might not be using all the digits that the calculator had stored and so were using a value less than the calculated value.
- To improve the fit of the Simpson's approximation to the actual area, the large area of Iowa, we might therefore:
    - Choose more intervening points to use in Simpson's Rule (keeping an even number, always).
    - Fail to round for ease of calculation convenience.
    - Be careful of the order in which operations are performed on the calculator (or move to a desktop computer) in order to maximize the number of fully stored digits used in each operation.
- The estimate is an underfit in the triangular area because:
    - The size of that area will vary depending on the amount of detail in the map used.
    - The assumption that we use a right triangle shaved off some of the land to the east.
    - There may be issues with rounding as there were for estimating the larger area.
- To improve the fit of the triangular area, we might therefore:
    - Choose a map that shows more detail.
    - Not use a triangle to estimate the area but instead use Simpson's Rule, once again, with the strips running N–S, from the southern boundary, extended from the long southern boundary of the entire State, south to the curvy boundary enveloping the rest of south-eastern Iowa.
    - Consider rounding issues as we did for estimating the area of the large part of the State.

Methods for improving the approximation are clear; individuals will make choices depending on how much benefit they see coming from extra time and effort expenditures. There is a balance that needs to be achieved.

This activity, while somewhat complex, draws together many of the concepts of this book. Indeed, one might say, why bother with using Simpson's Rule—just look up the area of Iowa. There are, however, good reasons to engage in this activity. First there is the opportunity to reinforce many of the concepts in this book and to gain great insight into how different concepts work together to forge a single answer. Some of these concepts are listed below.

- Educational Opportunity: integrate concepts to come to real-world solutions.
    - Earth measurement
    - Earth shape
    - Trigonometry
    - Geometry
    - Percentages
    - Decimal length, rounding, and communication
    - Approximation
    - Area
- Fieldwork Opportunity: a person wishing to survey an area and measure it, may now do so armed with only a GPS unit, a smartphone, and Simpson's Rule. Measure any shape of land or water. Gaining access to the land is the only major challenge. Collect coordinates, hopefully but not necessarily, on sets of arcs of circles of the same latitude, dissect the region to ease calculation, and use Simpson's Rule. We all become surveyors!

Do you see any others? Engaging in thought about how one topic relates to another not only reinforces concepts but also brings power to subsequent analysis.

Representing Numbers as Isolines

As you are seeing in this book, many ways exist to represent and analyze data and numbers using maps—choropleth maps, dot density maps, graduated symbol maps, and more. Another way to represent the world through maps is through isolines (iso = "same"). These are lines that connect variables having the same value. You have seen weather maps, which often show isolines that connect areas that currently have the same temperature or pressure. You may have gone hiking and taken topographic maps so you know how steep your trail will be—on such maps, contours are isolines that show land with the same elevation.

Isolines can also show frequency of crime in a city, types of diseases in a region, or earthquakes around the world. The manner in which these isolines are symbolized affects the user's perception of how that variable changes across a particular area of the Earth. Think about that as you experiment with changing topographic isolines, typically referred to as contours, or contour lines. Think about contours in relation to inequalities.

Symbolizing Contours

In a new web browser tab, open the linked Axis Maps (https://contours.axismaps.com/#12/37.2341/-113.0180) contour symbolizing tool.

The map (Figure 7.8) is centered (a clue is in the latitude-longitude coordinates in the above URL) on Zion National Park in Utah. On this map, the isolines are contour lines; each contour line represents and connects elevations that are the same. Areas where contour lines are close together represent landscapes where elevation changes more rapidly if you were to walk from one contour line to another on the surface of the Earth; that is, steeper areas. Areas where contour lines are farther apart represent landscapes of flatter terrain (such as in the northeast section of this map). Use Edit Style to experiment with changing the contour interval, line weights, background style, and colors. As you do, think about how your perception of the landscape changes as you change these elements. Use Search for places>to change the location from Zion National Park to a flatter or steeper area. Observe your resulting map. Think about how the perception could be influenced by similar changes of isolines on maps that show phenomena other than elevation—crime incidents, health incidents, natural hazards, and others. Symbology in mapping is powerful! It greatly affects the map-reader's understanding of a phenomenon.

FIGURE 7.8 Contours shown on a map covering an area inside Zion National Park using the Axis map contouring tool. Contours are a specific type of isoline that connect areas that have the same elevation. Axis Maps.

Investigating the Depths of the Great Lakes

In this activity, you have the opportunity to examine isolines and filter different types of data to foster understanding of numbers and geographic patterns.

Open the linked Great Lakes bathymetry map (https://www.arcgis.com/apps/mapviewer/index.html?layers=269f4cdba1704c1f86916f9f844da653) (Figure 7.9). This dataset is a cooperative effort between investigators at the NOAA National Geophysical Data Center's Marine Geology & Geophysics Division (NGDC/MGG) and the NOAA Great Lakes Environmental Research Laboratory (GLERL) showing the five Great Lakes plus Lake St Clair. Make the legend visible. Zoom in until you clearly see the individual contours—each of these lines is an isoline. These lines are similar to the contour lines you examined earlier, but here, they represent the depth of the water. They are called bathymetric contours, showing the bathymetry (depth) of the water bodies. Each is a line representing equal depth, in meters. The lake floor is steeper where the bathymetric contours are closer together and flatter where they are farther apart. Expand the layer to see the sub-layer underneath>click to make it active>on the right,>enable popups>click on contours to view the depth in different parts of the lakes (as shown).

Just like land contours, the shape of bathymetric contours in lakes and oceans may reveal plains, canyons, hills, and other features across these floors—even volcanoes!

Suppose you are assigned the task of identifying the deep places in the lakes where shipwrecks may have occurred, at specified depths. Open the data table behind the contours via the ellipsis to the right of the data layer>click on the Depth (*m*) field>sort descending. Each of the 119,813 records in the table represents one bathymetric contour on the map. Scroll down in the table, looking at the deepest contours. In which lake are all these deepest places? Maps are rich sources of information, but sometimes they contain too much information to be easily understood. Filtering maps or data tables can make specific data clearer (Figure 7.10). Use the Filter tool to the right of the map>Add an Expression>Depth is at least 100 m (as shown).

Examine the resulting patterns: Which of the Great Lakes do you believe has the highest percentage of its depths at 100 m or more? Which one(s) of the five Great Lakes have no depths more than 100 m? Where is the deepest part of Lake Huron? Open the table. How many records exist in the newly filtered table? Jot this number down. Then, clear the filter and examine the original table to be able to compare the number of records in the original table versus the filtered table.

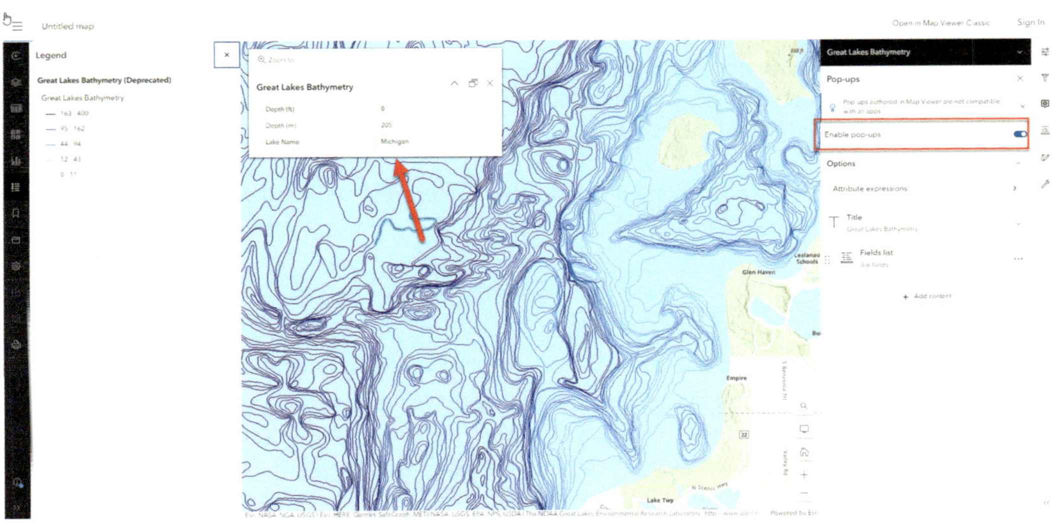

FIGURE 7.9 Adding popups to examine bathymetry for a section of Lake Michigan. NOAA (NGDC/MGG and GLERL) data. Mapped in ArcGIS from Esri.

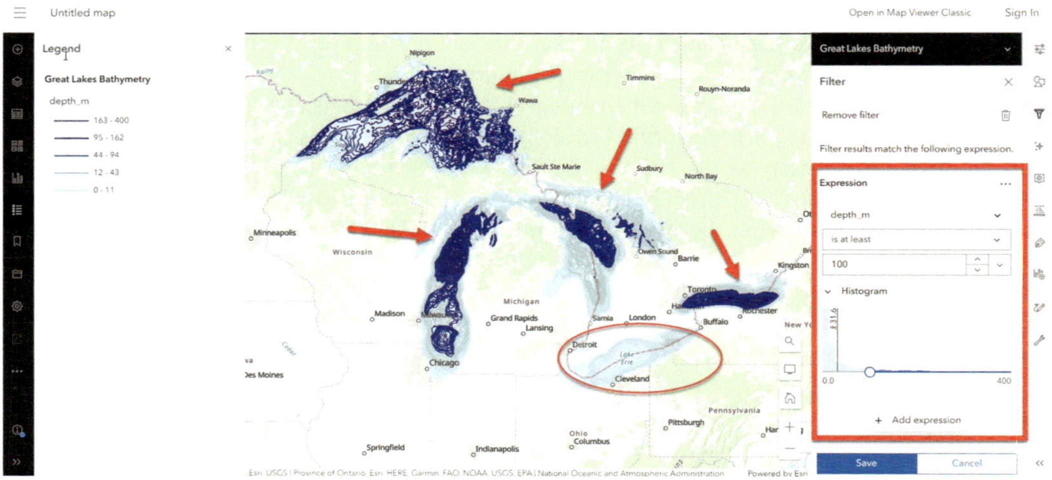

FIGURE 7.10 Filtered bathymetry for the Great Lakes. Mapped in ArcGIS from Esri.

What percentage of the original data set (as measured by the number of records) do you have in the filtered table of only the deep contours?

Probably the most famous Great Lakes shipwreck of all, that of the sad fate of the SS Edmund Fitzgerald on 10 November 1975, occurred at a longitude and latitude of −85.11017, 46.998500. Use the search tool to search for this location and add it to the map as a sketch. Zoom to this location. In which lake did this occur? Is the shipwreck located in waters over 100 m deep? Notice how close it actually was to the shoreline. Do some research on the shipwreck and find out why it might have sunk despite being so close to shore.

Suppose you just received word of new budgetary constraints on your shipwreck project that mean that you can now only focus on the lake where you found the deepest contours and where you located the above wreck. Add an expression where the Lake Name is the name of the lake that meets both criteria. Open the table. How many records exist in the newly filtered table? What percentage of the original data set (as measured by the number of records) do you now have? Using mathematical expressions in filtering aids in analysis and decision-making. Mathematical expressions can also be used in interactive maps to customize the appearance of the popups, for labeling features, for symbolization, for spatial analysis, and much more. Again, think about inequalities.

For further exploration, consider another rendering of Great Lakes bathymetry using a technique that results in an etched wood block effect, here: https://adventuresinmapping.com/2017/12/12/papercut-lake-map/

Or, consider the project from a nautical perspective captured by animating 3D maps (Arlinghaus and Blake, 2004).

DIMENSION

Overview

In the style of Edwin Abbott and his residents of Flatland, we noted earlier one difficulty in moving from 2D to 3D involved whether one could get "outside" the space in order to visualize it. Abbott's citizens lived in the plane and had trouble understanding the third dimension (Abbott, 1884). We live in three-dimensional space; what are our difficulties in thinking "outside our box?"

The study of geometry helps one to imagine, from a slightly different vantage point, what might be the difficulties in switching from one dimension to another: consider how polygons change as one switches dimension. In the plane, one might create infinitely many regular polygons simply by adding one more side. In three-dimensional space there are five regular polyhedral (singular, polyhedron), called Platonic Solids. They are: tetrahedron (with 4 triangular faces), the cube (with 6 square faces), the octahedron (with 8 triangular (equilateral) faces), the dodecahedron (with 12 pentagonal faces), and the icosahedron (with 20 triangular (equilateral) faces). We cannot keep adding faces to create arbitrarily more polyhedra. There is a problem when doing so with getting the polyhedron to close correctly. Real estate developers, architects, and construction workers are familiar with the problem of ensuring that a house closes properly when rounding the last corner to be built. There are similar issues in higher integral dimensions with only a handful of polytopes (generalized polyhedral forms) in each (Coxeter, 1947 and later).

Beyond the Usual—Fractional Dimension

So far, we have dealt only with integral dimension. More contemporary mathematics permits the visualization of fractional dimensions, a ratio that compares how the detail in a pattern changes with measurement scale. The fractal dimension measures the extent to which a pattern fills the space in which it is embedded (Mandelbrot, 1983).

One classic example of a fractal is the Koch Snowflake. It appears to have Euclidean dimension 1: however, the length of the pattern between any two points on it is infinite, composed of an infinite number of line segments joined at different angles. To generate the pattern, begin with an equilateral triangle and on the middle third of each side segment, add another triangle scaled to fit, and delete the base of the added triangle. Figure 7.11 shows the first few stages of this process that is to be carried out infinitely to create the full Koch Snowflake which will never fill the entire two-dimensional area in which it is embedded. From an intuitive standpoint, this pattern is too detailed to be one-dimensional but too simple to be two-dimensional. Using the ratio referred to above to capture this idea, its fractional dimension is calculated to be between 1 and 2, specifically about

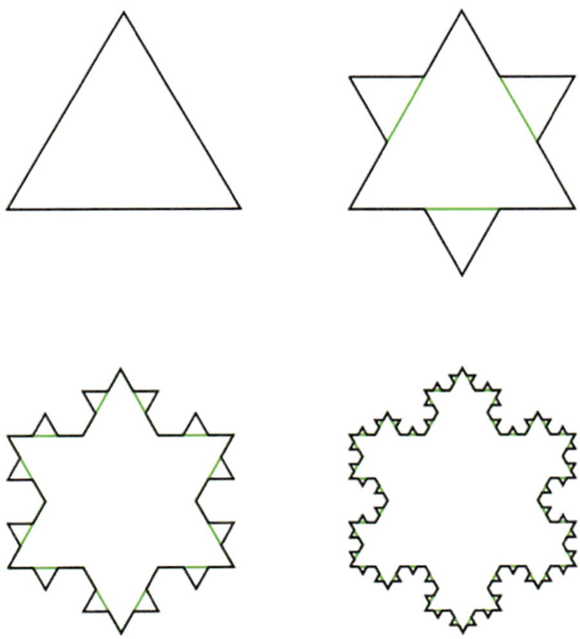

FIGURE 7.11 Koch Snowflake construction. Source. Original: Chas_zzz_brown, Shibboleth Vector: The original uploader was Wxs at Wikimedia Commons. - Own work based on: KochFlake.png, CC BY-SA 3.0, https://commons.wikimedia.org/w/index.php?curid=1898291.

1.2619, thinking of that as extending the integral Euclidean dimension into the space of rational numbers. Thus, we see transformations in thought process not only in moving from one integral dimension to the next but also in thinking about in-between dimensions as well: matching numbers to geometry is not easy.

Testing the Coastline Paradox: Fractals

In this activity, you have the opportunity to test the coastline paradox and hone your skills in measurement, scale, and mapping.

How long is the coastline of Great Britain (Mandelbrot, 1983)? The answer is more challenging than it may seem. The coastline paradox is an observation that the coastline of a land mass does not have a well-defined length. This is because coastlines have a fractal dimension: As the scale gets larger—that is, showing more and more detail—the measurement of the coastline begins to include every cape and bay. Therefore, the measurement lengthens as the scale gets larger. What is the smallest feature that should be taken into consideration when measuring? Should it include every rock outcrop? Therefore, there is no single well-defined perimeter to any landmass.

Access ArcGIS Online (https://www.arcgis.com/apps/mapviewer/index.html)>and change the basemap to National Geographic. Then,>Map Tools>Measurement>Measure Area (using square kilometers for area and kilometers for perimeter), and measure the coastline of Great Britain, clicking along the coastline until you circumnavigate the entire island (Figure 7.12).

Next, zoom in to a larger scale>New Measurement, and repeat the process to again measure the area and perimeter at this new, larger scale. You can use your mouse or touchpad to pan the map as you circumnavigate the coastline and delete any wayward points as you move. At this larger scale, you are considering more and more of the irregularities of the coastline (Figure 7.13) and the process should therefore take you a little longer.

Compare your two perimeter measurements. Which is larger? Why? How *much* larger is one compared to the other?

Coastlines are not the only mapped features with this challenge—it could be any irregular border. In fact, around 1950, Lewis Fry Richardson, researching the possible effect of border lengths on the probability of war, noted that Portugal reported their border with Spain to be 987 km, but Spain reported it as 1,214 km. This, the beginning of the "coastline problem," was expanded upon

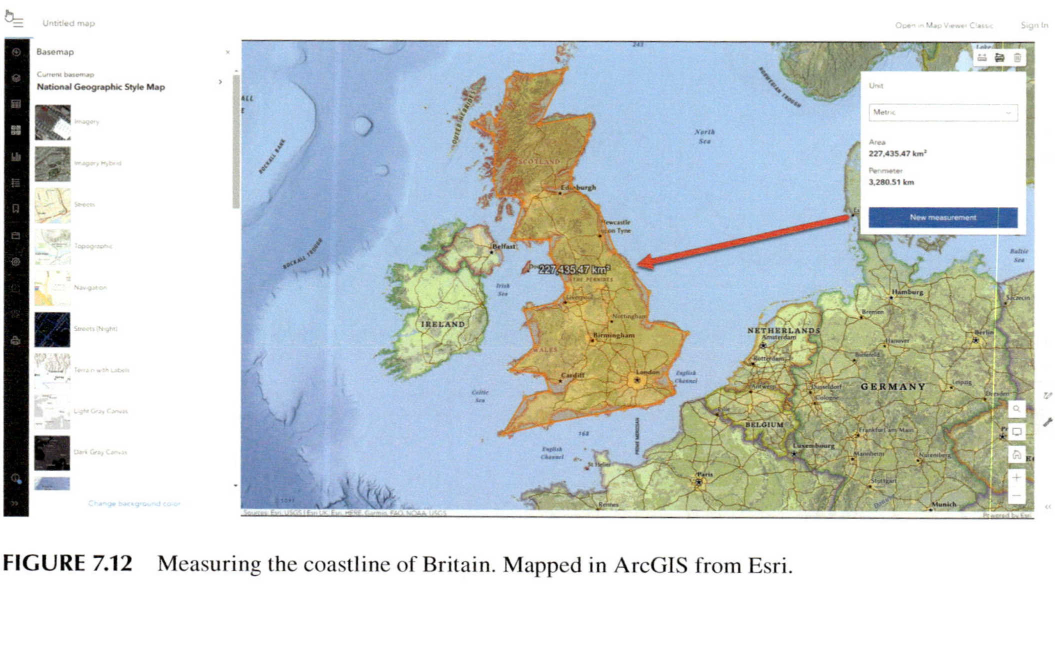

FIGURE 7.12 Measuring the coastline of Britain. Mapped in ArcGIS from Esri.

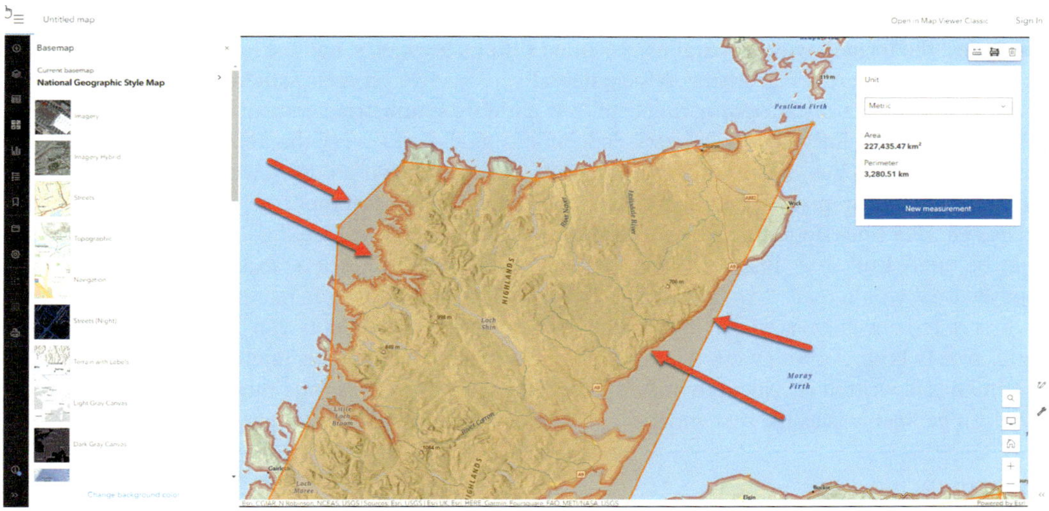

FIGURE 7.13 A closer look at Figure 7.12 at a larger scale showing the differences between a trace along the coastline and the coastline itself. Mapped in ArcGIS from Esri.

by Benoit Mandelbrot, who developed the visualization of earlier mathematical work from the 19th century. He called it fractal geometry (fractal, from Latin *frangere*, "to break").

Dig deeper: use ArcGIS Online to measure the boundary between Spain and Portugal and compare it to the two measurements given in the preceding paragraphs.

Why does this matter? Think back to earlier discussions involving the concept of uncertainty. It is something to be managed and understood, not necessarily *solved*. Maps are *representations* of reality; very useful representations to be sure, but they have been projected, generalized, and symbolized to represent the Earth. Even your large-scale measurement in this activity was dependent in part on the portrayal of the coastline of Great Britain, which does not show every inlet and cape of the coast: it is merely a *representation* of that coast. Measurement as performed here is probably sufficient for numerous applications: percentage of urban land in Great Britain, percentage of agricultural land, and so forth. Other applications might require greater precision and accuracy and a higher degree of certainty.

NON-EUCLIDEAN GEOMETRY

Euclid's Parallel Postulate remained a fixture in geometry until late in the 19th century (Coxeter, 1965). Think about it! Might it not be natural to conceive of a geometry with no parallel lines? After all, we cannot see parallel lines, given that a line is infinite in extent. Even railroad tracks, which we know to be parallel, appear to converge as we look to their vanishing point along the horizon. Artists have worried about this phenomenon for centuries—about the issue of representing three-dimensional views of the world on two-dimensional canvases. The concept of a world without parallelism is a natural one in a number of regards.

The Parallel Postulate says that through a given point, *P*, not on a line, *L*, there is exactly one line through *P* that is parallel to *L*. We see that the idea of no parallel makes sense. But what about the idea of more than one line through *P* that is parallel to *L*? Here, think of the conic sections: in this case, the hyperbola. Consider a branch of a hyperbola in the xy-plane that passes through a point p not on the x-axis and suppose that that branch of the hyperbola does not intersect the x-axis (but is asymptotic to it). There might be an infinite number of choices available for such a branch of a hyperbola. And, that branch of the hyperbola satisfies the idea of parallelism and the Parallel Postulate.

Both of these geometries that violate Euclid's Parallel Postulate are called non-Euclidean geometries: the former, with no parallels, is called Elliptic geometry, and the latter, with multiple parallels, is called Hyperbolic geometry. More generally, projective geometry studies alignment of points but does not consider distance or parallelism. It is a highly symmetric geometry employing a remarkable Principle of Duality, in which every concept has a dual and every theorem has a dual (Coxeter, 1955, 1961, 1965). Thus, for example, in the projective plane, two points determine a line (as is the case in the Euclidean plane) and, dually, two lines determine a point (not true in the Euclidean plane because parallel lines do not intersect). The terms "point" and "line" are dual concepts. What others might be dual? Let's try "intersect" and "join"; consider "two lines intersect in a point." Then the proposed dual would read: "two points join in a line." So, yes, the words intersect and join are duals. The first statement above is true in the projective plane but not in the Euclidean plane; the second statement is true in both. As with natural human languages, understanding the language in which a form of mathematics is written is critical to understanding its meaning. One of the most famous uses of non-Euclidean geometry is as the mathematics that serves as the foundation for general relativity theory.

ERROR ISSUES

What are the possible sources or error that can result from aligning mathematical concepts with real-world patterns? Can the user control them? There are many possibilities; we offer a quick view below.

Digitizing

Scanning maps and images is a good way to insert visual field evidence into mapping software. However, scanning is not always possible. A digitizer can convert pattern to number. When using a digitizer, a drafting table underlain with wire mesh that interacts with a computer mouse with multiple buttons to track position in relation to the wire mesh, there are a number of possible traps to fall into. Nowadays, most digitizing is done in a "heads-up" mode while tracing a feature in a GIS seen on a computer monitor, such as tracing the outline of a building as seen on a satellite image to create a vector layer representing buildings. Still, the principle is the same—assigning real-world coordinates so newly mapped features will "overlay" onto existing mapped features. Nevertheless, the following guidelines are still useful:

Generally, trace the outline of the largest polygon that contains all others that are to be digitized. If the process proceeds the other way around, greater error may be introduced. That is, trace the outline of Canada first, then subdivide that into polygons for each province, rather than assembling the outline for the full country from the boundaries of individual provinces.

When digitizing building footprints from aerial photographs, do not trace the rooftop and assume that the building footprint will be accurately represented. In the real 3D world, given how most buildings are actually built, it is indeed the case that the rooftop lies directly above the footprint. However, the eye of the camera is not directly overhead of most buildings. So there is some offset in angle and that needs to be compensated. Does this discussion remind you of Eratosthenes? It should—remember that he could only make his measurement on a day when the sun was directly overhead a certain well—where he knew that there was a 90° angle. The lessons of history endure; knowing them is critical!

Jordan Curve Theorem

A simple closed curve is one that is equivalent, by stretching, to a circle: it is like a circular rubber band. The Jordan Curve Theorem (Jordan, 1887) says that a simple closed curve J in the plane separates the plane into two distinct domains, each with boundary J. Think of this as "inside" and "outside" regions determined in relation to the curve. An example of a curve that is not a Jordan curve is a figure 8 shape. Do you see that when you "walk" along the curve, after you walk across the curve's crossing point that which might have intuitively been viewed as the inside has now become the outside? As with dimensions, more generally, viewer vantage point is important!

It is important for mapmakers to know what kinds of curves they are working with. If they do not, they might assign an address to the wrong side of a street. If, for example, all even-numbered addresses are assigned to the "inside" and all odd-numbered ones to the outside of a street pattern that can be deformed to a figure 8 shape, then the even-numbered addresses might appear within the bounded lower half of the figure 8 shape and on the unbounded side of the upper half of the figure 8 shape, contrary to the real-world situation.

This problem is easily overcome. Simply split any complex curve, such as the figure 8 shape, into simple closed curves: view the figure 8 shape not as a single curve with a crossing point but as two simple closed curves with a common point. When all complex curves are split apart so that all curves on the map are simple closed curves, then no errors of this sort arise. Contemporary software typically deals effectively with this issue; however, if unexpected assignment problems arise, consider looking at the underlying mathematics involving the Jordan Curve Theorem.

GEOMETRIC MODELS OF THE REAL WORLD?

There are assumptions about the underlying geometry that produce an associated world. But are the geometry and its concepts real? When applying these concepts, the user needs to keep that thought in mind and control for any lack of fit in modeling by being clear about where concepts do not fit reality, rather than trying to "bend" the analysis to fit a preordained agenda.

Are there, instead, other geometries that might be considered, and what sorts of trigonometric functions do they lead to and do these yield useful models of real-world activity?

If one considers geometry/trigonometry in higher dimensions, what sorts of error might be introduced in associated real-world interpretations and associated activities?

If we consider the three-dimensional measurement of the Earth, in various contexts, what are the difficulties? Is the Earth a sphere? What problems arise if it is not and can the user compensate for them?

The latter comment, involving the curvature of the Earth, is one that is richer in activity content than the more abstract comments that lead up to it. It thus opens the door to more activity questions derived from this conceptual consideration.

What is the actual shape of the Earth, as a geometric solid?

Compounding the challenge of measuring the shape of the Earth is that it not a solid—the molten part of the Earth that is beneath the crust and outside the core occupies a large part of the volume of the Earth, and it has a certain *plasticity* to it—after major earthquake events, such as off of Japan in 2011, the Earth's shape changes—it may swell, shrink, or crack—usually only a few centimeters, but these are enough to change the overall measurement of the Earth's shape.

When considering the Earth as a whole, should we also include mountain ranges, gorges, and other topographic features as well? What sorts of scale problems are there with including more detail?

How do we measure topography? How are topographic maps made?

Indeed, how do we measure the whole Earth?

In this sequence of concepts, we were led from a specific conceptual visualization to think a bit more broadly, in terms of measurement issues of the Earth as a whole. We went back and forth, jumping from one scale of observation to another as we attempted to fuse geometric and numerical patterns to create useful models of part, or all, of the Earth and its surface.

8 Proximity and Adjacency
Measurement

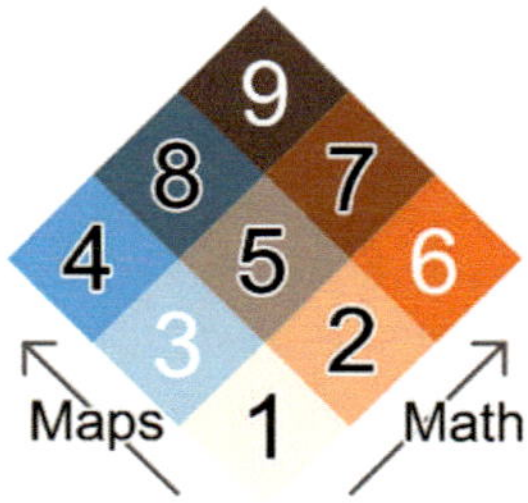

The line of images above is a visual abstract of this chapter designed to foster spatial thinking. From the chapter numeral, to the book structure, to the real world, the reader is offered gentle guidance to develop spatial intuition about what might be coming. Those thoughts are then reinforced with a detailed text outline of chapter content below. The images and outline forge as an abstract of chapter content.

CHAPTER OUTLINE

- Overview
    - Graph Theory
    - Topology
- Measuring Proximity
    - Graph Theory: Tracing a Walk through Königsberg
    - Proximity: Tracing Downstream
    - Topology: One-point Compactification of the Line
    - Proximity Zones and Wildfire
- Measuring Adjacency
    - Graph Theory: Coloring and Adjacency
        - The Mediterranean Basin Countries Sharing a Common Water Resource
        - Four Color Problem
        - Four Color Theorem
    - Topology
        - Continuity and Color
        - Color Ramps
        - Color: Fundamental Concepts
        - Color Vectors
    - Terrestrial Surface and Green Infrastructure
        - Agricultural Application
        - Recreational Application
    - Map Projection
        - Overview
        - Stereographic Projection: One-Point Compactification of the Plane
- How Many Colors Suffice to Color a Map on a Sphere's Surface?

DOI: 10.1201/9781003305613-8

OVERVIEW

Nearness, closeness, neighborliness, and more, are all concepts that are related to each other but that vary slightly in meaning, possibly as moving targets with geographical and cultural variation as driving forces. The words "proximity" and "adjacency" are two such words that also seem to embrace many others, including nearness, closeness, and neighborliness. To be clear, in the real world, entities:

- that are near each other, but not physically contiguous, will be said to be "proximate."
- that are physically contiguous will be said to be adjacent.

In the chapter on geometry, we saw that Cartesian coordinates serve as one way to measure distances. We also saw that geometry without coordinates (synthetic geometry) had many uses, as well. The situation with measuring proximity and adjacency is somewhat similar. One can use a ruler (coordinates) of various sorts to measure these entities. There are, however, powerful tools without coordinates to probe these ideas and come up with ways to model, or visualize, elements of structure. We take a brief look at two languages for structural models: graph theory and topology.

GRAPH THEORY

Graph theory is the study of mathematical structures that model pairwise relations between objects. Pairwise simply refers to things that are in pairs. The objects are represented as vertices (nodes or points) and the relations are represented as edges (links or lines) joining vertices. Two countries sharing a common boundary might be represented as two vertices—a "pair"; one for each country, with an edge representing the boundary to show adjacency. Or a country might be represented as a node and all other countries within 500 miles, as proximate neighbors, might be represented as nodes with edges to any country within 500 miles of them, creating a network of proximate neighbors on a continent. As interactive maps and decision-making is increasingly linked to Artificial Intelligence and Deep Learning models, encoding topology into these models to enhance decision-making is critical. It is especially important given the big data sets that are being processed by these tools, such as climate models, electrical wires, switches, and substations in an electrical network serving millions of customers, and the complex migration paths of geotagged bird and other species.

TOPOLOGY

General topology (point-set topology) is based on the following concepts: continuity, compactness, and connectedness. It is a broad and complex discipline. From an intuitive standpoint, however, one might learn to think in general about various patterns in the world. Continuous functions are ones that transform nearby points to other nearby points. Think of transforming one cultural community in one region of Europe to another in the USA, maintaining the cultural relations from the old world into the new world. Compact sets are sets that can be covered by finitely many sets of arbitrarily small size. A sphere is compact; the plane is not compact. You can hold the whole sphere in your hand if it is sized appropriately; you can never hold a whole plane in your hand. A connected set cannot be split into two sets that are far apart. The global railroad network is not connected; a local rail network might be connected.

A linking concept is that of topological equivalence. Intuitively, it means that one object can be stretched into another without crossing over, or cutting into, itself. Thus, a wiggly line stretches to a straight line; a doughnut stretches into a coffee mug.

Of these concepts, connectedness is generally the easiest to grasp intuitively and compactness the most difficult. However, as we shall see, compactness is a critical concept behind the scenes in mapping of all kinds.

Topological relationships underlie all interactive maps in a GIS environment, and enable all spatial analysis. Questions such as "how many water wells are within 1 km of a specific landfill," "how much land is under 5 m in elevation along this stretch of coastline," or "how many cottonwood trees are found within 100 m of this river, that is, in a river's calculated riparian zone" can only be resolved in a GIS if each of those mapped features "knows" where it is in relationship to the other features; that is, if they have topology.

MEASURING PROXIMITY

Graph Theory: Tracing a Walk through Königsberg

In the 18th century, citizens of Königsberg challenged themselves to take a walk around the city so that they crossed each of the seven bridges over the River Pregel exactly once and returned to their point of departure (Figure 8.1). After several attempts, they discovered that the task seemed impossible, but all they had was empirical evidence (Harary, 1969).

When "land areas" and "bridges" are represented respectively as "nodes" and "edges," a simplified graph emerged that captured structural elements of adjacency of landmasses across the bridges (Figure 8.2). Thus, it became possible for Leonhard Euler to show (1736) why the challenged walk-tracing was impossible.

In any sequence of nodes and edges representing a walk, an edge enters, and an edge exits each node. Thus, the total number of edges at each node must be an even number (this number is called the degree of that node). In the graph of the Königsberg bridges, all the nodes have odd degree: three of them have degree 3 and one of them has degree 5. Therefore, the challenge of tracing out a walk across all the bridges is impossible. Euler generalized his solution as a theorem and in so doing, the formal mathematical discipline of graph theory had its beginnings: Euler is regarded as the father of graph theory. To interact with a modern map of the same city, now a part of Russia, in ArcGIS

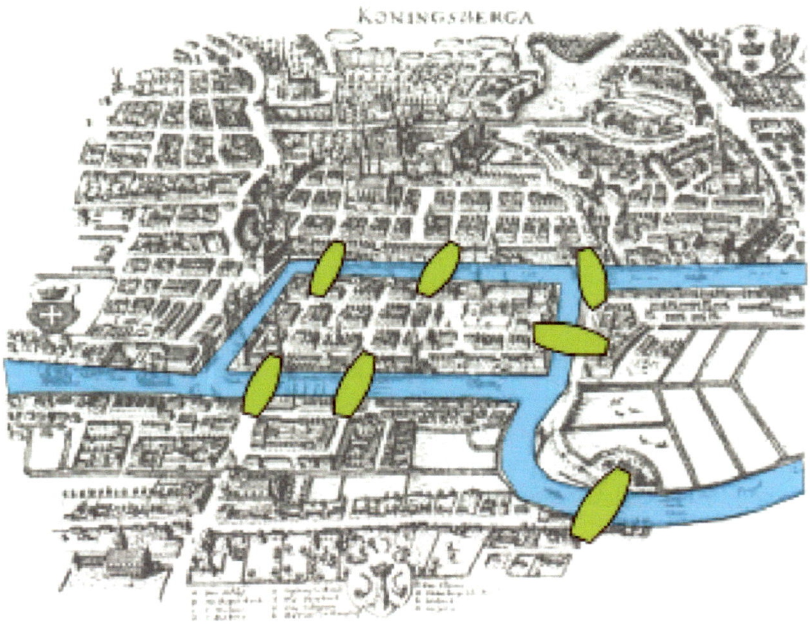

FIGURE 8.1 Map of a portion of Königsberg from 1890 showing the seven bridges. Source: Bogdan Giuşcă, CC BY-SA 3.0 <https://creativecommons.org/licenses/by-sa/3.0/>, via Wikimedia Commons, https://commons.wikimedia.org/wiki/File:Konigsberg_bridges.png.

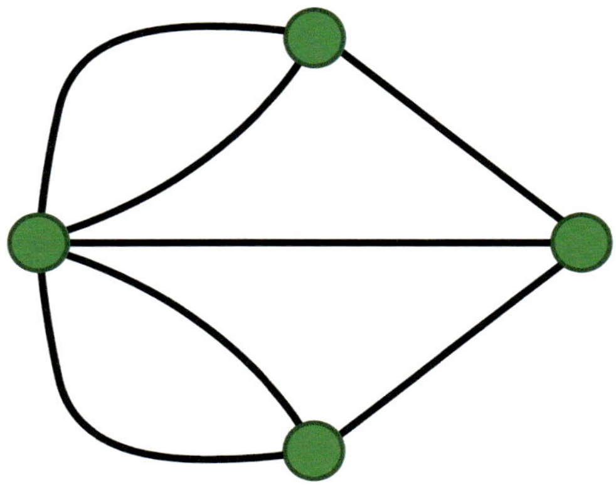

FIGURE 8.2 Graph of Königsberg Bridge problem: green nodes represent land areas on the map; edges joining nodes represent bridges joining land areas. Source: Original: Mark Foskey and Booyabazooka Vector: Riojajar, Public domain, via Wikimedia Commons. https://upload.wikimedia.org/wikipedia/commons/9/96/K%C3%B6nigsberg_graph.svg.

Online, start here: https://www.arcgis.com/apps/mapviewer/index.html?webmap=e961d42952a44535b67c16d54e023e5b.

Proximity: Tracing Downstream

Our world is full of networks, both natural and human-constructed. For thousands of years, rivers have served to join regions as physical landscapes and watersheds, and culturally, around common agricultural practices and through the cities that they gave rise to along their courses. Similarly, roads and railroads enabled swifter movement of people and goods, tying people together from lands even farther away. Bridges can hinder water flow but also serve as transmitters of human interaction. Two locations that are not proximate from one vantage point might become so from another vantage point. When they are viewed as connectors, bridges at strategic locations permit controlled interaction so that regions or towns that might not have been close, now become close—just across the bridge. The next activity suggests how that closeness might be achieved by moving along a network rather than across it.

In this activity, you will also have the opportunity to develop skills in mathematics and map interpretation as you work with river data. Rivers, along with their tributaries, represent a specific type of network, a hydrologic network, where water and all sediment carried from the natural and human-built landscape flows. The hydrologic network is bound by ridges and hills that make up the natural boundaries of a watershed, and is impacted by snow, rainfall, proximity to ocean currents and air flows, atmospheric pressure, types of soils, landforms, and much more. However, most hydrologic networks have also been modified, sometimes for thousands of years, by human activity. Canals, drainage ditches, dams, reservoirs, pumping stations, and other structures that divert river flow for irrigation, domestic water use, power, industry, or other uses, may take that water into completely different drainage basins.

Modeling flows through hydrologic networks is important to supply the needs for all these uses, for flood monitoring and control, and for many other reasons.

Open the linked map (https://www.arcgis.com/apps/mapviewer/index.html?webmap=079349bc1ebb438d9d4c37de8061c91b) in ArcGIS Online. The map opens with a shaded relief base map of the world and major world rivers. At the initial scale of the map, you should be able to pick out the Nile in Africa, the Amazon in South America, and the Ganges in Asia. Use bookmarks>Europe, and observe the pattern of rivers in Europe, noting that many of the rivers (the Drava, the Donau (Danube), the Po, and others) arise out of mountainous areas of Switzerland and Austria. The Europe bookmark zooms you in to a larger scale, so you should see more rivers and tributaries now than when you were at the smaller world scale. Investigate several of the courses of the rivers more closely, using the pan and zoom tools, noting how rivers flow from mountainsides, between valleys, and onto and forming wide plains.

Next, use bookmarks>Iowa, which moves you to an even larger scale: Which directions do the major rivers of Iowa flow? Why do they flow in these directions? One clue is the dotted brown lines that appear at this scale—these are the watershed boundaries. Zoom in closer to the watershed boundary in western Iowa (shown here, Figure 8.3). As you do so, note how rivers on the western side of the boundary flow southwesterly toward the Missouri River, while rivers on the eastern side of the boundary flow southeasterly toward the Mississippi River. Relate this partition to the idea of a half-plane.

Zoom to another part of the USA and observe the watershed boundary where you have lived or traveled, comparing it to the landforms that you observe on the shaded relief basemap.

Use bookmarks>48 States USA, to see most of the lower 48 states of the USA. Note the thick blue line that represents a trace downstream from a point in Boulder, Colorado (Figure 8.4). This trace represents the route of a cup of water as it would travel from Boulder (its source) to where it would enter the ocean (its mouth). Zoom to the beginning of this river trace, in Boulder, and pan the map along the entire trace, from source to where the trace enters the ocean. How many river names are represented by the water as it moves downstream? What is the name of the body of water where the water enters the ocean? On the right>Map Tools>Measurement>measure the length of the entire stream segment from source to mouth, clicking at specific points where the river changes direction to obtain the most accurate measurement of the distance (that you can in a short amount of time). What is the total distance? If the water flowed at a rate of 10 km/hour, how long would it take for a cup of water to flow from Boulder Colorado to the point where it would enter the ocean? Name three reasons why this rate of movement would not be uniform during the entire course of its route. Name one reason why floating along a river from its source to its mouth, even on a fast-moving segment flowing down a mountainside, would usually involve a greater distance and a longer time span than traveling on a highway along a similar route.

To create your own stream trace, use this tool: https://txpub.usgs.gov/DSS/streamer/web/. This tool will calculate the trace either upstream or downstream from a point you are interested in.

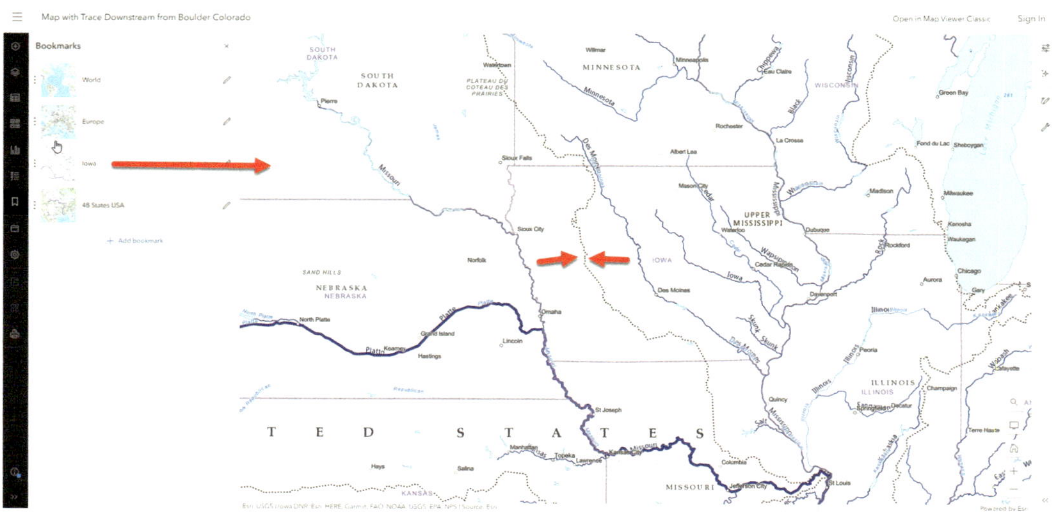

FIGURE 8.3 The watershed boundary in western Iowa. Mapped in ArcGIS Online from Esri.

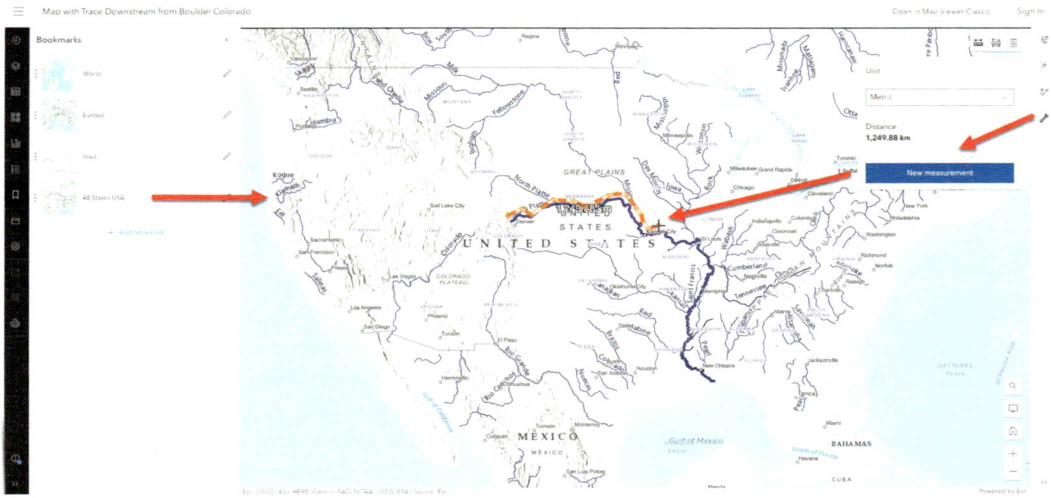

FIGURE 8.4 Measuring along the downstream trace from a point in Boulder, Colorado, to where the water from Boulder would empty into the ocean. Mapped in ArcGIS Online from Esri.

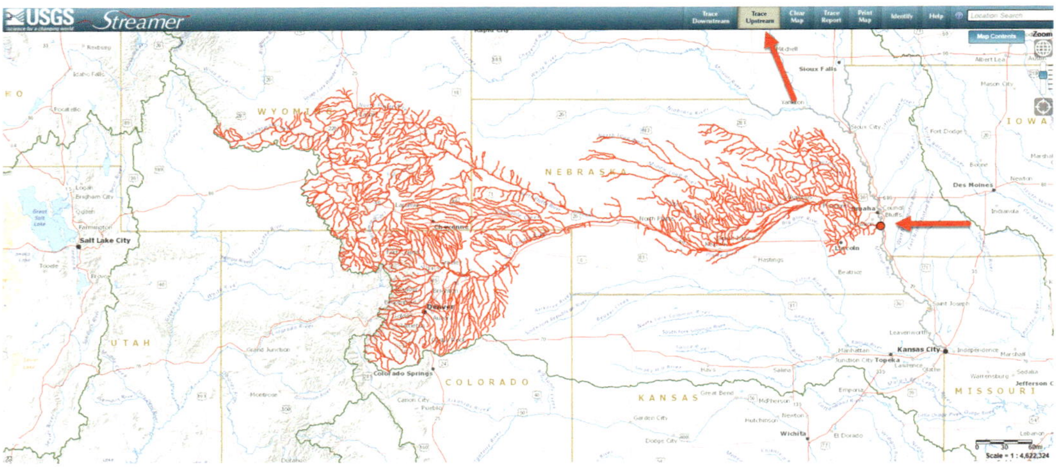

FIGURE 8.5 Upstream traces show all the rivers in red that drain *into* a point that you specify on the map. USGS.

Upstream traces show all of the rivers in red that drain *into* a point that you specify on the map—in this case, Plattsmouth Nebraska (Figure 8.5).

Consider how the amount of water at any given point along a river reflects the influence of the entire watershed that drains into that point. Try the River Runner tool (https://river-runner.samlearner.com/) and (https://river-runner-global.samlearner.com/), which allows you to follow the trace of a river while simulating flying above satellite imagery from any point in the USA or the world, respectively.

Topology: One-Point Compactification of the Line

When a stream is mapped, only a portion of its full extent is generally shown. One might think of it as topologically equivalent to a straight-line segment; then, imagine that when the stream leaves the map, beyond what we see, that it continues as a straight line forever—as a non-compact object.

Little extra insight comes from that approach. Instead, imagine that a single point is added to that straight line: the point at infinity where the straight-line wraps back around on itself to form a circle. Now, the idea of a stream in that form suggests the stream flowing into a larger river which then flows into a bay, then the ocean, then around the world in that ocean and back to where it all began as water drops. The added point at infinity made the non-compact line into a compact circle (Kelley, 1955), and in so doing, inserted the valuable suggestion of the hydrological cycle in association with the stream pictured in a single-snapshot map!

Proximity Zones and Wildfire

Hazards offer natural ways to look at spatial issues associated with the concept of proximity. No one wants to be adjacent, right next to, a hazard. But we all want to know how close we are to a hazard. As the hurricane approaches, we need to know how far away it is so that we might plan an evacuation route that will maintain at least a minimal safety buffer. So it is with other real-world hazards.

Perimeters, areas, and perspectives: In this activity, you have the opportunity to develop skills in measuring, calculation, estimates, percentages, and 3D perspectives. Thus, you will create zones based on the ideas underlying proximity. Open the linked 3D scene (https://www.arcgis.com/home/webscene/viewer.html?webscene=8b208cebeb3140179fbf6fd8d265a869) of the perimeter of the Camp Fire 2018 wildfire in Paradise, California (Figure 8.6). The Camp Fire was the deadliest and most destructive wildfire in California's history. Named after Camp Creek Road, its place of origin, the fire started on Thursday 8 November 2018, ignited by a faulty electric transmission line. High winds drove the fire and created a firestorm in the foothill town of Paradise.

The 3D scene shows the perimeter and area of the Camp Fire wildfire and the surrounding terrain. Use the rotation and perspective tools to rotate and tilt the area to get a sense of the terrain. Use the slice tool to help you get a sense of the elevation of the area. Each slice represents a horizontal plane of equal elevation or "cake layer" that could be sliced off the terrain. If you get stuck, refresh the tab containing the map in the web browser, as shown.

Use search > Enter Poe Dam and zoom to it. This is where the first 911 call was made, at 6:25 am. Based on the location of Poe Dam, and your observation of the total area burned, which direction(s)

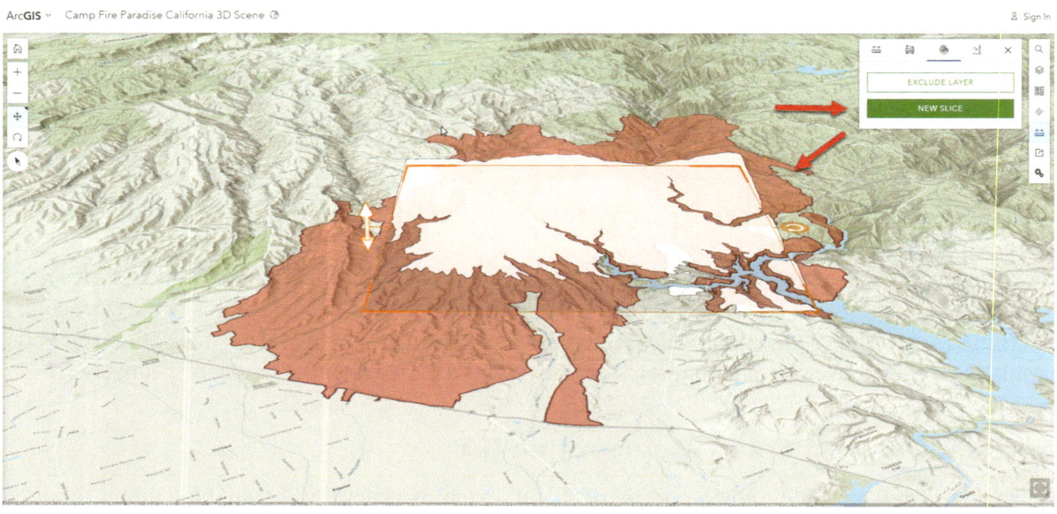

FIGURE 8.6 3D scene of the perimeter of the Camp Fire 2018 wildfire in Paradise, California. Note the use of the slice tool. (From Mapped in ArcGIS from Esri.)

did the fire spread? From which direction was the wind blowing? Based on your observation of the terrain, did the fire primarily spread uphill from the origin point or downhill from the origin point?

At 7:44 am, the first fires began burning in the town of Paradise. Compare this time to the 6:25 am time of the first 911 call, and the distance between Poe Dam and Paradise (12 km) to determine how fast the fire was traveling between those two points in kilometers per hour. Is the speed the wildfire was moving faster than you can walk? Run? Drive a car?

Zoom the map to the entire extent of the fire. Under analyze > use the profile tool > draw a line from the northeast side of the wildfire to the southwest side > generate an elevation profile to show the steepness of the terrain over your line (Figure 8.7). Because the terrain profile graph must be compressed to fit into the screen, it gives the impression that the area is steeper than it actually is on the ground. Therefore, zoom and pan your map along the profile line to obtain a more accurate sense of the terrain profile on the land. Consider the impact that the terrain had on the ability of the wildfire to spread, the ability for people to evacuate, and the ability for rescuers to access the area. From the terrain profile, calculate the average degree of slope for the line.

Next, analyze the extent of the wildfire. Use the tool on the left side of the map to reset map orientation so that north is at the top. State Highway 99 is a major highway in central California. Did the wildfire cross the state highway? If so, at how many points did it cross? Use Analyze > Measure Distance > measure the farthest distance from Highway 99 that the wildfire spread. How close did the wildfire come to the downtown of the next major community to the northwest, Chico, California?

Next, use Analyze > Measure Distance > measure the distance spanned by the wildfire from its most northerly extent to the most southerly, in kilometers. Note your result. Measure the distance spanned by the wildfire from its most easterly extent to the most westerly. Note your result. Use these two sides to calculate the area that would be covered as a rectangle bounded by these two sides, noting your result. Next, do a different kind of measurement: Use Analyze > Measure Area > measure the area covered by the Camp Fire, tracing the perimeter of the fire as best you can (shown here). Note that you can snap your line, to the perimeter line, as you are measuring. Note your result (Figure 8.8).

Convert your two figures to hectares. Compare your two figures to the following figure: the Camp Fire burned 62,053 hectares before it was finally put out. Which of your figures was closer to the actual figure of the burned hectares? What are a few reasons for the differences between your two figures and the actual "burned" figure?

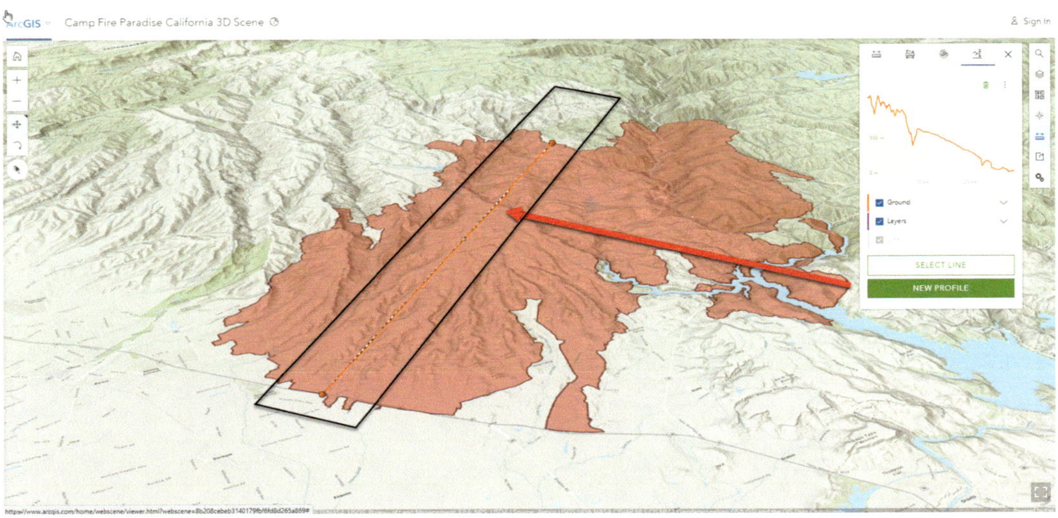

FIGURE 8.7 Analyzing the terrain to gain a sense of its relationship to the spread of the wildfire. Mapped in ArcGIS from Esri.

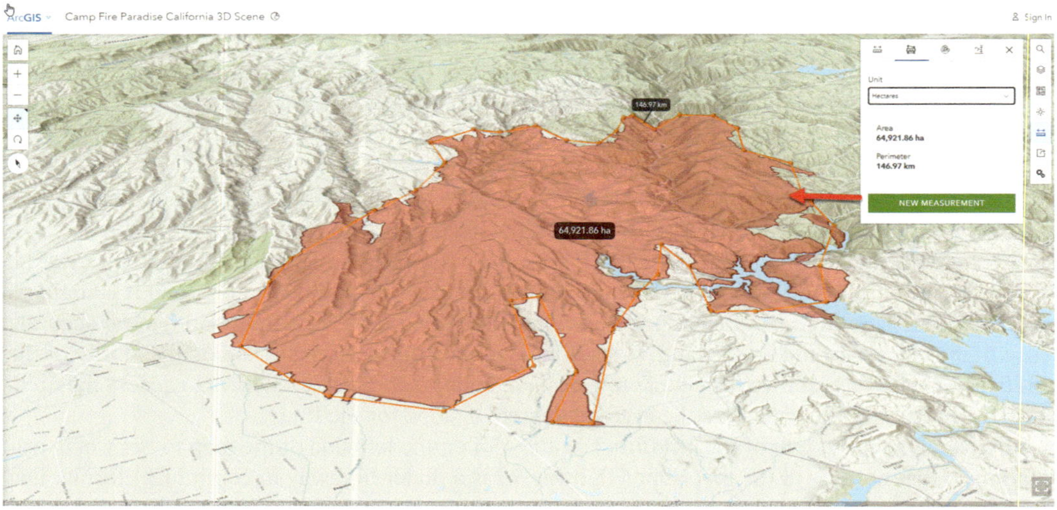

FIGURE 8.8 Determining the area covered by the wildfire. Mapped in ArcGIS from Esri.

A circle is the shape with the smallest perimeter to area ratio. A circle has a perimeter of $2\pi r$ and an area of πr^2. Convert your measured area to square kilometers. Note that the area burned by the Camp Fire has a much larger perimeter than if the burned area been a perfect circle, in part because of the varied terrain and the reservoir along part of its southern boundary that occupied (and in some ways impeded) its forward progress.

The fire caused 85 deaths and burned nearly 19,000 structures (Figure 8.9). To see more clearly the devastation on the ground in Paradise, (1) turn on the 2018 UAV imagery, collected by a drone, (2) change the base map to imagery, and (3) use the layer tool to turn off the 2018 Camp Fire wildfire perimeter layer. First, zoom to the hills above the town of Paradise. What sort of vegetation is covering them? What effect do you think this type of vegetation had on the spread of wildfire? Next, zoom in to a section of the town of Paradise of your choice. Observe the large number of housing lots that

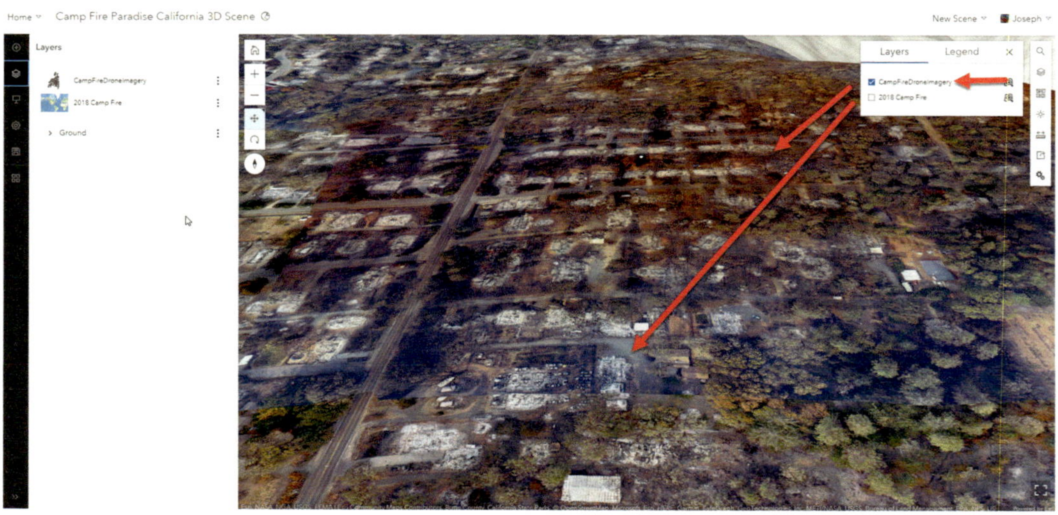

FIGURE 8.9 Examining UAV imagery to understand the extent and severity of the burned area. Mapped in ArcGIS from Esri.

do not have any houses on them. Even after several years, these homes have not been rebuilt, as evidenced by the UAV imagery.

In your chosen area, by comparing the 2018 imagery to newer (but coarser in resolution) satellite basemap imagery, count the number of structures that have been built. Then count the number of lot sites that have *not* been rebuilt. Compute the percentage of rebuilt structures as compared to the total number of structures in your neighborhood. Pan across the map to one more part of town and repeat the process. Compare the numbers and the percentages of the housing in your two neighborhoods.

The world is changing. As the new construction described by this article indicates (https://www.ktvu.com/news/town-of-paradise-sees-housing-boom-after-catastrophic-2018-camp-fire) (Vacar, 2022), homes *are* being rebuilt here, so your percentage of the number of constructed buildings will no doubt change if you repeat the process a year from now.

For more information about the Camp Fire, examine these two story-maps, 1 (https://storymaps.arcgis.com/stories/4d08a97e87664c66b051f97df09a1107) and 2 (https://storymaps.arcgis.com/stories/7f7602261edf4d6185fdbe2991d57492).

MEASURING ADJACENCY

Graph Theory: Coloring and Adjacency

When carefully used, color can be used to extract meaning from a graphic, a visualization, an image, or a map. When is color necessary rather than merely decorative and, when color is necessary, how many colors are required to make the desired distinctions? We consider questions of this sort below and tie them to graphical and to geographical evidence.

The Mediterranean Basin: Countries Sharing a Common Water Resource

Often one sees maps that use colors to distinguish one region from another. But, how many separate colors are *required* to make the needed distinctions? The Mediterranean Sea and the countries that have at least some coastlines along it form a natural environmental grouping sharing direct access to a common water resource. Getting along with one's neighbor is always a good idea; even more so, perhaps, when adjacent countries share a critical resource.

The map in Figure 8.10 shows a set of countries in the Middle East, some of which are adjacent to the Mediterranean Sea (colored blue/cyan). Clearly, adjacent countries must be colored something other than blue. Use new colors only when forced to do so. Color Turkey green. Color Syria red. When the Golan Heights is not included as a part of Syria, Israel, or elsewhere, it is not adjacent to the sea and so just assumes the background gray coloring as not part of the set because it is not adjacent to the water. Thus, Syria and Israel are separated by gray space and can be colored the same color. Color Israel red. Lebanon can be colored green because it is only adjacent to red countries (Syria and

FIGURE 8.10 top. Golan Heights not included.

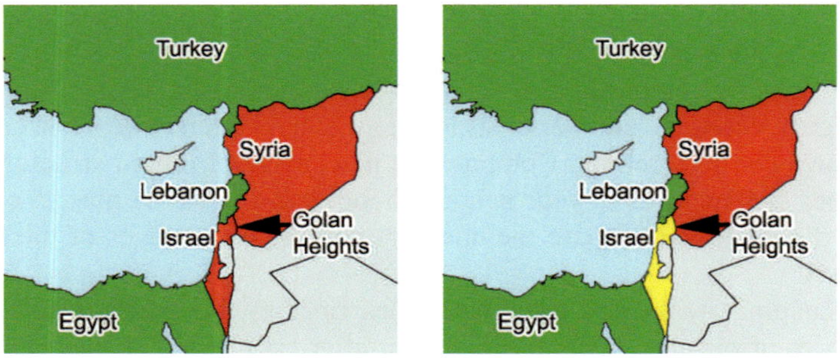

FIGURE 8.11 (a) left. Golan Heights included in Israel. (b) right, Israel requires a fourth color. Source: S. Arlinghaus, using tool at MapChart.net together with Adobe Photoshop.

Israel) and blue water. Continue coloring in the natural manner. Three colors, blue, green, and red were enough to color the map and make all adjacency patterns clear because no two touching entities were colored the same color (revised from the original in Arlinghaus, Arlinghaus, and Harary, 2002).

Now suppose, however, that the Golan Heights is appended to Israel (Figure 8.11a). Then it is colored red, along with the rest of Israel and now has direct access to the Mediterranean Sea. Then Israel, which is already adjacent to both the sea and to Lebanon, has now become adjacent to Syria and can no longer be colored with the same color as Syria: red. Color Israel (including the Golan Heights) yellow: a necessary fourth color.

The political boundary issues surrounding the Golan Heights, Israel, and Syria are reflected in the seemingly simple task of coloring a set of nations. Possible political hot spots may emerge as those nations requiring the fourth color, regions with relatively complex political boundary adjacency patterns, offer the potential to fuel boundary conflict. Tools that can simplify structural complexity, such as coloring, can offer insight into other issues, as well.

What happens, however, when the coloring strategy is no longer obvious? Graph theory may help with that. Shrink each country to a point and join two points with an edge if they share a common boundary; we saw that approach with the Königsberg bridge problem. How many colors will be enough to color a more complicated mapping situation whether it is viewed as a graph or as a set of regions? The next section focuses on the latter question and employs methods based in graph theory.

Four Color Problem

As we saw in the previous section, we could create a map that forced the use of four colors: that is, four colors were necessary. It is well-known that there are maps that require four colors—that four colors are necessary; it has been conjectured, but remained unsolved for a long time, that indeed four colors suffice. In 1941, Courant and Robbins summarized the history of the four-color problem as follows.

> The problem of proving [Four Color Theorem] seems to have been first proposed by Moebius in 1840, later by DeMorgan in 1850, and again by Cayley in 1878. A "proof" was published by Kempe in 1879, but in 1890 Heawood found an error in Kempe's reasoning. By a revision of Kempe's proof, Heawood was able to show that five colors are always sufficient…It has been proved that five colors suffice for all maps and it is conjectured that four will likewise suffice. But, as in the case of the famous Fermat theorem…, neither proof of this conjecture nor an example contradicting it has been produced, and it remains one of the great unsolved problems in mathematics.

It was not until the last half of the 20th century, aided by the capability of contemporary computing equipment to examine large numbers of cases, that the age-old "four color problem" became the "four color theorem" when Appel and Haken (1976) showed that four colors suffice. The careful map-reader, when viewing a map of many colors, should be able to see a good reason why many colors were used or perhaps question the need for so many colors.

The fact that four colors suffice to color any map in the plane seems to stimulate a challenge to show it false. Often people draw planar maps that appear to need five or six colors; such maps, when the coloring scheme is re-examined, can always be colored more efficiently with four colors. Another challenge arises when some people believe they have successfully proved the four-color theorem when they rediscover a result first proved by Augustus DeMorgan.

DeMorgan's Theorem. No five regions in the plane can be mutually adjacent.

This fact does not suffice to prove the four-color theorem, however. For example, we can find a map in which no more than three regions are mutually adjacent but that nonetheless needs four colors (Figure 8.12).

Courant and Robbins made reference to Kempe's 19th century (1879) response to the four-color challenge and noted that his result was in error. Kempe's ideas, however, were what led Appel and Haken, in the late 20th century, to a proof of the four-color theorem employing computer enumeration. Fundamental conceptual material from Kempe endured the test of time and its validity provided a platform for an eventual new style of proof employing current technology—much as mapping has generally evolved from well-tested concepts in a pen and ink world into a more modern digital environment.

Four Color Theorem

The Four-Color Theorem (Appel and Haken, 1976). Every planar graph can be colored with four, or fewer, colors so that no two adjacent nodes have the same color.

To understand that this theorem applies to geographical maps, simply reverse the process of representing a country or other region in the plane by a node and a boundary between adjacent regions by an edge. Simple ideas often lead to powerful results when they are creatively integrated with other fundamental ideas. Historic preservation is important in mathematics as well as in a variety of more familiar contexts!

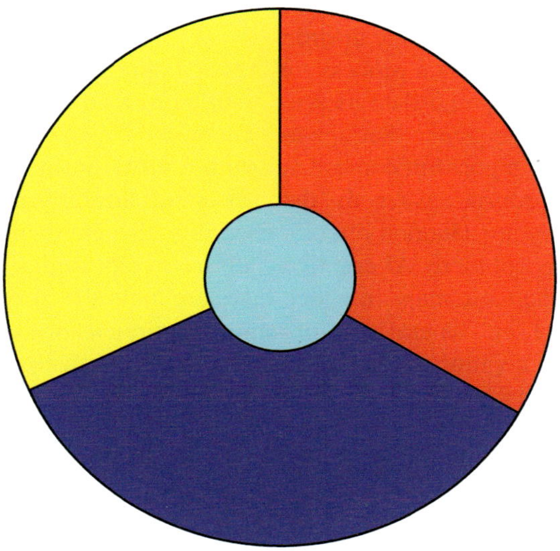

FIGURE 8.12 Map requiring four colors. Source: Original, Vector, Germo, Public domain, via Wikimedia Commons. https://commons.wikimedia.org/wiki/File:Fourcolorsmap.svg.

Topology

Continuity and Color

We have seen several cases in which subsets of mapped data sets are assigned separate colors representing separate ranges of data. In very simple cases, shades of gray suffice to make distinctions: the lowest quarter of the data might be colored white, the next quarter light gray, the next quarter a darker gray, and the final quarter, black. This "gray scale" is sort of like a linear scale; everyone will see it pretty much the same way and will infer the correct pattern from it. When the data sets are clearly organized to correspond to being depicted in gray scale, that is a fine choice for optimizing communication of spatial data pattern. With more complicated data set patterns, care in selection to communicate the intended data pattern must be exercised. Here is where visual continuity, a topological concept, enters. A single color might be introduced, from white through light shades and then dark shades: a monochromatic scale—a linear scale equivalent in nature to the gray scale. A red monochromatic scale, for example, might go from a very faint pink to a deep burgundy color. The intensity of the color reflects the increasing density (or other characteristic) of the underlying data. In the diamond-shaped pattern of the book, we intersected two monochromatic scales, producing a "bivariate" pattern: see if you can figure out what these two monochromatic scales are (the answer is given at the end of Chapter 9).

Not all maps with color require concern with the topological concept of continuity. The example above, involving coloring of countries surrounding the Mediterranean, made a deliberate color selection that violated the principle of color continuity: why? Because we wished to emphasize distinction among position of countries, there are no underlying data sets and hence no continuity in the underlying data. The emphasis was only on the graph-theoretic concept of adjacency whose presence was emphasized with contrasting blocks of color.

Color Ramps

In the material above, we briefly discussed the concept of continuity and color as they might be represented by gray scale and by monochromatic color scale. Beyond those, things get complicated quickly. Color exists only in the eye of the beholder; thus, it is highly subjective. What appears to be a house painted light green may appear to you as a house that is painted light blue. Several factors enter into variation in perception and these factors offer difficulty in trying to capture color in a systematic fashion. Nonetheless, both computers and logic have been employed to offer some sense of order in what otherwise might be mere disorder.

Computer software often offers a process for creating appropriate sets of color, designed to optimize communication of spatial data mapped with more than one color, using a smoothing algorithm to create a color sequence that appears continuous, as a color "ramp." In it, the user selects two colors, and the algorithm comes back with a visually smooth color pattern. A continuous ramp is often appropriate when the data set is not naturally partitioned into separate, discrete, subsets; thus, any introduction of color used in a discrete manner might imply separation within the data that is not there. Alternatively, discrete color ramps may also be created and these we see often with thematic maps in which there are naturally appearing breaks or classes within the underlying data set. Generally, color ramps, continuous or discrete, are easily created by all levels of user, from beginner to expert and they offer an effective way to simultaneously advance the personal color knowledge of the user as well as to create a respectable map that clearly communicates spatial information. There are many behind-the-scenes decisions that create an effective, and perhaps seemingly effortless, map; and these decisions are, ultimately at least, all based in mathematics.

Color ramps create an "assembly line" product. If one wishes to go beyond that, and make a map with color that matches the underlying data with a customized fit, care must be taken to get a good fit. Cartographers, printers, and others worked on this topic long before computers existed. As we have seen elsewhere with spatial objects or data linked to mathematics, a fundamental approach with color has been to discover basic concepts as fundamental principles to introduce systematic order into complexity. The Color Brewer tool (https://colorbrewer2.org/) can be effectively used to determine colors and the number of classes that a map user can easily determine.

Color: Fundamental Concepts

Color, in theory and practice, has long been a topic of concern to scholars in various disciplines, yielding no unique agreement on how best to characterize it. From the times of Newton and Goethe, great minds have puzzled on the complexity of using color to advantage and capturing it both mathematically and linguistically. Much contemporary software is based on a set of primitive terms designed to capture color in relation to elements of physics. This categorization rests on capturing the reactions of a typical observer to fundamental concepts of hue, saturation, and luminosity (HSL).

Hue describes the basic colors of white light broken up by a prism into the rainbow color spectrum: they form a discrete set of values, between which continuity might be introduced to produce logical infill. Associated numerical values vary from 0 to 255. Why is that a natural value to choose? Think about the powers of 2. What is 2^8? How many colors are there in a standard rainbow? Definitions may vary but one common categorization is, in addition to the white light, seven others: red, orange, yellow, green, blue, indigo, and violet—for a total of eight.

Saturation, or intensity, describes the amount of hue in a color. Burgundy is more intense than pink; royal blue is more intense than light blue; forest green is more intense than light green. Similarly, numerical range to capture saturation goes from 0 to 255 (there are 256 numerals in this set). Notice a difference here from other axiom sets: saturation is not independent—it depends on the definition of hue for its existence.

Luminosity describes relative lightness or darkness of a color. It functions as a gray scale to make comparisons of lightness or darkness. Again, luminosity varies from 0 to 255.

How many colors are there? There are 256^3, or 16,777,216, different colors that can be described using this pattern, each with its own unique ordered triple of coordinates. Figure 8.13 shows an example as it appears in the Color picker for MS Word. The coordinates for a particular shade of magenta are shown.

Color models other than HSL are also typically available in color pickers. You might also see the RGB (Red-Green-Blue) description of color, again expressed as three coordinates based on percentages of red, green, and blue as primary colors to make other colors. Based on the RGB scheme, you

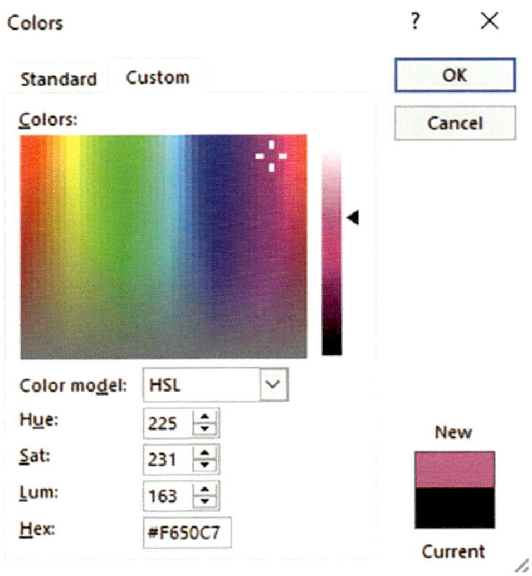

FIGURE 8.13 Color picker, magenta, focused on HSL color model. Source: MS Word.

will likely also see the Hex code, where Hex, refers to "hexadecimal"—a number system that goes from 0 to 9, as does the decimal system, and then extends beyond that using A, B, C, D, E, F as other basic digits (think back to Chapter 2). There are a total of 16 basic numerals in this system. Why 16? Well, 16 is 2^4! A typical Hex value for a color model is #RRGGBB where RR denotes a value specifying color intensity using hexadecimal integers from 00 to FF (255) for red, GG from 00 to FF for green, and BB a value from 00 to FF for blue.

Notice the reliance on powers of 2 again; because computers are based fundamentally on switches, with an "on" and an "off" position, powers of 2 are important in developing systems that work well on them. Here, colors expressed in terms of Hex values work in all Internet browsers. In the case of the HSL and the RGB color models, both return values of ordered triples to describe requested colors. However, the triples are not the same from one system to the other. The Hex value provides a convenient pivot to execute the conversion. Thus, for a particular shade of magenta shown in Figure 8.13, the HSL triple was (225, 231, 163). Its Hex value was #F650C7. To translate the HSL triple to an RGB triple, open a new color picker window and enter the Hex value in the appropriate slot (Figure 8.14). Then, after doing that, pull down the menu so it reads RGB for color model. Then, in Figure 8.14 read off the coordinates: (246, 80, 199). Notice that they are different from those for the HSL model for the same color. Notice that the Hex value is the same for both models. Are there any colors for which the ordered triples are the same for both the HSL and RGB models? What does this mean? Perhaps that the use of concepts to describe colors is not the same as using the colors themselves, even though the colors are derived from those concepts? There are interesting philosophical questions associated with color that have puzzled minds for centuries.

Other color models are in use to describe color. One is the printmakers color separation into layers of CMYK (Cyan-Magenta-Yellow-blacK). There are various color wheels available that can help users to change and to mix colors. For example, to decrease magenta, either subtract magenta or add cyan and yellow (opposite from magenta on a color wheel, Figure 8.15). Notice that there appear to be three axes emanating from the center void of white: one in a cyan direction, one in a yellow direction, and one in a magenta direction. If the magenta axis were continued on through the central void, it would bisect the angle between the cyan and yellow axes.

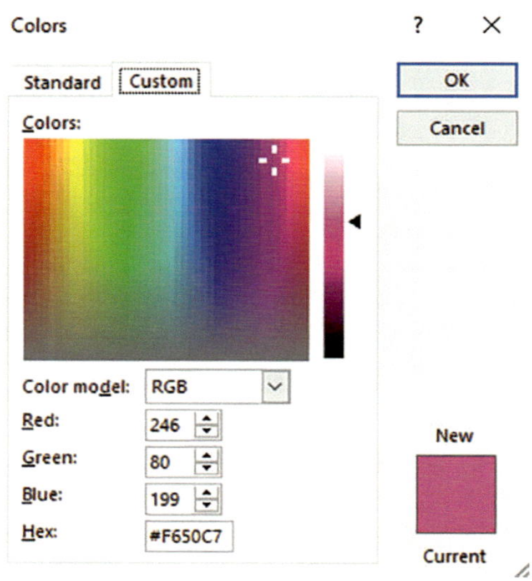

FIGURE 8.14 Color model RGB used to describe the same color as in Figure 8.13. The Hex value is a pivot used to transform from the HSL coordinate system to the RGB coordinate system. Source: MS Word.

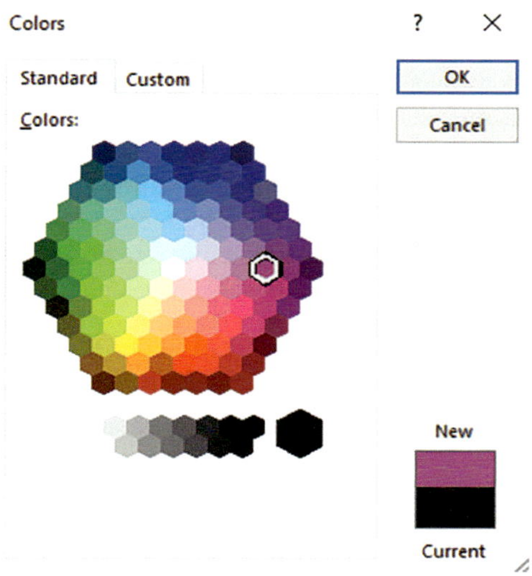

FIGURE 8.15 Cyan, Magenta, and Yellow axes emanating from a central void. If the magenta axis were extended across the void, it would bisect the angle between the cyan and yellow axes. Source: MS Word.

FIGURE 8.16 Venn diagrams of primary and secondary colors: RGB on the left, CMYK on the right. Sources: Left: By Wiso – Own work, Public Domain, https://commons.wikimedia.org/w/index.php?curid=3106363. Right: CC BY-SA 3.0, https://commons.wikimedia.org/w/index.php?curid=113383.

Different professionals may employ different color wheels, in part depending on whether their approach to what they are doing is additive or subtractive in nature. In Figure 8.16, we see Venn diagrams of three circles to represent different sets of primary colors and associated secondary colors (as set intersections of primary colors). On the left, the diagram displays the RGB color model and on the right the CMYK color model.

Color Vectors

Another conceptually natural way to look at color is as an ordered triple in Euclidean three-space as has been done in various applications in the past. However, because the values of coordinate triples are all integral, this cube has become discretized; there are gaps between successive values. The cube is not a solid Euclidean compact space but is rather a cubical shape based on coordinate triples of integral values. To work with this, hue might be measured across a horizontal x-axis as a horizontal color vector. Saturation might be measured along a vertical y-axis as a vertical color vector. The resultant of these color vectors is a square or rectangular color space with discretized vertical strips of color corresponding in order to the pattern on the color wheel. A third axis of luminosity (a gray scale), a z-axis, matches the selected color from the two-dimensional color space, generated by hue and saturation vectors and then extruded into the third dimension, against light/dark values. In that regard, visualize color as a cube of 256 units on a side. Color space is formed from a set of volume pixels making up the cube. Since $256 = 2^8$, there are therefore $2^8 \cdot 2^8 \cdot 2^8 = 2^{24} = 16{,}777{,}216$ volume pixels within the color cube matching the number of colors generated in HSL and RGB. Vector arithmetic within this color cube with colors viewed as tips of vectors, may suggest other color ramps. The focus here is to try to take advantage of the large, but finite, number of options available in a discrete arrangement of colors to create new approaches and new color ramps. Here, we have viewed the cube as composed of a finite number of volume pixels; in the true sense of a Euclidean cube, there are an infinite number of points—a difference once again between continuous and discrete.

In the following activity, you will have the opportunity to consider how color can communicate effectively on maps. Open the linked ArcGIS Online map (https://www.arcgis.com/apps/mapviewer/index.html?layers=d431d945a68f40988eda26353d1c01bd) that shows the number of marine mammal species, by marine ecoregion, from the Atlas of Global Conservation (Figure 8.17).

Notice how darker colors were used to signify where more of a species were identified, and lighter colors used where fewer of a species were identified. Making sure that the mammal layer is active or highlighted on the left side of the map > to the right of the map, use Style > Options > Counts and Amounts Color > Experiment with a few different color ramps (the continuous tones of color at your fingertips here), one at a time (Figure 8.18).

Note the "flip ramp colors" option. Generally, more of something is usually represented by a darker color, which might be the most useful option here. Which color ramp do you think best communicates the number of species? In mapping, there are few hard-and-fast rules; more often, there are guidelines of best practice. None of the color ramps are "wrong" but a few can be misleading, if the colors are meant to signify a number. One good choice is to vary the saturation in a single color, such as a light green for a smaller number, and a dark green for a larger number. One choice that is usually misleading is to choose the rainbow color scheme from blue to red, with pale yellow in the middle, which is close to the last choice in the color chooser style window. Why is this misleading? Is blue "more" of a species or is red "more" of a species? The rule of thumb is—rainbows are nice in the sky, but usually make poor color choices for maps.

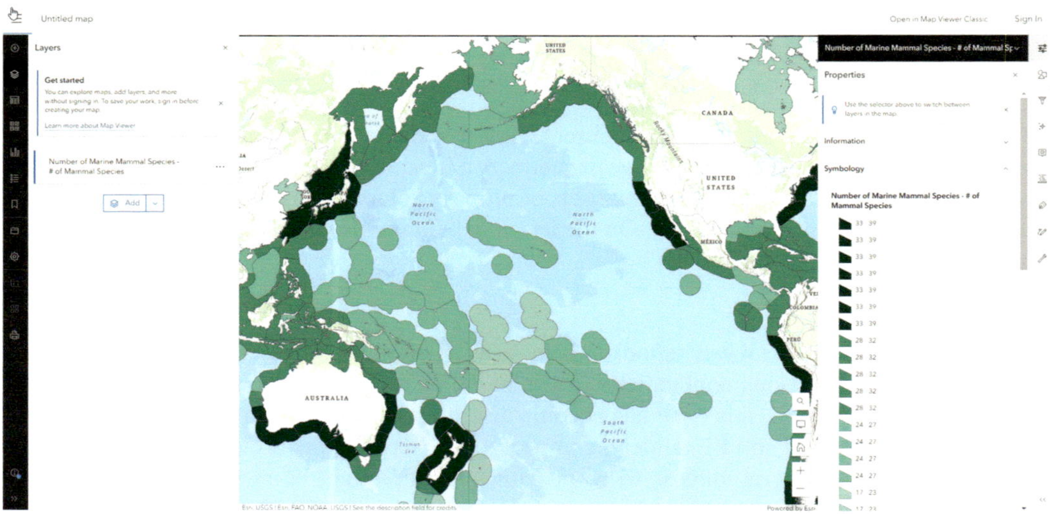

FIGURE 8.17 Examining the number of marine mammal species in different parts of the world. Mapped in ArcGIS Online from Esri.

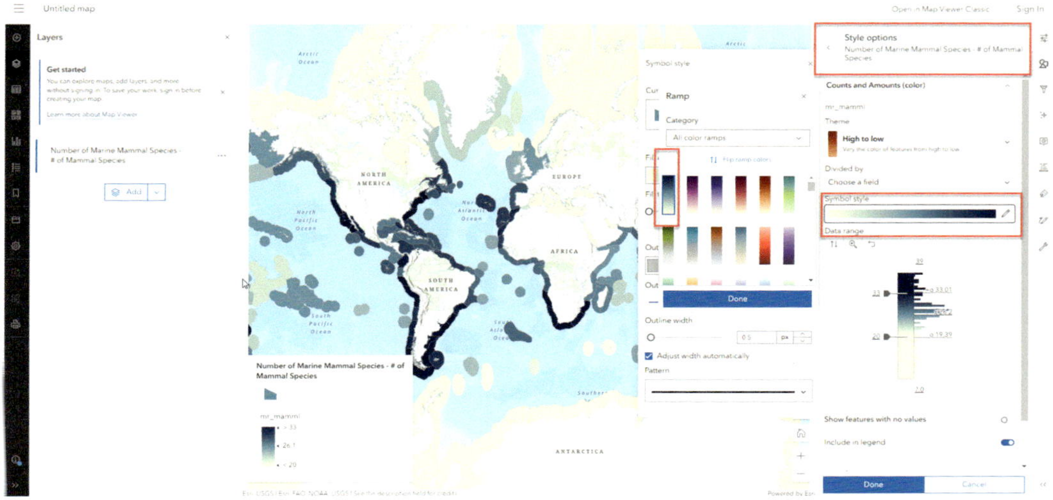

FIGURE 8.18 Choosing colors to represent the numbers of marine mammal species. Mapped in ArcGIS Online from Esri.

Terrestrial Surface and Green Infrastructure

In this activity, you will have the opportunity to work with different components of the land surface using percentages (weights) in an overlay operation. In the case of topological surface, we looked at ways to deform, or stretch the surface to reveal different structural aspects of it. On a terrestrial surface, we look instead to layering and weighting data to reveal patterns and to make decisions about the land. Overlaying data layers to assess the current situation and plan for the future across multiple themes or aspects of the environment, each of which is stored in a layer, is central to many GIS applications.

Open the following Green Infrastructure web mapping application: https://green-infrastructure.esri.com/LandCover/index.html.

You will see a USA map on the right, and on the left, a set of map layers. To each layer you can assign a weight, which will perform a weighted overlay on your selected map layers and display the result on the map on the right. Move a slider all the way to the left to remove a layer from consideration. Slide it to the right to include it. Up to 10 variables may be evaluated at any one time. Click a detailed settings box to expand it and weight the individual sub-items for that variable.

Click a location on the map to see how much each selected attribute contributed to its final score. While doing this, do not move your cursor or mouse, or it will pan the map instead. If you click to identify and then hover over one of the pie chart wedges, you'll see statistics, including the value of the pixel clicked and the minimum and maximum values for that attribute within the screen's extent.

Agricultural Application

Overlay operations are routinely done in city and regional planning to determine the most suitable land for the development of cropland, residences, reservoirs, wind farms, transportation, or other uses. Suppose you are tasked with finding high ground and flat ground to test specific types of crop seeds to determine their resiliency to cold, high-elevation winters. To determine suitable areas, move all the slider bars to the left. Next, move the elevation and slope sliders all the way to the right. Notice that when you consider two factors equally, they are each at 50%. Why? If you added aspect and moved its slider all the way to the right, each of the three factors now weighs in at 34%, 33%, and 33%. All factors must sum to 100%.

Move aspect back to the left (0%) so you are only considering elevation and slope (Figure 8.19). Expand the detailed settings under elevation and move 2,501–3,500 m all the way to the right, and

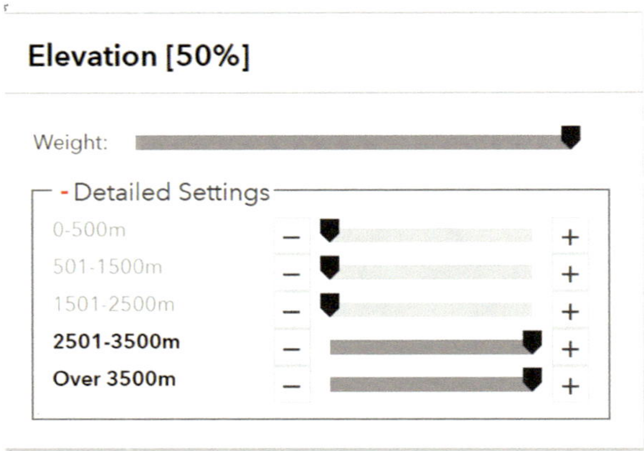

FIGURE 8.19 Elevation settings for green infrastructure overlay operation. Esri.

Over 3,500 m all the way to the right, setting the other three elevation zones to the far left (that is, so they will not be considered).

Under slope, because you only wish to consider flat areas, expand the detailed settings > move the 0%–2% slopes, 2%–4% slopes, and 4%–6% slopes all the way to the right, and the remainder of the slope values to the left. When done, in the upper center of the pane to the left of the map > you should see "2 selected" > Apply.

Once the weights are applied and the map redraws, observe the spatial pattern on the map (Figure 8.20). The dark green areas on the map represent areas of the Lower 48 USA states where the elevation is high and yet the slopes are flat or nearly flat. That is, they represent the high mesas, plateaus, and valleys. You will be able to see that the results include North Park and South Park in Colorado (which are high valleys), the Grand Mesa (a high flat-topped mountain), and some areas near the Grand Tetons and the Wind River Range in Wyoming. Why do no places in the central or eastern USA meet these two criteria? Per the color discussion in this chapter, note how effective the color is with this mapping tool combined with the base map: The dark green contrasts well with the light shaded relief base map, clearly communicating to the map user the areas that are currently under consideration.

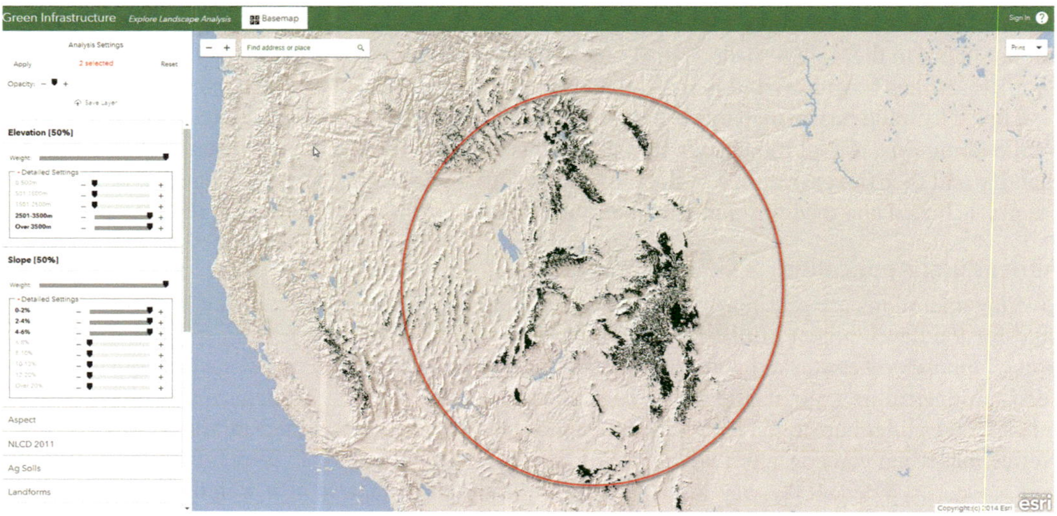

FIGURE 8.20 Results of spatial analysis showing areas of high elevation with simultaneously relatively flat slopes. Mapped in ArcGIS from Esri.

Zoom in to a small area on your map so that you can see the individual pixels (Figure 8.21). The cells meeting the criteria are shown as individual cells, or pixels, because the data layers are served as raster files. In these raster files, each cell of a certain width and height receives a value, such as land cover, a biodiversity index, or, in the case of the layers you are examining, elevation and slope. These values were originally stored as paper maps and converted to digital form by scanning, which resulted in raster files.

Click on the map inside one of your selected areas to see more information about each pixel. You will see the weight you assigned to that layer and the raw value of that pixel (Figure 8.22). The

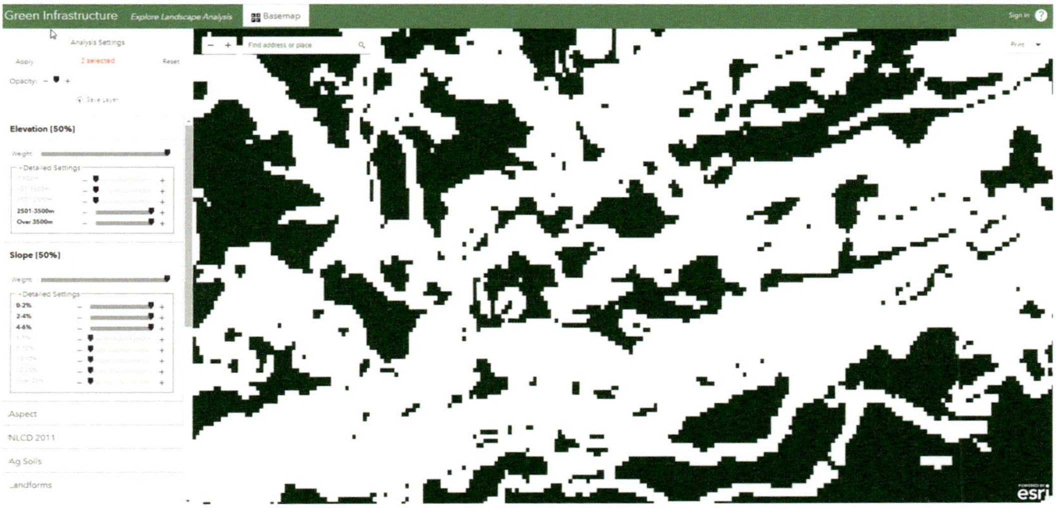

FIGURE 8.21 A large-scale view of the map from Figure 8.20 shows individual pixels evident because the data layers are represented as raster grids. Mapped in ArcGIS from Esri.

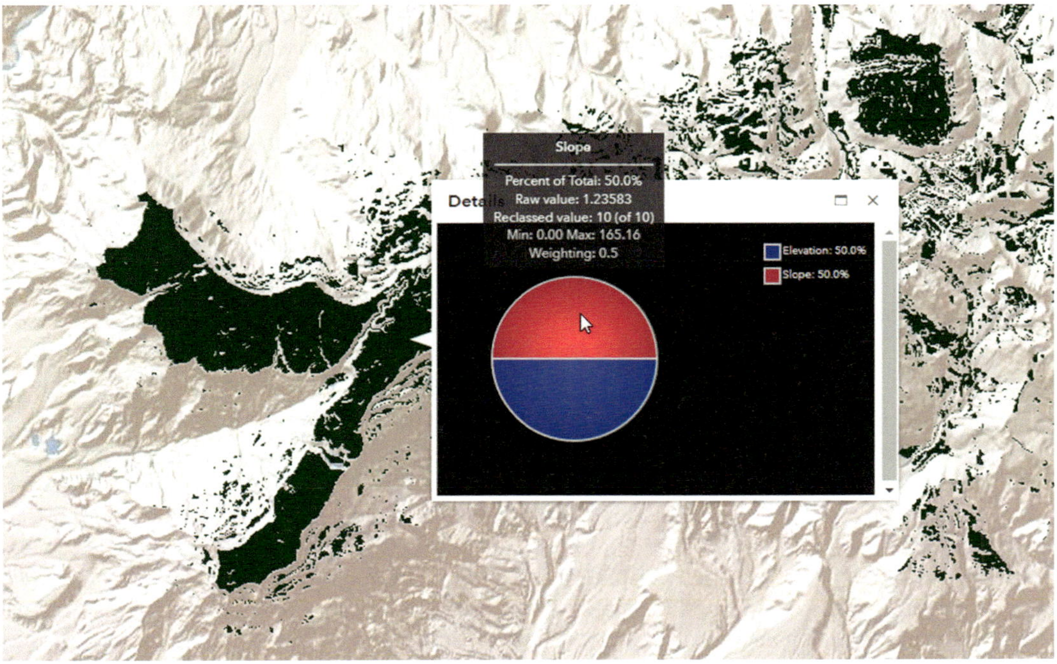

FIGURE 8.22 Weights assigned to pixels reflect the 50/50 assignment of weight equally to elevation and slope. Mapped in ArcGIS from Esri.

weighting should reflect your 50–50 assignment of elevation and slope. Think about the discussion in this chapter on color and note the colors that enable you to make a clear distinction between the colors in the pie chart. When you create your own maps, charts, infographics, and other visualizations in the future, keep clarity of interpretation at the forefront. These include color blindness considerations that are available for most modern graphics and GIS software.

You know that all factors must sum to 100%, but each of the factors does not have to be weighted evenly. Experiment with weighting the slope at 67% and the elevation at 33%. Does the resulting map show more land under consideration than with each factor at 50%, or less? Switch the weights so that the elevation is weighted at 67% and the slope at 33%. Again, compare the amount of land now under consideration than under the other two scenarios you just used. Add the 1,501–2,500 m elevation zone to the elevations under consideration and observe the difference. Add the 6%–8% slope zone to the slopes under consideration and observe the difference. You can see that the factors considered and the weights for each have a great deal of influence on the final sites under consideration. Therefore, weighting and overlay operations, and the data quality of the layers themselves, must be considered carefully.

Recreational Application

Let us say that you are now in charge of planning a new ski area. Change the weighting so that you are examining high ground, high (steep) slopes, and aspects (direction of slopes from 0°–45° to 315°–360°, that is, the **north**-facing slopes). Apply. You will now have the ideal locations for ski areas. Indeed, in central Colorado, some of the most famous ski areas of the state are located in the dark green areas—Breckenridge, Keystone, Arapahoe Basin, and Loveland. Think spatially: high elevations ensure skiing well into the springtime, and steep terrains are fun and challenging. But why are north-facing slopes important? Clue: North-facing is only important for ski areas in the Northern Hemisphere. Think about: what other layers would be important in locating ideal sites for testing seeds, as you did in this activity? What other layers would be important in locating ski areas? Do some of those layers exist in the Green Infrastructure web mapping application? If so, try them out!

If one thinks of regions as representing physical territory, it makes intuitive sense to consider adjacency as meaning only sharing a common edge (and not merely a common vertex). There are, however, real-world scenarios, such as tracking animal movement through natural habitats by superimposing grid cells in which to take counts, in which it may make sense also to count regions as adjacent that also touch only at a node. A challenge for the future may lie in understanding what the parallel set of theorems might be if "adjacency" also permits the idea of regions being considered geographically adjacent if they share either a common edge or a common node. Will extra colors become necessary?

Map Projection

Overview

The term map projection describes a set of transformations that sends the two-dimensional curved surface of the globe to the two-dimensional surface of the plane: coordinates on the globe, such as latitude and longitude, are transformed to coordinates in the plane. In earlier work in this book, we have introduced projections as needed rather than within a broad context. Here, we introduce a few broad elements of projections and offer a variety of references for further reading. The topic of map projections is one that has engaged cartographers, and others, for centuries: it is a deep and rich subject and one fundamentally tied to mathematics. Indeed, the very shapes that map projections are cast onto are based on geometry—cones, cylinders, and others.

Any projection of the surface of a sphere on a plane necessarily distorts the resulting map: think of peeling an orange and trying to flatten the peel completely. Something has to tear or stretch. An

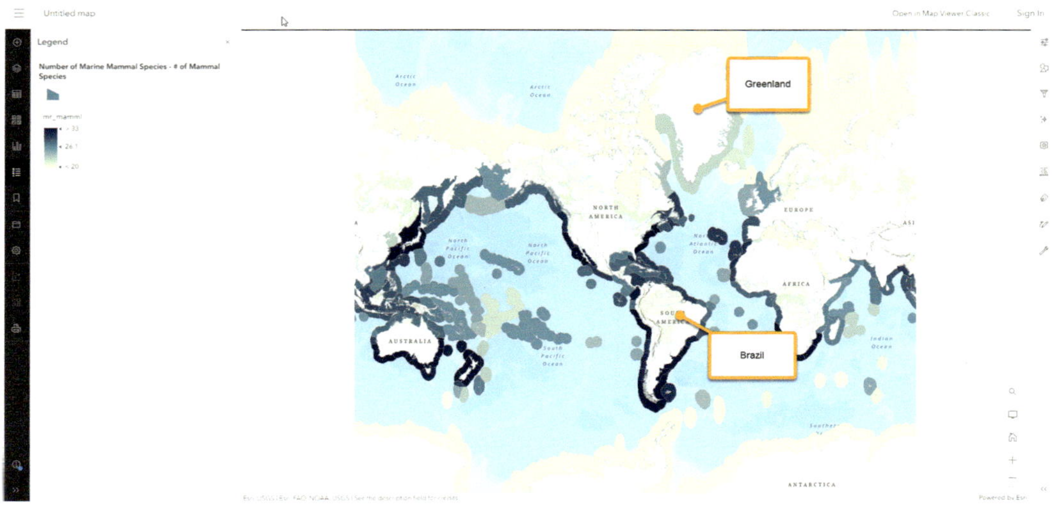

FIGURE 8.23 The web Mercator Projection for the marine mammals map you were analyzing earlier shows land distortions particularly for areas near the poles. Mapped in ArcGIS Online from Esri.

infinite number of map projections are possible; this fact may seem at first to make life difficult. However, the good news is that depending on the purpose of the map and which distortions are acceptable, a map exists to suit the purpose (although it might not be easy to find).

There are broad classes of map projections. Some common ones are perspective projections in which a shadow of the Earth's landmasses is cast on a screen. The screen may be flat, as is a plane, or it may be cylindrical and unrolled to a plane, or conical and unrolled to a part of a plane. Projections of this sort are geometric in nature. Many other projections are algebraic in nature and focus on optimizing the transformation of one coordinate system to another.

One projection in common use is the Mercator projection; it is conformal (angle is preserved locally). However, size is not preserved, and Greenland usually appears to be larger than Brazil (Figure 8.23).

The Mercator projection is the default projection in Google Earth, ArcGIS Online (that you have examined above), and in other common mapping tools for its ease of use. However, other projections can and should often be used instead because of the distortion evident at the global scale.

Mercator is not a good choice when comparing land areas. It has been a good choice as a chart for navigation at sea, which traditionally relies on local angle preservation using odd-looking instruments to attempt to calculate position. It is an easy one to use for the default for modern web-based mapping tools because coordinates can easily be mapped onto it.

Equal-area projections, in contrast to the Mercator, do show the correct areas of countries in relation to each other. They distort angles. Many contemporary atlases choose some sort of "compromise" projection as its base map where the compromise is between angular and areal distortions. Some in common use are the Robinson, the sinusoidal, and the Gall-Peters projections. One projection will be explored in more detail in the next section because it can offer insight into the coloring of maps on surfaces other than the plane: that is, it will enable us to wrap back around on the Four-Color Problem!

Stereographic Projection: One-point Compactification of the Plane

A stereographic projection (Figure 8.24) is a map projection obtained by projecting points of the sphere's surface from the sphere's North Pole, at the top, to a plane tangent to the sphere's opposite South Pole (Coxeter 1969, p. 93). All points of the sphere map to points in the plane, except the sphere's North Pole. This is the best we can do; the map on the sphere fails, at this one point,

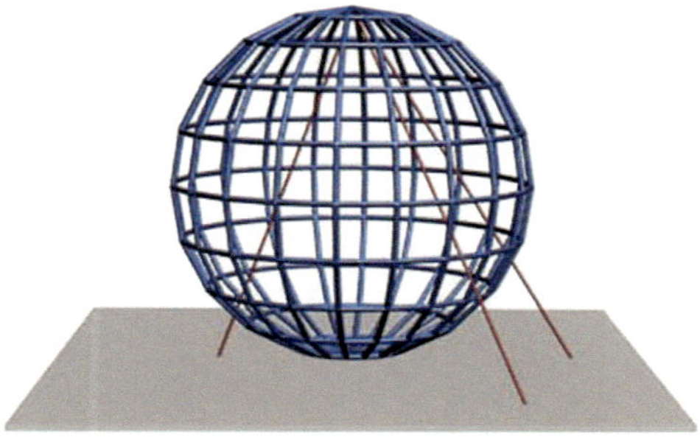

FIGURE 8.24 Stereographic projection. Source: CC BY-SA 3.0, https://commons.wikimedia.org/w/index.php?curid=85234.

to lie within the plane. There is one point that is left out of the transformation. The reason the North Pole does not map to the plane is because of Euclid's Parallel Postulate: the ray emanating from the North Pole that would carry it to the plane is parallel to the plane and therefore does not intersect it. The reader might find it interesting to speculate what would happen in this case if Euclid's Parallel Postulate did not hold.

The sphere is a compact surface; the plane is not. A stereographic projection tells us how to make the plane compact. Reverse the mapping and send every point in the plane to its associated point on the sphere. That process covers every point on the sphere except one: the North Pole. So, to make the plane compact, we need to add a single point (corresponding to the North Pole). Hence, a "one-point compactification of the plane."

How Many Colors Suffice to Color a Map on a Sphere's Surface?

It is startling that the number of colors that suffice to color a map in the plane will also suffice to color a map on the surface of the sphere (note that we are not specifying what that number is). To see why this is the case, we use stereographic projection. Consider any map on a sphere. Remove a point, N, in the interior of any region (bounded by a simple closed curve) of the map on the sphere and use N as the projection pole. Use stereographic projection to project the map on the sphere from N into the plane tangent to the sphere at the point, S, antipodal to N. The plane extends forever in all directions. Points close to N project to points extremely distant from S, the point of contact of sphere and plane. The map projected from the sphere into the plane can be colored using X colors. Project the plane map back to the sphere using the same configuration for stereographic projection. Thus, all points but one (N) on the sphere have been colored. Because the added point N is a point that is interior to a region of the map, color that point, N, the same color as that of its surrounding region. Hence, X colors also suffice for the sphere.

As with coloring and other issues in this chapter, careful alignment of geographical and mathematical elements can penetrate in many different directions. It is such alignments and their consequences that are the thrust of this book. As you move on to Chapter 9, the final chapter in this book, keep looking for such alignments and reflect on how the activities and concepts in that chapter align not only with each other but also with other concepts throughout the previous chapters. And, keep the scaffolding logo of the book in mind, as you will see it unfolded as the last element in this work!

9 Measuring Hierarchies and Patterns

The line of images above is a visual abstract of this chapter designed to foster spatial thinking. From the chapter numeral, to the book structure, to the real world, the reader is offered gentle guidance to develop spatial intuition about what might be coming. Those thoughts are then reinforced with a detailed text outline of chapter content below. The images and outline forge as an abstract of chapter content.

CHAPTER OUTLINE

Overview

The Dot Density Map

- Interactive View
- Conceptual Principles
- Nested and Non-nested Hierarchies
- Breaking a Principle

Non-Nested Hierarchy: The Cartogram

- Non-contiguous Cartograms
- Contiguous Cartograms

Hierarchical Organization Chart for This Book

OVERVIEW

Perhaps the most pervasive theme throughout this book is that of "hierarchy." A hierarchy is an arrangement of items that are represented as being "above," "below," or "at the same level as" one another. It is also a system that organizes or ranks things, often according to power, importance, or size. There are many examples of hierarchies in society: In political geography, countries are the first order of national hierarchies, then regions, then cities and other local governments. In commodities, one could consider electronics, and then inside electronics, computer software, and under that, photo editing and design tools. In a university, one hierarchy is schools or colleges, and inside those, departments. In mapping, hierarchical rules must be used or else a map would be too cluttered or confusing. When highways cross each other, which type should take precedent and be symbolized—the major highway, or the minor arterial? Usually the major highway is symbolized at such intersections. All mapping is a series of choices about hierarchies. Consider the following map layout example (Figure 9.1). The assembly of elements into an effective communication tool is dependent on horizontal and

DOI: 10.1201/9781003305613-9

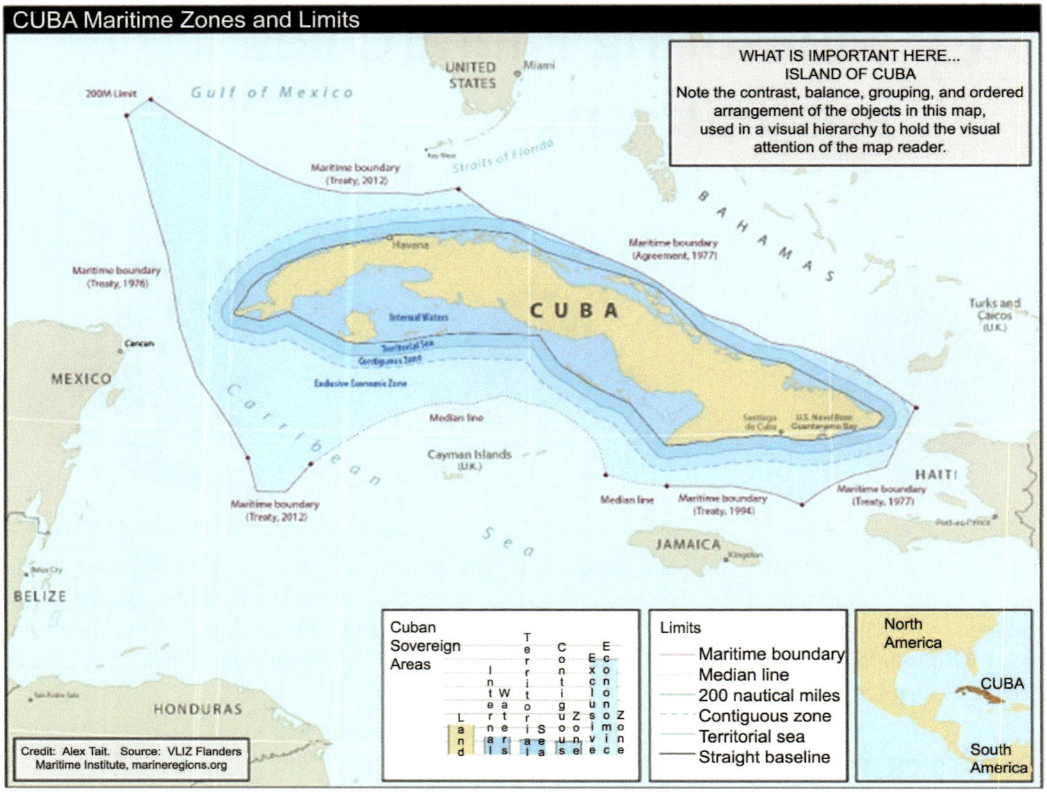

FIGURE 9.1 A map layout showing visual hierarchy. Source: UCGIS & T Body of Knowledge, Chapter CV-07-Visual Hierarchy and Layout, Tait 2018. Used with permission from Alex Tait and UCGIS.

vertical balance and ordering of mapped elements. Contrast, balance, grouping, and ordered arrangement of objects are used in a visual hierarchy to gather and hold the visual attention of the map-reader. In Figure 9.1, the map title in the upper left has the highest contrast in the image (white on a black background) which raises it in the visual hierarchy, but its position in the layout at the top edge makes it attract lesser attention than the island of Cuba in the center of the map.

On the Earth, hierarchies abound, such as biomes, inside of which are ecoregions, and inside those, sub-ecoregions. In river systems, small tributaries flow into larger streams, which flow into larger rivers, which eventually flow into the oceans. The land areas containing river systems are called watersheds. The watersheds can also be grouped into areas that drain one small tributary to large watersheds that cross national boundaries or even occupy major portions of a continent, such as the Mississippi-Missouri watershed of much of the USA and part of Canada. In the constructed, human-built world, hierarchies also are numerous: Metropolitan areas contain several or even dozens of individual municipalities, and inside those municipalities are individual neighborhoods. This latter example in mapping shows that some hierarchies have definitive boundaries, such as a municipal boundary, but some have indefinite boundaries, such as a neighborhood. Even in the physical environment, a lakeshore is a definitive boundary dividing land from water, but a biome may have an "ecotone"—a zone where it gradually, not suddenly, transitions to another biome, such as from the Great Plains of the USA to the Central Lowlands.

Another common mapping hierarchy is that of statistical enumeration areas. In the USA, these census geographic units increase in detail from state, to county, to census tract, to block group, to block (Figure 9.2).

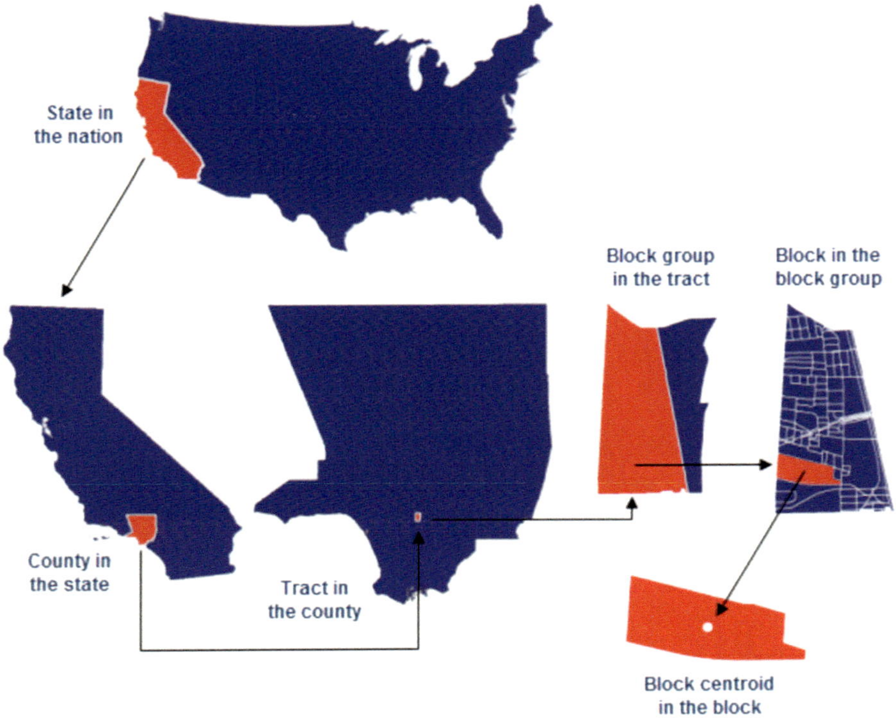

FIGURE 9.2 Map of Census statistical areas. Source: US Census Bureau.

Data may be collected hierarchically; measurements may be made hierarchically; maps may be layered hierarchically; knowledge may be organized hierarchically. Often, the hierarchy involved was a "nested" one. Nested hierarchies are an efficient organizational tool. It makes sense to have data that is collected for small areas become assembled as a large area whose boundary exactly contains those of the smaller area: there are no problems with overlap in set intersections when the hierarchy is nested. Axiom sets, are in some sense, nested hierarchies. The theorems that derive from them contain them as part of their underlying structure.

As we saw, however, in the case of Euclidean geometry, great discoveries can be made by breaking the nesting structure: when Euclid's Parallel Postulate was "broken" by altering the fundamental concept of parallelism, brand new geometries arose that permitted the consideration and solution of significant real-world problems. Simply breaking the way of thinking did not automatically lead to new achievements. New achievements were only possible from thoughtful reflection of advancements made and what remained to be done. In this chapter, we present material that epitomizes the hierarchy: We will include a nested example in the form of the census statistical boundary map, the dot density map, and a not-nested example, breaking with tradition, in the form of the cartogram.

You now have the opportunity to explore a map with a hierarchy as follows: Open the linked map (https://www.arcgis.com/apps/mapviewer/index.html?layers=a6549bd25a0e42bbab8758efd49a54b4) in ArcGIS Online of Housing Units. Scroll in or use the+sign in the lower right corner of the map to zoom in to a larger scale, a small amount at a time. As you do so, you will move from the state-level of geography, to the county level, to the census tract level. Search and zoom to Sioux City, Iowa (Figure 9.3). Note that with this particular set of housing variables, the largest scale is the census tract scale. Toggle between the tract and the county level, using Properties>and adjust the visible range and the transparency so that you can see how the census tracts nest within counties.

FIGURE 9.3 Interactive map showing housing unit characteristics in US Census statistical areas to illustrate nested geographic units. Mapped in ArcGIS Online from Esri.

No census tract boundary crosses a county line. Next, toggle between the county and the state level, using Properties>and adjust the visible range and the transparency so that you can see how the counties nest within states. No county boundary crosses a state line.

THE DOT DENSITY MAP

A dot density map is a type of thematic map where dots or points indicate the relative distribution of something. The key adjective here is *relative:* The dots do not show the exact location of a phenomenon. Rather, each dot may represent either a single instance of a sheep, person, acre of soybeans, or a representative group of those things. A typical dot map may have a title of "2020 Population over the age of 65 in Wisconsin" accompanied by a legend that says, "1 dot=1000 people." Or it may be "Soybean Cultivation in Kossuth County Iowa" with a legend that says, "1 dot=10 acres."

Dot density is an effective method to visualize concentrations of qualitative data. The points in a dot density map are all the same size, so more of something in a specific geographic area receives more dots and looks "denser" to the eye than another geographic area with less of something.

One might see a map in a newspaper showing concentrations of voters of different political persuasion, by ward, prior to an election. Often, though, these maps might appear to be (or actually be) in error: a glance at one's own home ward generally elicits reactions such as "no one lives over there where that dot is!" Dot density maps should not be confused with pushing pins in a wall map to indicate position; dot density maps show pattern, not specific position tied to address or latitude/longitude. They show proximate locations rather than exact ones: relative rather than absolute location. When the dot density map is properly constructed, it can serve as a tool offering valuable insight into clustering.

Interactive View

Open the linked interactive map (https://www.maps.arcgis.com/apps/mapviewer/index.html?webmap=393184cc34524fb293226f75413c066c) of 2020 population of the USA represented as a dot density map. The map opens showing the Chesapeake Bay as a large estuary running roughly north to south with many side inlets in the vicinity of Baltimore Maryland (Figure 9.4). The red dots provide an immediate sense of the population density in the area.

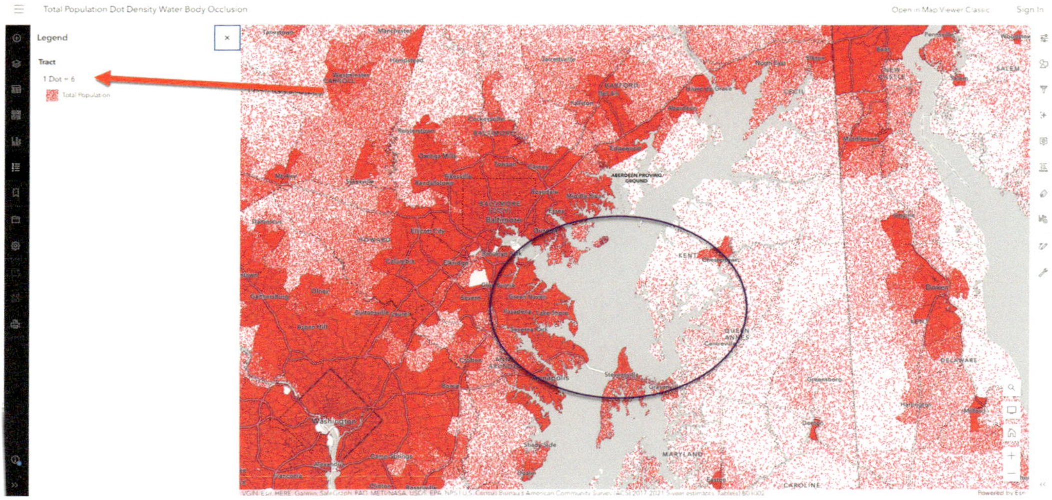

FIGURE 9.4 Population of the Chesapeake Bay region shown as a dot density map. Note that the map shows no dots in the water. Mapped in ArcGIS from Esri.

Notice the legend indicating 1 dot=6. This means 1 dot represents six people. Pan around the area and make observations about the population density—including how the density is lower on the eastern shore of the Chesapeake Bay and how it is higher from Baltimore all the way to Washington DC, part of the first "megalopolis" of the USA that extends all the way to Boston.

At least three characteristics on maps affect your interpretation of any phenomenon—here, the phenomenon is population. First, the size and shape of the area that is being mapped affects interpretation. In this example, the unit mapped is the census tract. Because they aim to include about 4,000 people in each, and because of the great differences in the population patterns of the USA because of historical settlement and landforms, census tracts vary widely in size and shape. A much larger area is needed for a tract to include 4,000 people in a rural area than an urban one. Click on the census tract in Chestertown (Tract 9503) on the eastern shore of the Chesapeake Bay and note how small its area is. Click on a few census tracts outside of Chestertown and note how large they are in area, but like Tract 9503, even these rural tracts have around 3,000 or 4,000 people living in them. Thus, because of the different sizes of the census tracts, the dots are much more densely distributed in Chestertown versus the surrounding area. Across the Bay to the west, most census tracts have even more dots in them because of the density of the population in the city of Baltimore.

The other two characteristics that affect perception are the size of the dot or point that is chosen, and how many people each dot represents (Figure 9.5). Note how the dots on this map are small and red. As always on these interactive maps you have the power to change the map symbols. On the left side of the map, make sure that Tract is the layer selected. Then, on the right, go to>Styles>Dot Density Style Options, and experiment with adjusting these two characteristics, or "style options": (1) The number of people each dot represents, and (2) the size of each dot, as shown.

The objective in mapping is for the map to enable the user to gain understanding. Therefore, feel free to experiment. After doing so, think about: Which value and size do you find most useful? Why?

You can even change the color of each dot by touching the pencil to the right of the symbol style color ramp. What color might be clearer or more understandable than the color red? Part of the answer depends on the basemap you are using. You can change the basemap on the left. The default basemap is the human geography map, so chosen because its simple style does not interfere with the population data being shown.

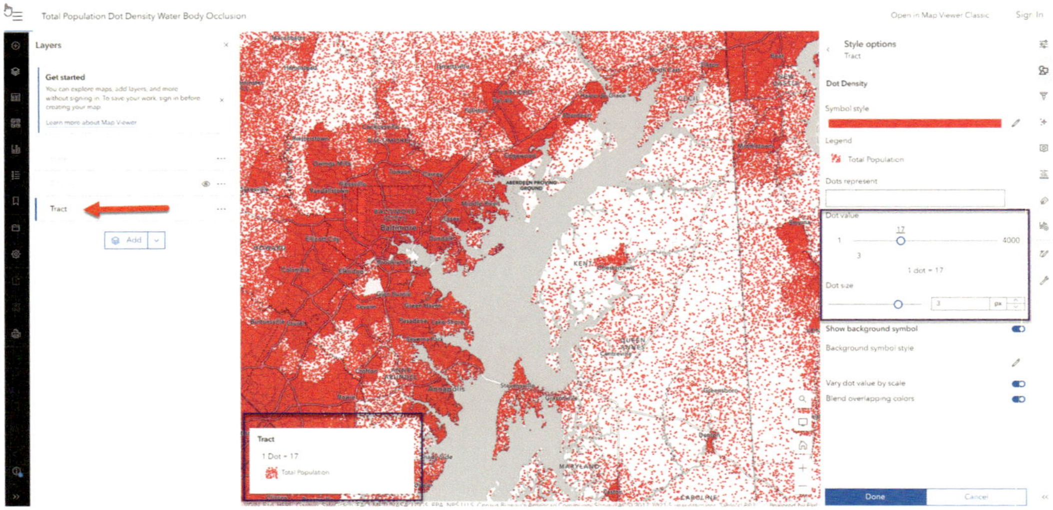

FIGURE 9.5 Changing the parameters of the original map in Figure 9.4: dot size and color. Still, no dots exist in the water but changing size and color can help the map-reader understand patterns more clearly. Mapped in ArcGIS from Esri.

Refresh the map to reset it and to ignore your changes. Next, examine two other areas. On the left side of the map, use Bookmarks>Memphis to pan the map to the area around Memphis Tennessee. Note the higher density of the population on the east bank of the Mississippi River, which is the higher elevation, versus the west bank (which is lower elevation and prone to flooding). Next, Bookmarks>Lake Elsinore to pan to this location in California, noting where people prefer to settle there. The preference is for settling around the lake, which makes sense! Have you ever wanted to live on a serene lakeshore and watch the sunrises or sunsets? Many people here feel the same way.

Use the bookmarks>Chesapeake Bay to return to the Chesapeake Bay. Zoom out and note how the dots disappear. Many types of map symbols, such as dot density, do not work for small-scale studies. This map has a visualization setting that shows the states as solid colors rather than dot density. Zoom in and note how even the areas that looked like solid colors at a small scale are, at a larger scale, composed of a series of dots. Zoom in to the largest scale, until the legend says 1 Dot=1 (person). Recall the definition of a dot density map: The dots do not, even at a scale of 1 dot for 1 person, represent *exactly* where that person lives in a neighborhood, but rather, that *somewhere* in that geographic area, one person lives, according to the data tabulated by, in this case, the US Census Bureau. Remember that all maps and visualizations hinge on the data they represent, so a good phrase to keep in mind in mapping and mathematics is always, "According to this data, we can say that ___." Zoom in on the map in an urban area until you can see buildings shown as light gray. Note how some dots fall within the gray building outlines but many fall outside the buildings, in backyards and on streets, again reflecting the random scattering of the dots within each census tract (Figure 9.6).

Dot density maps can also be effective tools to understand issues surrounding data quality and visualizing mathematical data. Pan the map so that you are looking at part of the Chesapeake Bay. Note how there are no dots in the Chesapeake Bay itself even though census tract boundaries extend into the water so that all land 9 and water territory 0 within the USA is covered by a tract.

In a new browser tab, open the following map (https://www.arcgis.com/apps/mapviewer/index.html?webmap=460f1a605b964ef981c636160cad43ad) (Figure 9.7). This map shows the same population density by census tract as you were examining below but with a different water algorithm applied. On this new map, note that the dots appear **in** the Chesapeake Bay. Use the bookmarks to go to Memphis and Lake Elsinore. Note in these places that dots appear in the Mississippi River and

FIGURE 9.6 Dot density map at the neighborhood scale. Mapped in ArcGIS from Esri.

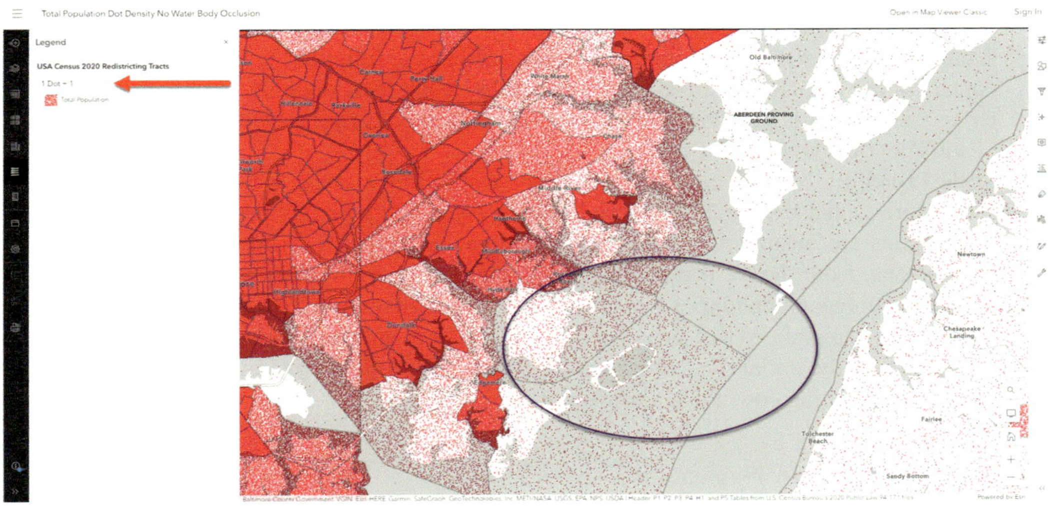

FIGURE 9.7 Population mapped as dot density with a different water algorithm applied; now, dots appear in the Chesapeake Bay and in nearby rivers. Mapped in ArcGIS from Esri.

in Lake Elsinore, respectively. Are these "aqua people"—a different human species perpetually on water skis or in scuba gear? No, and they are not people living on houseboats, either: Even if they live full time on a houseboat, the Census Bureau counts them where the boat is docked. Why then are these points shown on the water?

The points are in the water because the dots are randomly assigned to the statistical area that was used. In both of these maps, the statistical areas are census tracts. Randomly distributing the points means that some points fall into the water. Random distribution was applied to the first map you examined as well, but an algorithm was applied to cookie cut **out** the water, hence the first map's name "water body occlusion," or "blocking," and the second map's name "no water body occlusion."

Let's dig one level deeper: Would a dot density map more accurately represent the data underlying if it is portraying if the occlusion algorithm simply covered up the water bodies, or would the map be more accurate if those points that would have fallen into the water be added to that part of the census tract that was on land? Why?

Understanding differences in how data are represented and displayed are an important part of understanding mathematics through maps and offer data-rich discussion points in class. Issues of data collection, algorithms used in making the map, the symbols, colors, and sizes chosen, and the geographic unit by which the data was collected—all of this impacts our understanding and interpretation, and hence, these issues *matter.*

Conceptual Principles

One set of conceptual principles, often useful in making these maps, is enumerated below. They do not all apply to every map; indeed, the reader with experience may find shortcuts or other useful elements to add to this set.

- Projection Principle: A dot density map should be based on an equal area projection for comparisons between areas to make sense. At the outset, choose an equal area projection from the list offered.
- Nested Hierarchy Principle: A nested hierarchy of layers provides optimized assignment of dots to more global layers. When creating layers for the map, choose polygonal layers that will nest inside each other; for example, census block groups and census tracts rather than census tracts and school districts.
- Randomizing Principle: Randomize the dot scatter at a scale that is local relative to the content of the map. Randomization provides relative dot location, only. No single dot scatter is "correct"; it is random rather than absolute. Randomization is like throwing grains of sand out on a table.
- Scale Principle: To have the randomized dot scatter make sense, it must be viewed at a scale more global than that of the randomizing layer. Clusters that form in the local level of polygons are meaningless as they are random; those that form in more global layers of polygons, from the local randomization, reveal meaningful clusters at the local level. Removing polygon boundaries, used in creating the map, may help to reveal clustering.
- Dot Clutter Principle: When a change in scale produces dot clutter from the density of the dots, return to the initial randomization and select a smaller dot size or thin out the number of dots by changing how much a single dot represents.

Nested and Non-Nested Hierarchies

Dot density maps offer one way to synthesize information derived from point sources to suggest information about areas by clustering patterns of dots. The linked interactive map (https://deepblue.lib.umich.edu/handle/2027.42/58253) shows a map constructed on all four of the principles above (Arlinghaus, 2005). It offers both vertical and lateral movement within a hierarchy, including state boundaries (heavy white lines), county boundaries (yellow lines), Census tract boundaries (pink lines), and dot scatter at the 1:800 level. Use the zoom feature to shift hierarchical levels in a scale transformation. Within a given hierarchical level, click on single polygons to view associated database information. Or, search the underlying database using the "search" feature to find polygons with attributes of special interest.

The most common error in making dot density maps is to show the dot scatter within the polygons used for randomizing the scatter; hence, the common, but incorrect, complaint that these are "not accurate." In addition to their value for showing clusters of data and pattern within nested hierarchies, dot density maps have outstanding value in showing clusters when data boundaries do not mesh into nested hierarchies. Thus, data can be reasonably mapped as a dot density map that relates school district boundaries to census tracts or ZIP code.

There are a number of ways of extracting information from Census units about, for example, zip code areas. One way involves using computer software to calculate the centroids of Census units and count the Census unit as lying "within" the zip code area if the centroid of the Census unit lies within the zip code area. Zip code areas are often larger than tracts and block groups, especially in urban areas (as Census units are scaled in area according to population). Thus, in an urban setting, one might properly construct a dot density map randomizing the dot scatter at the block group level and then viewing that scatter through tract boundaries and zip code boundaries, creating a properly constructed dot density map that is not optimized because the hierarchy of spatial units on which it is based is not a nested hierarchy.

Breaking a Principle

The dot density map is important in overcoming incommensurable measurement issues derived from a non-nested hierarchy of polygonal layers—a "breaking" of the rules.

Violating the "nested hierarchy principle" to derive new information is a bit like what geometers did in violating the Parallel Postulate to derive new approaches in physics!

It is a breaking of established rules that is constructive and additive to tradition while solving new and previously unsolvable problems.

NON-NESTED HIERARCHY: THE CARTOGRAM

Another way to map a set of data is through cartograms. Cartograms map the size of an areal unit mapped by something other than the areal size of that unit; thus, there is no nesting. For example, a state could be sized by the number of a certain type of convenience store chains in that state or the number of universities there. A country could be sized by the number of hydroelectric dams in that country or by the number of mountains above a certain elevation.

The geometry, or areal represented by a cartogram, can be anything, such as census tracts in a city, states in a country, or countries around the world. However, cartograms work best when the map-reader has a mental picture of what the map is "supposed" to look like before the distortion is applied. Then the map-reader can more readily understand the resulting patterns. Therefore, small scale less detailed areas such as states in a country or countries around the world work better as cartograms than large-scale detailed areas such as neighborhoods.

Non-Contiguous Cartograms

In a web browser, examine variables by country around the world using the Bouncy Maps cartogram tool: https://www.bouncymaps.com/#!/bouncymaps/world/-2102779804

Set the tab under the map to "regular map" and observe the map symbolized not by population but by continent. Move the tab to "Bouncy Map" and compare Population 2019 to the regular map you examined a moment ago (Figure 9.8). The bouncy map is a cartogram, and the countries are still symbolized by continent, but are sized according to their population. What are two patterns that you notice? What happens to the relative sizes of India and China compared to other world countries between a "regular" map and as a cartogram? Why? Note the data table, the "I" part of interactive mapping, below the bouncy map. Make a few more maps of agriculture, economy, or another topic of interest to you and think about the resulting patterns. For example, the cartogram shows a relatively high number of sheep in China, Iran, Australia, and New Zealand (in fact note that the 27.5 million sheep in New Zealand is about five times more than the number of people there!).

These Bouncy Maps are non-contiguous cartograms: the countries no longer share common boundaries. Adjacency is not maintained. Shapes are retained. Area is modified.

What did you learn about one of the variables on the Bouncy Maps site in terms of that variable's spatial pattern and relative number by countries around the world? In your opinion, are these

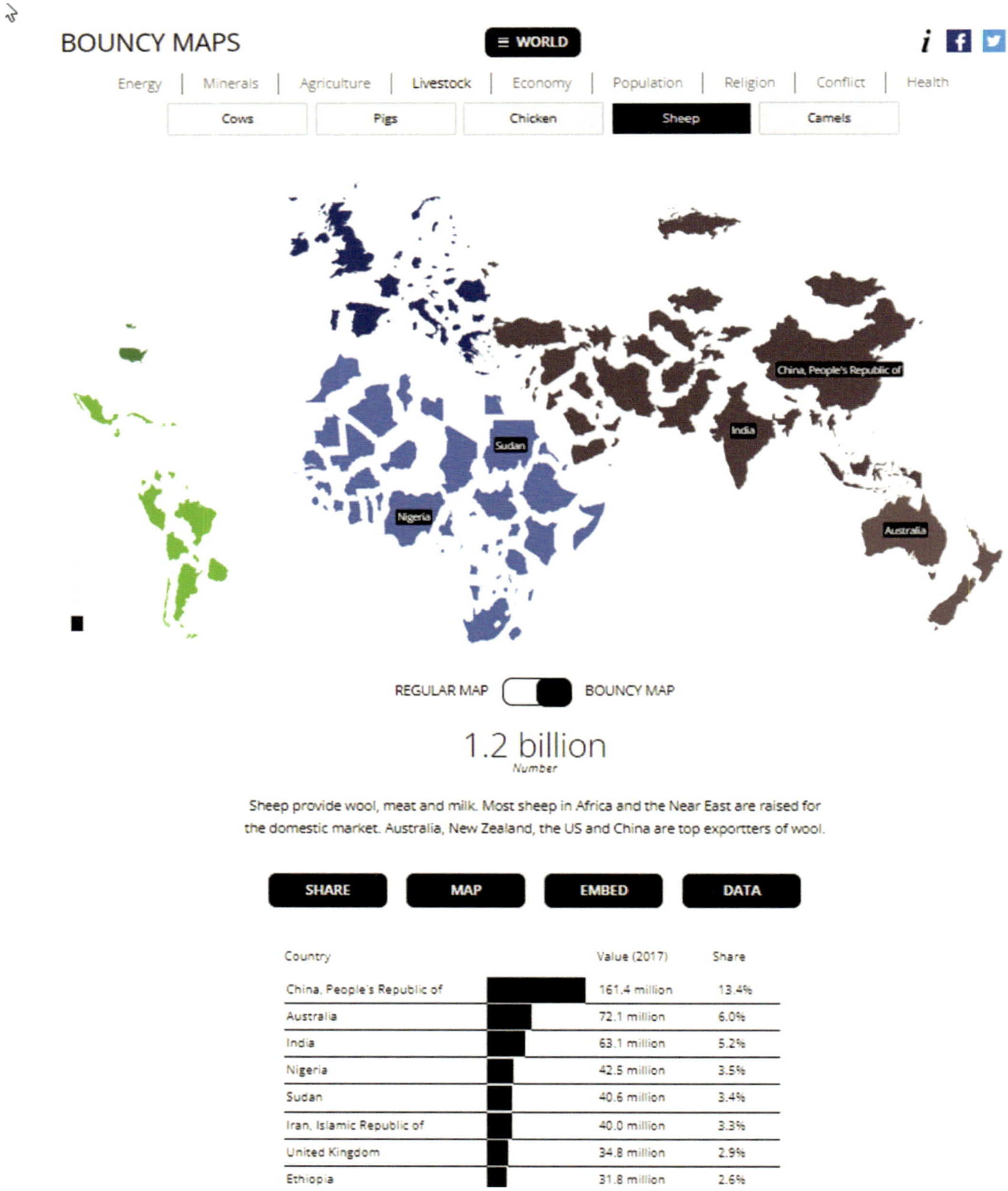

FIGURE 9.8 The number of sheep in 2019 by country. Source: Bouncy Maps.

non-contiguous cartograms a tool helpful in fostering understanding or are they confusing? Create a cartogram from Bouncy Maps on a variable of interest to you and assess the resulting patterns.

Contiguous Cartograms

Sometimes readers find the lack of adjacency in cartograms to be a source of confusion. Figure 9.9 shows an early cartogram of sorts (Johnson, 2008) by Levasseur (1876) and displays clearly two major issues associated with creating cartograms (true or otherwise): preservation of country

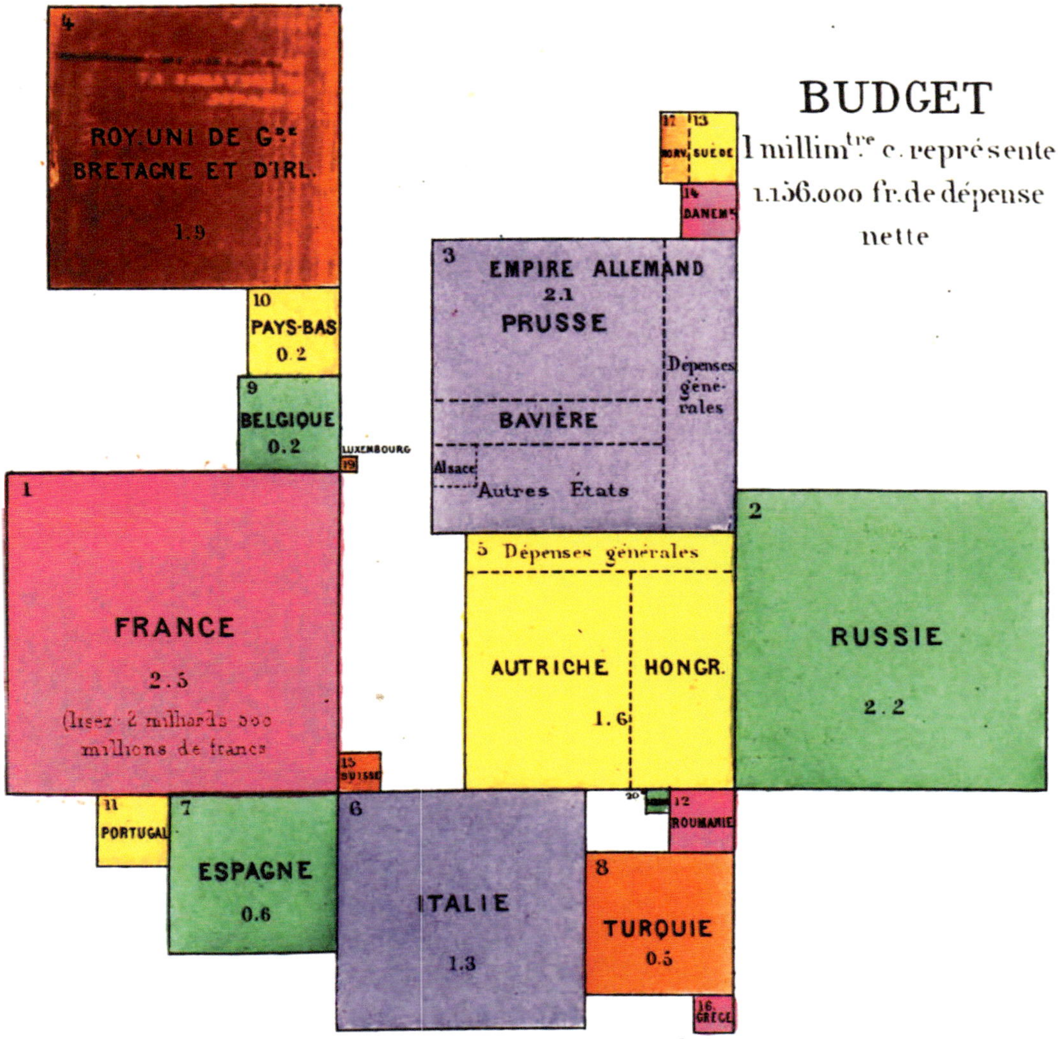

FIGURE 9.9 Levasseur original cartogram from 1876. Public domain.

shape and preservation of country adjacency pattern. Over the years, algorithms have been created that attempt to find a compromise as a balance between these competing values. One interesting approach, by Waldo R. Tobler (1963), involved warping the underlying space containing the map rather than the map itself. It is an abstract geometrical and topological approach that D'Arcy Wentworth Thompson (1917) employed in biological research.

Look around for cartograms on the Internet. One convenient starting point is, as it is with many topics involving digital mapping, the GIS&T Book of Knowledge; try reading this chapter on cartograms (https://gistbok.ucgis.org/bok-topics/cartograms). Look at the contiguous cartograms. What did you learn about the variables in terms of spatial pattern and relative number by countries around the world? In your opinion, are contiguous cartograms a tool helpful in fostering understanding or are they confusing? Do you find the non-contiguous cartograms in Bouncy Maps or the contiguous cartograms more understandable? Why? Maps are powerful because they may confirm our prior understanding of how the world works and the variability around the world; they may also challenge our prior understanding of how we "thought the world worked."

Cartograms can also be created for other areas. Consider population from 1900 to 2010 for a state, such as the state of Kansas. You could create standard choropleth maps showing the population each census year and even an animation to help visualize the changes. But creating cartograms of the population in each county provides additional insight. Consider the output from selected years (Figures 9.10–9.14). The cartograms show the settlement of the high plains (western Kansas) from 1900 to 1930, followed by population loss that continues in some counties all the way to 2010. Coupled with that is the rise of the urban centers of Wichita (south central Kansas) and Topeka, Lawrence, and Kansas City (northeast Kansas). The combination of these trends, brought about by

Kansas Map, Albers Projection, Cities and Counties

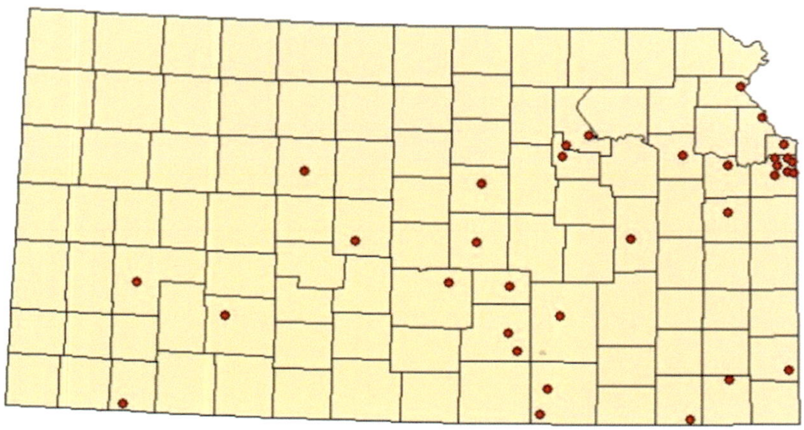

FIGURE 9.10 State of Kansas—Counties and cities—in a projected coordinate system. Source: Esri ArcMap software, data from US Census Bureau, map by Joseph Kerski.

1900 Cartogram

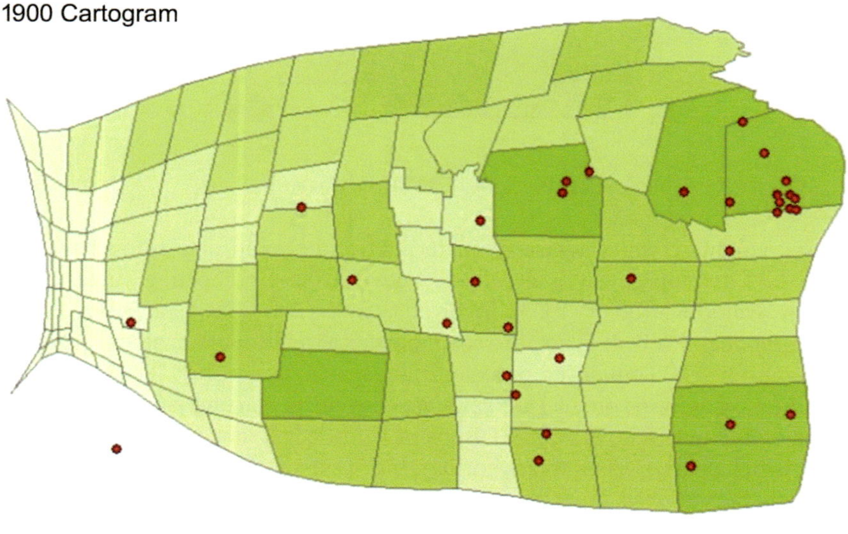

FIGURE 9.11 Population in 1900 shown as a cartogram by county in Kansas. Many of the eastern and central Kansas counties had similar population totals. Source: Esri ArcMap software, data from US Census Bureau, map by Joseph Kerski.

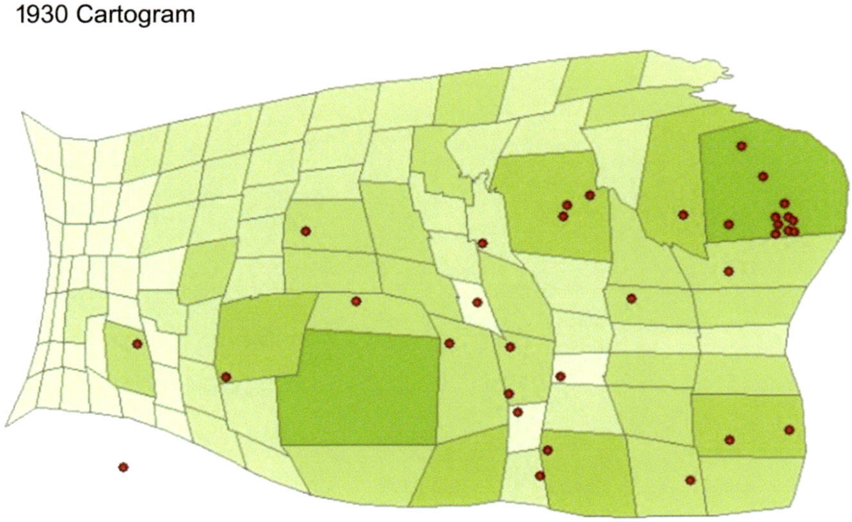

FIGURE 9.12 Population in 1930 shown as a cartogram by county in Kansas. The western third has now been settled, but the east continues to increase at a more rapid rate. (From Esri ArcMap software, data from US Census Bureau, map by Joseph Kerski.

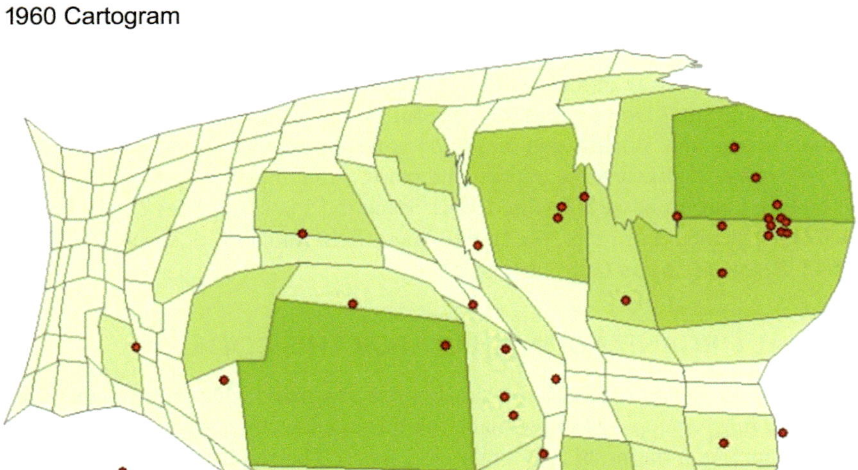

FIGURE 9.13 Population in 1960 shown as a cartogram by county in Kansas. Many western counties have lost population or gained very little compared to the rise of the major urban centers of Wichita (south central), Topeka (east central), Lawrence (east central) and Kansas City (northeast corner). Source: Esri ArcMap software, data from US Census Bureau, map by Joseph Kerski.

social, physical, and economic forces, squeeze some of the northern and western counties so much that they are almost invisible by 2010 (even though they still contain some vibrant communities!).

Consider doing this for your own area of interest—population change in your own state over time, water quality or river flow differences by watershed, or crime rates or median age by neighborhood in your own community. If you do this, again, it is advantageous if the readers of your cartograms know what the areas that you are analyzing look like as a standard map for comparison

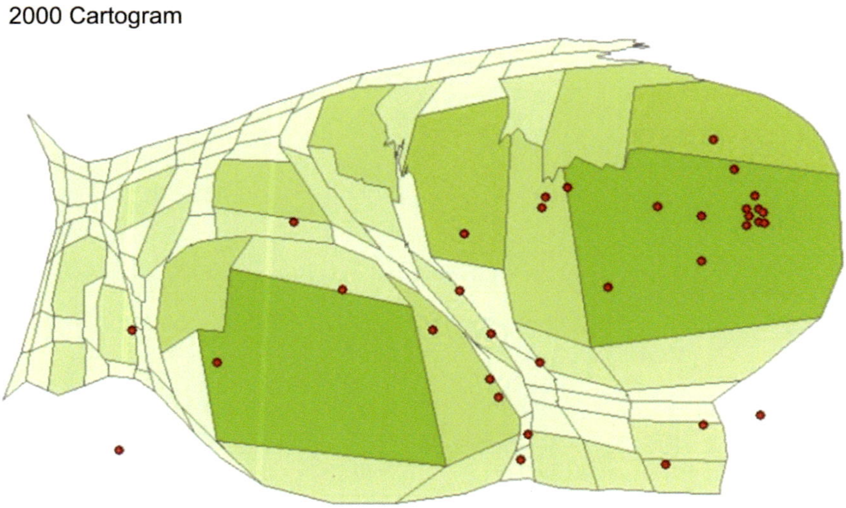

FIGURE 9.14 Population in 2010 shown as a cartogram by county in Kansas. Urbanization continues, particularly Johnson County (suburban Kansas City, northeast corner) which has become larger than Kansas City itself. Selected rural counties are gaining population as well but are usually those adjacent to large urban areas. Source: Esri ArcMap software, data from US Census Bureau, map by Joseph Kerski.

purposes. Thus, consider providing such a standard map at the front of your set of cartograms, as is done below. That way, your audience will more readily understand how the variables you are mapping distort the "standard" way of looking at that area.

Do you like cartograms? Do you find them useful? Why? Many people, including some cartographers, dislike cartograms. What is one reason do you suppose they might not like cartograms?

The goal with cartographic representations and maps is that they are useful if they help us understand something in a different way. To dig deeper: Even the shape of tables can be distorted by the variable each cell contains. For more, see this linked article (https://link.springer.com/chapter/10.1007/978-3-642-40450-4_36) on table cartograms.

HIERARCHICAL ORGANIZATION CHART FOR THIS BOOK

Is the image (Figure 9.15) that suggests how this book is organized, chapter by chapter, based on the percentage of how much mapping or how much math is in each chapter and the degree of difficulty of the math or mapping, a hierarchically organized chart? The reddish colors along the math axis suggest that chapters adjacent to the math axis emphasize math (Chapters 1, 2, and 6) and that the degree of difficulty of that math corresponds directly to the saturation of the color. The same idea is present with the cyan chapters adjacent to the maps axis (Chapters 1, 3, and 4) while those in between the two axes reflect more of a balance in content. Chapters 1–3 with lower intensity form a group of "easier" chapters; Chapters 4–6 with moderate intensity in color form an intermediate level, while Chapters 7–9 create a more advanced approach—all independent of the balance between math/maps content (reflected in proximity to the axes). One might also see some sort of imagination vector, based on Chapter 1, rising through Chapters 5 to 9.

Did you see all that from looking at this apparently simple image? Often we need to think carefully about what we see around us; it may have much deeper roots than is suggested by a superficial view. Indeed, that is true—"superficial" is very different from "simple." The former suggests a shallowness of thought while the latter should suggest a depth of thought resulting in a construct that is elegant conceptually.

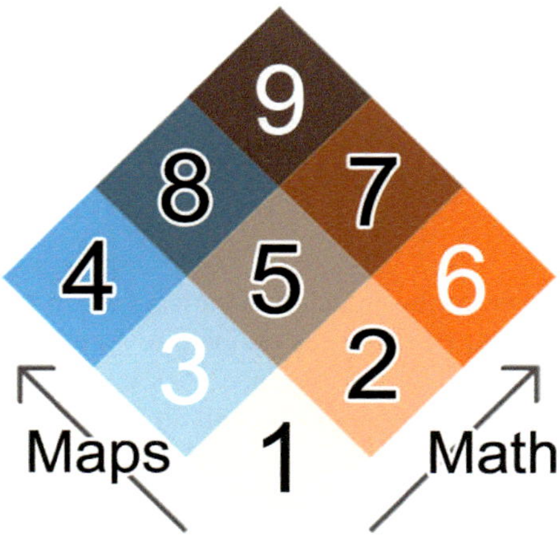

FIGURE 9.15 Organizational image for the content of this book, chapter by chapter, showing both balance in math/maps content (measured by proximity to axes) and degree of difficulty of content (measured by color saturation).

FIGURE 9.16 Two independent color hierarchies, layered, create the book logo.

Take another look. Is there a hierarchy, or not? How was the image made? It was made by superimposing two images, shown in Figure 9.16. From that image, we could say that the final image is composed of two-color hierarchies: one along the cyan color scale and the other along the orange/red color scale. When superimposed, those cells adjacent to the axes retain the true colors of the merged underlying hierarchies while those near, but not adjacent, to the axes have blended colors: reveal them by layering the two squares on the left, in Adobe Photoshop (or another graphic design software), and then in the Blending Option, choose to "Darken" or "Multiply" the top layer—watch the amazing result emerge! This simple method does yield a color pattern similar to, but not identical with, the logo for this book. For that logo, we chose to use an established Esri color scheme that has been tested to be "colorblind-safe" (a tweaked version of the technical Blending Option result).

So, the answer is "yes"—the notion of hierarchy as well other concepts within this book, are all embodied in the single simple organizational image. Evidently, one cannot judge a book by its cover, nor an eBook by its logo! Look around you everywhere you go and think objectively, constructively, positively, and creatively about all that you see—do not underestimate even those items that appear simplest. They may be full of pleasant, and unforeseen, surprises!

References and Further Reading Suggestions

A number of the readings listed below are much more advanced, in terms of mathematics, than the content of this book. That apparent mismatch is deliberate. Another skill in motivating people to move forward with mathematics is to build confidence. Pick up a book at any level of mathematical knowledge. Page through the whole thing. Some of it will no doubt be covered with notation that may be unintelligible to most readers. However, much of it will also be covered with natural language that explains generally what the author is trying to work on. Read the background; understand the context; master the general gist of the problems at hand. If you do that, at least you will build appreciation for the kinds of problems and world-views that mathematics can deal with. That alone builds confidence. Find particular areas that seem interesting and use those to help guide your further detailed study of mathematics. Books in mathematics that are well-written have clear text explanations and are aided with esoteric notation only when needed; they tend to endure over decades. Don't be intimidated by notation that you do not yet understand; read around it and get to the point of what is really being done by reading explanations given in natural language to set the context.

The readings below offer a variety of examples for you to sample in your leisure. Entries without page numbers are presented as general readings of this sort. Entries with specific page ranges are materials to which specific reference is made within the text.

Abbott, E. A. 1884. *Flatland: A Romance of Many Dimensions*. London: Seeley & Company.

Appel, K. and W. Haken. 1978. "The Four Color Problem." In: Lynn Arthur Steen, eds *Mathematics Today Twelve Informal Essays*, 153–180. New York: Springer-Verlag.

Appel, K. and W. Haken. 1989. *Every Planar Map Is Four Colorable*. Providence: American Mathematical Society.

Aristotle. 1952. "Metaphysics." In: Hutchins, R. M., Ed., *Great Books of the Western World*. W. D. Ross (trans.), Vol. 8. Chicago: Encyclopædia Britannica, Inc.

Arlinghaus, S. 1985. "Fractals Take A Central Place." *Geografiska Annaler*. 67B, 83–88.

Arlinghaus, S. 1986. *Essays on Mathematical Geography*. Ann Arbor: Institute of Mathematical Geography, Monograph #3. The University of Michigan Persistent Online Archive, Deep Blue: https://deepblue.lib.umich.edu/bitstream/handle/2027.42/58266/Mongraph03.pdf?sequence=2&isAllowed=y

Arlinghaus, S. L. 2005. "Spatial Synthesis. The Evidence of Cartographic Example: Centrality and Hierarchy." *Solstice: An Electronic Journal of Geography and Mathematics*. XVI(1). Ann Arbor: Institute of Mathematical Geography.

Arlinghaus, S. L. and W. C. Arlinghaus. 1989. "The Fractal Theory of Central Place Hierarchies: A Diophantine Analysis of Fractal Generators for Arbitrary Löschian Numbers." *Geographical Analysis: An International Journal of Theoretical Geography*. 21(2), 103–121. Columbus: Ohio State University Press.

Arlinghaus, S. and W. C. Arlinghaus. 2005. "Animated Prime Number Sieve." *Spatial Synthesis, Volume I: Centrality and Hierarchy, Book 1*. Ann Arbor: Institute of Mathematical Geography. https://deepblue.lib.umich.edu/handle/2027.42/58264

Arlinghaus, S., W. C. Arlinghaus, and F. Harary. 2002. *Graph Theory and Geography: An Interactive View*. Hoboken, NJ: John Wiley & Sons, Wiley Interscience Series. Link.

Arlinghaus, S. and B. Blake. Winter 2004. "Two Rivers Ridge: Capturing Art." *Solstice: An Electronic Journal of Geography and Mathematics*. XV(2). Ann Arbor: Institute of Mathematical Geography. https://deepblue.lib.umich.edu/handle/2027.42/58220

Arlinghaus, S. and J. Kerski. 2013. *Spatial Mathematics: Theory and Practice through Mapping*. Boca Raton, FL: CRC Press, Inc.

Arlinghaus, W. C. 1985. *The Classification of Minimal Graphs with Given Abelian Automorphism Group*. Memoirs of the American Mathematical Society 57. Providence: American Mathematical Society.

Arlinghaus, W. C. 2021. *Calculus II Workshops*. Monograph 24. Ann Arbor: Institute of Mathematical Geography.

Ball, K. M. 2003. "Fibonacci's Rabbits Revisited." *Strange Curves, Counting Rabbits, and Other Mathematical Explorations*. Princeton, NJ: Princeton University Press.

Baumgaertner, E. January 30, 2023. "Students Lost One-Third of a School Year to Pandemic, Study Finds." *New York Times*. https://www.nytimes.com/2023/01/30/health/covid-education-children.html

Beck, M. and R. Geoghegan. 2010. *The Art of Proof: Basic Training for Deeper Mathematics*. New York: Springer.

Berge, C. 1966. *The Theory of Graphs*. Alison Doig (trans.). Hoboken, NJ: John Wiley & Sons.

Birkhoff, G. and S. Mac Lane. 1969 (3rd Edition). *A Survey of Modern Algebra*. New York: MacMillan.

Bittinger, M., N. Brand, et al. October 18, 2005 (1st Edition). *Calculus for the Life Sciences*. New York: Pearson.

Birkhoff, G. 1948 (Revised Edition). *Lattice Theory*. Providence: American Mathematical Society.

Bourbaki, N. 1998. "Foundations of Mathematics & Logic: Set Theory." *Elements of the History of Mathematics*. Berlin: Springer.

Buckley, A. March 24, 2011. Tissot's Indicatrix Helps Illustrate Map Projection Distortion. https://www.esri.com/arcgis-blog/products/product/mapping/tissots-indicatrix-helps-illustrate-map-projection-distortion/#:~:text=Tissot's%20indicatrix%20was%20first%20developed,%2C%20i.e.%2C%20a%20map%20projection

Campbell, J. 2001. *Map Use and Analysis*. New York: McGraw-Hill.

Cantor, G. 1874. "Ueber eine Eigenschaft des Inbegriffes aller reellen algebraischen Zahlen." *Journal für die Reine und Angewandte Mathematik*. 77, 258–262. doi:10.1515/crll.1874.77.258. S2CID 199545885. Archived (PDF) from the original on October 7, 2017.

Cantor, G. 1955, 1915. In: Jourdain, P., Ed., *Contributions to the Founding of the Theory of Transfinite Numbers*. New York: Dover Publications.

Cartograms. https://link.springer.com/chapter/10.1007/978-3-642-40450-4_36

Comparing Subtraction and Addition of Integers. Last accessed April 2023. https://www.youtube.com/watch?v=YB6zcw-2Ses

Courant, R. and H. Robbins. 1941. *What Is Mathematics*? London: Oxford University Press.

Coxeter, H. S. M. 1955 (2nd Edition). *The Real Projective Plane*. New York: Cambridge at the University Press.

Coxeter, H. S. M. 1961, 1969 (and later editions). *Introduction to Geometry*. Hoboken, NJ: John Wiley & Sons.

Coxeter, H. S. M. 1965. *Non-Euclidean Geometry*. Toronto: University of Toronto Press.

Coxeter, H. S. M. 1947, 1973 (3rd Edition). *Regular Polytopes*. New York: Dover.

Danzig, T. 1988 (Revised). *Number: The Language of Science*. New York: Dover.

Demana, F. and J. R. Leitzel. January 1, 1990 (Revised Printing). *Transition to College Mathematics*. Boston: Addison-Wesley.

Demana, F. and B. Waits, et al. January 27, 2018 (10th Edition). *Precalculus: Graphical, Numerical, Algebraic*. New York: Pearson.

Descartes, R. 1954. *The Geometry of René Descartes: With a Facsimile of the First Edition*. Smith, D. E., Latham, M. L. (trans.). New York: Dover.

Dorling, D. 1996. *Area Cartograms: Their Use and Creation*. Concepts and Techniques in Modern Geography Series No. 59. Norwich: University of East Anglia.

Duke, B. and J. J. Kerski. 2022. Geography in Everyday Life in 2022: How Geography Is Embedded in Our Everyday Lives. https://www.directionsmag.com/article/11897

Eves, H. 1963. *A Survey of Geometry Volume One*. Boston: Allyn and Bacon.

Euler, L. 1736. "Solutio problematis ad geometriam situs pertinentis," *Commentarii academiae scientiarum Petropolitanae*. 8, 128–140.

Freedman, D., R. Pisani, and R. Purves. 1978. *Statistics*. New York: W. W. Norton & Company.

Gamow, G. 1954. *One, Two, Three, Infinity*. New York: Mentor.

Gastner, M. T. and M. E. J. Newman. 2004. "Diffusion-Based Method for Producing Density-Equalizing Maps." *Proceedings of the National Academy of Sciences*. 101, 7499–7504.

Geach, P. T. 1980. *Logic Matters*. Berkeley: University of California Press.

Getis, A. and J. K. Ord. 1996. "Local Spatial Statistics: An Overview." In: Longley, P. and Batty, M., Eds., *Spatial Analysis: Modeling in A GIS Environment*, 261–277. Hoboken, NJ: John Wiley & Sons.

Gewin, V. 2004. https://www.nature.com/articles/nj6972-376a

Graham, R., D. Knuth, and O. Patashnik. 1989 (1st Edition). "Binomial Coefficients." *Concrete Mathematics*, 153–256. Reading, MA: Addison-Wesley.

Griffith, D. A. and B. Li. August 3, 2022. *Advanced Introduction to Spatial Statistics*. Elgar Advanced Introductions Series. Northampton, MA; Cheltenham: Edward Elgar Publishing.

Halmos, P. R. 1964. *Measure Theory*. Princeton: D. Van Nostrand.

Harary, F. 1969. *Graph Theory*. Reading, MA: Addison-Wesley.

Hardy, G. H. and E. M. Wright. 2008 (6th Edition). *An Introduction to the Theory of Numbers*. Oxford University Press.
Hausdorff, F. 1914. *Grudnzüge der Mengenlehre*. Leipzig: Veit.
Heath, T. L., Ed. 2013. "Apollonius of Perga." *Treatise on Conic Sections*. Cambridge: Cambridge University Press.
Heath, T. L. 1956. *The Thirteen Books of Euclid's Elements*, Vol. I. New York: Dover.
Heawood, P. 1890. "Map-ColourTheorem." *Quarterly Journal of Pure and Applied Mathematics*. 24, 332–338.
Hennig, B. D. 2021. "Cartograms." *International Encyclopedia of Geography: People, the Earth, Environment and Technology*. Hoboken, NJ: John Wiley & Sons.
Hennig, B. D. 2013. *Rediscovering the World: Map Transformations of Human and Physical Space*. Berlin, Heidelberg: Springer.
Herstein, I. N. 1964. *Topics in Algebra*. New York: Blaisdell (A Division of Ginn).
Higgins, K., D. Miner, C. N. Smith, and D. B. Sullivan. 2004. "A Walk through Time" (Version 1.2.1). https://physics.nist.gov/time [2010, July 12]. Gaithersburg, MD: National Institute of Standards and Technology.
House, D. H. and C. Kocmoud. 1998. "Continuous Cartogram Construction." *Proceedings of the IEEE Conference on Visualization 1998.*
Howse, D. 1980. *Greenwich Time and the Discovery of the Longitude*. Oxford University Press.
Jacobs, H. R. 1974. *Geometry*. San Francisco: W. H. Freeman.
Jacobson, N. 1951. *Lectures in Abstract Algebra, Volume I, Basic Concepts*. Princeton: D. Van Nostrand.
Jenks, G. F. 1967. "The Data Model Concept in Statistical Mapping." *International Yearbook of Cartography.* 7, 186–190.
Johnson, Z. F. 2008. Early Cartograms. Indiemaps.com/blog.
Jordan, C. 1887. *Cours d'analyse.* Reprinted by Cambridge University Press, 400 pages, 2013
Kasner, E. and J. R. Newman. 1940. *Mathematics and the Imagination*. New York: Simon & Schuster.
Kelley, J. L. 1955. *General Topology*. Princeton: D. Van Nostrand.
Kempe, A. B. 1879. "On the Geographical Problem of Four Colors." *American Journal of Mathematics*. 2, 193–204.
Kerski, J. J. 2014. "Earth Science Inquiry with Web Mapping Tools." *The Earth Scientist* 31(4), 11–18.
Kerski, J. 2014. "Geo-awareness, Geo-enablement, Geotechnologies, Citizen Science, and Storytelling: Geography on the World Stage." *Geography Compass*. 9(1), 14–26. Wiley. https://dusk.geo.orst.edu/Pickup/Esri/Kerski-Geog-Compass.pdf
Kerski, J. 2016. *Interpreting Our World: 100 Discoveries That Revolutionized Geography*. ABC-CLIO. https://www.amazon.com/Interpreting-Our-World-Discoveries-Revolutionized/dp/161069919X
Kerski, J. October 1, 2019. Al Idrisi: Map of Silver, Cartographer, Geographer. https://www.josephkerski.com/al-idrisi-map-of-silver-cartographer-geographer/
Kerski, J. April 15, 2020. Teaching GIS as a Combination of Its Letters: G-I-S. https://community.esri.com/t5/education-blog/teaching-gis-as-a-combination-of-its-letters-g-i-s/ba-p/1164902
Kerski, J. 2020. Connecting GIS Education to Bloom's Taxonomy. Esri Community. https://community.esri.com/t5/education-blog/connecting-gis-education-to-bloom-s-taxonomy/ba-p/1011891
Kerski, J. Geography: A New Way of Seeing the World. Last accessed April 2023. https://www.josephkerski.com/geography-new-way-seeing-world/
Kerski, J. and J. Clark. 2012. *The GIS Guide to Public Domain Data*. Redlands: Esri Press. https://www.esri.com/en-us/esri-press/browse/the-gis-guide-to-public-domain-data
Kinkeldey, C., and H. Senaratne, 2018 (2nd Quarter 2018 Edition). "Representing Uncertainty." In: Wilson, J. P., Ed., *The Geographic Information Science & Technology Body of Knowledge*. Chesapeake, VA: University Consortium for Geographic Information Science. DOI:10.22224/gistbok/2018.2.3
Kleene, S. C. 1991 (10th Printing). *Introduction to Metamathematics*. Amsterdam: North-Holland Publishing Company.
Kneale, W. and M. Kneale. 1975 (Reprinted with Corrections). *The Development of Logic*. Oxford: Oxford University Press.
Kolmogorov, A. N. 1925 (Reprinted with Commentary). *On the Principle of Excluded Middle*, 414. van Heijenoort. https://www.cs.cmu.edu/~fp/courses/15816-s10/papers/Kolmogorov25.pdf
Kosko, B. 1993. *Fuzzy Thinking: The New Science of Fuzzy Logic*. New York: Hyperion.
Laskowski, P. 1989. "The Traditional and Modern Look at Tissot's Indicatrix." *The American Cartographer*. 16(2), 123–133.
Livio, M. 2003. *The Golden Ratio: The Story of Phi, the World's Most Astonishing Number*. New York: Broadway Books.
Longley, P. 2005. *Geographic Information Systems and Science*. Hoboken, NJ: John Wiley & Sons.
Mandelbrot, B. 1983. *The Fractal Geometry of Nature*. San Francisco: W. H. Freeman & Company.

Mansfield, M. J. 1963. *Introduction to Topology*. Princeton: D. Van Nostrand.

Masterworks Fine Art Gallery, Palo Alto, CA. Printmaking Techniques, Defined and Explained in Plain English. Last accessed April 2023. https://news.masterworksfineart.com/2021/01/20/printmaking-techniques-defined-and-explained-in-plain-english#photogravure

Math Is Fun: Number Line. Last accessed May 31, 2023. https://www.mathisfun.com/number-line.html

McCoy, N. H. 1961. *Introduction to Modern Algebra*. Boston, MA: Allyn and Bacon.

McMaster, R. B. and S. McMaster. 2002. "A History of Twentieth-Century American Academic Cartography." *Cartography and Geographic Information Science*. 29(3), 305–321.

Merserve, B. E. 1983. *Fundamental Concepts of Geometry*. New York: Dover.

Monmonier, M. 1991 and later. *How to Lie with Maps*. Chicago: University of Chicago Press.

Multiplication on the Number Line. Last accessed April 2023. https://www.youtube.com/watch?v=PLDfl6daajo&feature=youtu.be

Newman, J. R. 1956. *The World of Mathematics*. New York: Simon & Schuster.

New York Times. 2022. https://www.nytimes.com/2022/10/06/education/learning/mississippi-schools-literacy.html

Olmsted, J. M. 1956. *Intermediate Analysis*. New York: Appleton-Century-Crofts.

Palomo, M. 2021. "New Insight into the Origins of the Calculus War." *Annals of Science*. 78(1), 22–40.

Raisz, E. 1962. *Principles of Cartography*. New York: McGraw-Hill.

Reid, C. 1996 (Revised). *Hilbert*. Göttingen: Copernicus Publishing.

Richardson, L. F. 1950s work discussed in 2008. Fractals and the Fractal Dimension. Last accessed 30 January 2008. Vanderbilt University website. Archived 13 May 2008 at the Wayback Machine.

Rogers, L. 2011. The Four Colour Theorem. The University of Cambridge. https://nrich.maths.org/6291

Rudin, W. 1964 (2nd Edition). *Principles of Mathematical Analysis*. New York: McGraw-Hill.

Russell, B. 1997 (New Edition, first published 1912). *The Problems of Philosophy, with a New Introduction by John Perry*. New York: Oxford University Press.

Russell, B. 1974 (Edition, First Published 1968). *The Art of Philosophizing and Other Essays*. Totowa, NJ: Littlefield, Adams & Co.

Sigler, L. E. 2002. *Fibonacci's Liber Abaci: A Translation into Modern English of Leonardo Pisano's Book of Calculation, Sources and Studies in the History of Mathematics and Physical Sciences*. New York: Springer.

Smail, L. L. 1923. *Elements of the Theory of Infinite Processes*. New York & London: McGraw-Hill Book Company Inc.

Snyder, J. P. 1987. *Map Projections-A Working Manual*. *Professional Paper 1395*. Denver: USGS.

Steiner, J. 1867. *Vorlesungen über synthetische Geometrie*, published posthumously. Leipzig: C. F. Geiser and H. Schroeter.

Strunk, W. Jr. and E. B. White, 1959 (and Subsequent Editions). *The Elements of Style*. New York: Harcourt, Brace & Co. and others subsequently.

Teubner, B. G. 1887. *Jacob Steiner's Vorlesungen über synthetische Geometrie*. Leipzig (from Google Books: (German) Part II follows Part I) Part II.

Thomas, G. B. and R. L. Finney. 1979 (5th Edition). *Calculus and Analytic Geometry*. Reading, MA: Addison-Wesley.

Thompson, D. W. 1917. *On Growth and Form*. Cambridge: Cambridge University Press.

Tobler, W. 2004. "Thirty-Five Years of Computer Cartograms." *Annals of the Association of American Geographers*. 94, 58–73.

Tobler, W. R. 1963. "Geographic Area and Map Projections." *The Geographical Review*. 53(1), 59–79.

Tuan, Y.-F. 1977. *Space and Place: The Perspective of Experience*. Minneapolis: University of Minnesota Press.

Upton, G. and I. Cook. 1997. *Understanding Statistics*. Oxford: Oxford University Press.

Vacar, T. June 8, 2022. Town of Paradise Sees Housing Boom After Catastrophic 2018 Camp Fire. https://www.ktvu.com/news/town-of-paradise-sees-housing-boom-after-catastrophic-2018-camp-fire

Venn, J. 1880. "I. On the Diagrammatic and Mechanical Representation of Propositions and Reasonings." *The London, Edinburgh, and Dublin Philosophical Magazine and Journal of Science*. 5. 10(59), 1–18.

Vogt, H. G. 1895. "Leçons sur la résolution algébrique des équations." *Nony*. 1, 9.

Waerden, B. L. v. d. 2013 (Based on Earlier Publications). *A History of Algebra: From al-Khwārizmī to Emmy Noether*. New York: Springer Science & Business Media.

Wilder, R. L. 1952. *Introduction to the Foundations of Mathematics*. Hoboken, NJ: John Wiley & Sons.

Wilks, S. S. 1962. *Mathematical Statistics*. Hoboken, NJ: John Wiley & Sons.

SELECTED TOPICAL GROUPINGS

Eratosthenes

- English translation of the primary source for Eratosthenes and the size of the Earth by Roger Pearse. https://www.roger-pearse.com/weblog/2017/04/15/cleomedes-how-big-is- the-earth/
- How the Greeks estimated the distances to the Moon and Sun. https://galileoandeinstein.physics.virginia.edu/lectures/gkastr1.html
- List of ancient Greek mathematicians and contemporaries of Eratosthenes. https://aleph0.clarku.edu/~djoyce/mathhist/greece.html

Longitude at Sea Measurement

- Board of Longitude Collection, Cambridge Digital Library. https://cudl.lib.cam.ac.uk/collections/longitude
- British Museum, Babylonian Clay Tablet link: https://www.britishmuseum.org/collection/object/W_1882-0714-509
- Nova Online: Lost at Sea, the Search for Longitude. https://www.pbs.org/wgbh/nova/longitude/
- Royal Observatory Greenwich: John Harrison and the Longitude Problem. https://www.rmg.co.uk/stories/topics/harrisons-clocks-longitude-problem

Standard Deviation

Bozeman Science Presentation, video: Standard Deviation

https://www.youtube.com/watch?v=09kiX3p5Vek

David Barr Lithograph

SunSweep: lithograph, David Barr, 1985. The white arc represents the path of the sun above the USA/Canada border and the contoured hands represent the friendly relations between cultures. Source of art: Arlinghaus, S. and W. Time Is No Object: SunSweep Video, primary source.
https://www.youtube.com/watch?v=CtUnTb7-5Go

Miscellaneous

- Bivariate, 2D, legend display. https://pro.arcgis.com/en/pro-app/latest/help/mapping/layer-properties/bivariate-colors.htm
- Boundary zones. https://spatialreserves.wordpress.com/2012/11/05/maps-as-representations-of-reality-the-deciduous-coniferous-tree-line/
- Earthquakes and critical data. https://spatialreserves.wordpress.com/2015/11/22/understanding-your-data-it-is-critical/
- Prime Meridian on Mars. https://www.esa.int/Science_Exploration/Space_Science/Mars_Express/Where_is_zero_degrees_longitude_on_Mars

RESOURCES

Government and Business Sites

- Bouncy Maps. https://www.bouncymaps.com
- EPA EnviroAtlas. https://www.epa.gov/enviroatlas
- Esri. ArcGIS Online Map Viewer. https://www.arcgis.com
- Esri. ArcGIS Living Atlas of the World. https://livingatlas.arcgis.com
- Gapminder. https://www.gapminder.org/tools/
- Library Research Guides: Trinity College, Hartford, CT. Introduction to US Census Data. Last accessed April 2023. https://courseguides.trincoll.edu/Intro_Census_Data/census_geography#:~:text=Hierarchy%20of%20Geography%20Levels,-This%20diagram%20shows&text=The%20geographic%20types%20connected%20by,cannot%20cross%20a%20county%20boundary
- National Aeronautics and Space Administration: https://www.nasa.gov
- National Geographic Society, MapMaker Interactive. https://mapmaker.nationalgeographic.org/
- National Oceanic and Atmospheric Administration: https://www.noaa.gov
- Statistique Canada: https://www.statcan.gc.ca/
- United Nations: https://www.un.org
- United Nations Environment Programme: https://www.unep.org/
- United States Census Bureau: https://www.census.gov/
- United States Department of Commerce. The United States and the Metric System. May 1992. https://www.govinfo.gov/content/pkg/GOVPUB-C13-439a4837a9eab3ea058c00e40e0119b6/pdf/GOVPUB-C13-439a4837a9eab3ea058c00e40e0119b6.pdf
- United States Geological Survey: https://www.usgs.gov
- WorldMapper. https://worldmapper.org/maps/

Co-Author Resource Sites

- Institute of Mathematical Geography. Arlinghaus, S. L. and W. C. resources. https://deepblue.lib.umich.edu/handle/2027.42/58219.
- Joseph Kerski Resources: https://www.josephkerski.com/
- Arlinghaus, Sandra L. and Kerski, J. J. 2013. *Spatial Mathematics: Theory and Practice through Mapping*. Boca Raton: CRC Press.

Postscript

MAPS AND MATH: BENEFITS OF THIS INTERACTIVE APPROACH

Over the years, courses about how to use specific software usually appeared before courses that used that same software as a tool in a broader subject. For example, there was a time when instructors spent much time within a course teaching about how to use word-processing software and spreadsheet software. Now, they are simply part of the toolkit of courses across a wide variety of disciplines. So too, in today's world, almost all books using GIS are focused on teaching GIS or GIScience, rather than on teaching a content area. This book provides a clear method and approach to aid instructors in teaching *with* GIS, rather than teaching *about* GIS, and it does so in the context of the discipline of mathematics.

Indeed, a gap is widening in the workforce between the need for professionals skilled in data and GIS, and the availability of graduates with these skills. One reason is that most GIS-skilled graduates arise from geography and environmental science, and these programs have traditionally been small in enrollment. Furthermore, most modern career pathways that use GIS will not generate an employee that is a "GIS analyst" or "GIS manager" but rather, a statistician, city manager, wildlife biologist, public safety officer, health official, or other professional who will be expected to understand and use GIS for analysis. GIS will not be the focus of their profession, but a tool on their toolbelt. This book helps bridge that gap by providing students with spatial, data, and analytical skills grounded in mathematics that they can use in a variety of professions.

The teaching and learning approaches of the authors are well grounded in learning theory including problem-based learning (PBL), constructivism, inquiry, and service learning. The book firmly guides students beyond memorization and toward a focus on content, its assessment and evaluation, based on logical reasoning (Kerski, Bloom 2020).

Solving 21st-century issues in all disciplines requires students to be:

- skilled in mathematical principles and applications. Basic Mathematics courses in universities and community colleges are well established and required for most degrees. The courses tend to be large in enrollment independent of their specific focus; they are excellent ways to reach a large number of students and faculty in building their skills.
- to be fluent in working with many types of data. Emerging data science programs in universities are being created to help meet this need, but the expansion of data and the need in society are proceeding far more rapidly than these university programs are evolving.

Furthermore, maps and visualizations are rapidly emerging as a critical language across the planet for solving problems and for communicating the results of research and investigations.

Working with interactive maps, as done in this book, fosters a variety of important educational values: spatial thinking, critical thinking, project-based learning, geographic and scientific inquiry, data fluency, community connection, mobile workforces, career pathways content knowledge, and students as agents of change (Kerski, 2014).

The interactive linkage between Maps and Math is a clear match for many reasons. One final one is that the US Department of Labor, in 2004, identified three rapidly growing technologies as most relevant to the modern era: nanotechnologies, biotechnologies, and geotechnologies (Gewin, 2004). This book is firmly anchored in one of these: geotechnologies.

PERFORMANCE ON STANDARDIZED TESTS: TIPS

The best way to perform well on any method of evaluation is to know the material thoroughly. In addition, though, one key piece of information can be extremely useful. On many standardized mathematical test questions that are multiple choice in nature, the correct answer is already present in the set of choices for the answer. So, the student does not always need to solve the problem. When in doubt, simply test the answers offered.

> **KEY TIP**: On most standardized multiple-choice mathematical test questions, the correct answer is already in the question—as one of the choices listed.

There are numerous derivative ideas, depending on the structure of the test, that evolve from this important observation.

If the test is, for example, a five-part multiple-choice test, in which one may work the questions on the full test in any order:

- Go through and first answer all the questions to which you are immediately certain of the answer, so that no question is left out that would have given you credit.
- Go back through the test and substitute answer choices to find the right answer in questions that you did not know the answer to on the first pass, AND that have no choice for "none" or "all."
- Then go through and work on the questions that have "all" as a choice. Start substituting answer choices. If you find that one choice is not a solution, then you have eliminated "all" as a possible solution.
- Finally, work on any that have "none" as a choice. Substitute all of the numerical answer choices to determine if the correct answer is present.
- What is the grading strategy? You may still have some questions that are not covered by the strategy above. So, should you guess, or not guess?

If a percentage of wrong answers will be subtracted from right answers, don't worry if there are a few questions you do not answer. There may be greater harm in guessing (it depends on the extent of the penalty) than in leaving the question blank.

If there is no penalty for guessing, do not leave any answers blank. Random guesses are better than blank answers.

What is the grading history?

If the test designer sees a test as a learning tool, he or she may want the student to read all choices. If that is the case, the correct answer is likely to be closer to the bottom of the list than to the top (assuming most start reading the choices at the top).

Read the choices from the bottom and move up; you will likely get to the correct answer faster.

When guessing at random, guess an answer not at the top of the list to increase the odds of hitting the correct answer.

If the test designer simply tries to have each of the five parts be the correct answer one-fifth of the time then there is no benefit to be gained involving reading order or guessing pattern.

Experienced teachers who have used similar standardized tests from year to year may have an idea of which choices seem to turn up, as correct answers, more often than others.

With different structures of tests, different fallout from the Key Tip may emerge. Students should not spend much time on this as it is mastery of content that is truly important. However, they should know, and grasp the importance of, the fact that the correct answer is already present in many of the problems they will see in multiple choice mathematics testing. A few extra points can make the difference between a good score and a great score.

MAJOR ACTIVITIES, LINKS, AND QR CODES

The following list contains the major activities and maps we have included in this book, including activities created by Joseph Kerski. Other activities with shorter URLs are contained within the chapter text and are not listed here, nor do they have an associated QR code; for these links, we believe the reader can accurately type in the shorter URLs. The "Start Here" references are recommendations for the next logical step in your mathematics and mapping journey, then dig deeper into the resources in the sections that follow.

Start Here

ArcGIS Online guided tutorial: https://learn.arcgis.com/en/paths/try-arcgis-online/

An introduction to GIS as an open course with 10 activities, by Joseph Kerski: https://community.esri.com/t5/education-blog/an-introduction-to-gis-as-an-open-course-with-10/ba-p/1204626

Book: *Everything You Need to Ace Math in One Big Fat Notebook.* 2016. https://blog.workman.com/big-fat-notebooks *Workman,* 528 p.

What is spatial thinking? Essay. https://community.esri.com/t5/education-blog/a-working-definition-of-spatial-thinking/ba-p/892576 Video: https://youtu.be/LOAfrTQmHBA

Front Matter

Esri. Soil Chemistry, North Dakota. https://www.arcgis.com/apps/mapviewer/index.html?webmap=f3a47dd8234a4de1a3dc8413651c4e80

Esri. Story Map: Demonstration of the elements of GIS: https://storymaps.arcgis.com/stories/f130ea8dbc0349c9b79f36b9d934f975

North Dakota Soil Chemistry

Story Map

Chapter 1

Esri: Scale Bars. https://pro.arcgis.com/en/pro-app/latest/help/layouts/scale-bars.htm

Esri: Legends. https://pro.arcgis.com/en/pro-app/2.7/help/layouts/work-with-a-legend.htm

Esri: Map Viewer of ecoregions, hydrography, and cities: https://www.arcgis.com/apps/mapviewer/index.html?webmap=8c7d2d43d3ae40988c46519c6153716a

Esri: Global Population. https://www.arcgis.com/apps/mapviewer/index.html?webmap=f5e6469be6c34f2194758b7af02109f3

United Nations, Human Development Index. https://www.arcgis.com/apps/mapviewer/index.html?webmap=2c9f37a511a749adb8a820c3e9a9b299

Scale bars Legends MapViewer Global Population

UN HDI

Chapter 2

Esri. Precision and Rounding Activity, Eiffel Tower. https://www.arcgis.com/apps/mapviewer/index.html?webmap=fce1f5d41bbf4eafbbc5757a57a01cfe

Esri. Key Graticule Lines and France. https://www.arcgis.com/apps/mapviewer/index.html?webmap=b02ef0f9af5e47eeb6b4d5da0505155d

Esri. 3D view of key graticule lines. https://www.arcgis.com/home/webscene/viewer.html?webscene=037abe3b7e454a26823e05f3fa5126f1

Eiffel Tower, France, and Great Circles Key Graticule Lines 3D Graticule Lines

Chapter 3

Esri. Sentinel-2 Florida Land Cover. https://livingatlas.arcgis.com/landcoverexplorer/#mapCenter=-80.533%2C25.696%2C10&mode=swipe&timeExtent=2017%2C2022

Esri. San Diego Urban Litter Smart Mapping. https://www.arcgis.com/apps/mapviewer/index.html?webmap=f78dbbb432d946a58009763c68caa84f

Esri. San Diego Donut Map. https://www.arcgis.com/apps/mapviewer/index.html?webmap=60718b2751604ac0b70696d9df01b00f

Esri. UK Poverty Map. https://www.arcgis.com/apps/mapviewer/index.html?layers=ef8c3344e9f7497caa5be549a0003d82

Esri. Zebra Mussels. https://www.arcgis.com/apps/mapviewer/index.html?webmap=460bab410f7d4d67b67072a767740342

Esri. Global Population: Vector and Raster Visualization, UN Data. https://www.arcgis.com/apps/mapviewer/index.html?webmap=f5e6469be6c34f2194758b7af02109f3

Esri. Colorado. https://www.arcgis.com/apps/mapviewer/index.html?webmap=4168c7a43f6d423dad872796af9f9c9e

Esri. Lincoln, NB. Crime Analysis. https://www.arcgis.com/apps/mapviewer/index.html?webmap=cbe7a6bba2d24e06931ad93d8c6f123a

Esri. Florida Traffic Accidents. https://www.arcgis.com/apps/Cascade/index.html?appid=9a27635635c940539b96fb5ef954e4d5

Florida Land Cover
San Diego Urban Litter
San Diego Donut Map
UK Poverty Map
Zebra Mussels
Vector/Raster Visual
Colorado
Lincoln Crime
Florida Accidents

Chapter 4

Esri. Arc de Triomphe. https://www.arcgis.com/home/webscene/viewer.html?webscene=037cceb0e24440179dbd00846d2a8c4f

Esri. Bismarck, ND, Scale. https://www.arcgis.com/apps/mapviewer/index.html?webmap=39b4e135a17d487c92ef9ada99e2f7d4

Esri. Kansas Football Stadium. https://www.arcgis.com/apps/mapviewer/index.html?webmap=725c686f9aff407b837d722cd114832a

Esri. Mars Story Map. https://www.arcgis.com/apps/Cascade/index.html?appid=96cc9a6f8df447b4940e3ebca611faba

Esri. Craters on Mars. https://www.arcgis.com/apps/mapviewer/index.html?webmap=b164957a19544f55a8d6a001c952a590

Esri. Mars 3D Explorer. https://explore-mars.esri.com

Arc de Triomphe | Bismarck, ND, Scale | Kansas Football | Mars Story Map

Mars Craters | Mars 3D Explorer

Chapter 5

Esri. Iowa Public Libraries. https://www.arcgis.com/apps/mapviewer/index.html?webmap=56a9e68b8669430fb7b095b12959f48e

Esri. Dashboard of Geography in Everyday Life. https://www.arcgis.com/apps/dashboards/cf80c2d91102482782fc1c1bd81052c4

Esri. Water Tank and Trail Tracks. https://www.arcgis.com/apps/mapviewer/index.html?webmap=771b15f99bb24e96b5bc4d7617cc2f51

Esri. Earthquakes. https://www.arcgis.com/apps/mapviewer/index.html?webmap=167402e476f84e2e95b52e2b19cd10f7

Esri. Internet Access by Income. https://www.arcgis.com/apps/mapviewer/index.html?webmap=485dd7cbf0a4405094f7b1c53ef4a14b

Iowa Public Libraries | Dashboards | Water Tank | Earthquakes

Internet Access

Chapter 6

Esri. Boulder Hazards. https://www.arcgis.com/apps/mapviewer/index.html?webmap=28aa62c3d07b45b2a1b5f5b4ac36f5c3

Esri. Powerlines in Kansas. https://www.arcgis.com/apps/mapviewer/index.html?webmap=c97b1c3300ef4ad4be868c37d04eaa2d

Esri. Syria, Inequality and Global Population and Maternal Health Indicators: https://www.arcgis.com/apps/mapviewer/index.html?webmap=5a91ed9e44e84fbaaf0e84e4e61ddaff/

Esri. Oblate Spheroid. https://www.arcgis.com/home/webscene/viewer.html?webscene=5c3a2807b34841f2afa4c383fb57b98a

Boulder Hazards | Kansas Powerlines | Syria Inequality | Oblate Spheroid

Chapter 7

Esri. Powerline Triangle. https://www.arcgis.com/apps/mapviewer/index.html?webmap=c97b1c3300ef4ad4be868c37d04eaa2d

Esri. North Table Mountain Golden, Colorado 3D scene. https://www.arcgis.com/home/webscene/viewer.html?webscene=b6be327daad64537ab682fe16c2c5daa

Axis Maps. Isolines/contours. https://contours.axismaps.com/#12/37.2341/-113.0180

Esri. Bathymetry, Great Lakes. https://www.arcgis.com/apps/mapviewer/index.html?layers=269f4cdba1704c1f86916f9f844da653

Powerline Triangle | Golden, Colorado | Isolines/Contours | Great Lakes, Bathymetry

Chapter 8

Esri. Hydrology. https://www.arcgis.com/apps/mapviewer/index.html?webmap=079349bc1ebb438d9d4c37de8061c91b

Esri. Camp Fire California Wildfire. https://www.arcgis.com/home/webscene/viewer.html?webscene=8b208cebeb3140179fbf6fd8d265a869

Esri. Story Map 1, Camp Fire. https://storymaps.arcgis.com/stories/4d08a97e87664c66b051f97df09a1107

Esri. Story Map 2, Camp Fire. https://storymaps.arcgis.com/stories/7f7602261edf4d6185fdbe2991d57492

Esri. Marine Mammal Species. https://www.arcgis.com/apps/mapviewer/index.html?layers=d431d945a68f40988eda26353d1c01bd

Hydrology Wildfire Wildfire Story-Map 1 Wildfire Story-Map 2

Marine Mammals

Chapter 9

Esri. Housing Variables. https://www.arcgis.com/apps/mapviewer/index.html?layers=a6549bd25a0e42bbab8758efd49a54b4

Esri. Dot Density water body occlusion. https://www.maps.arcgis.com/apps/mapviewer/index.html?webmap=393184cc34524fb293226f75413c066c

Esri. Dot Density, No water body occlusion. https://www.arcgis.com/apps/mapviewer/index.html?webmap=460f1a605b964ef981c636160cad43ad

Housing Variables Dot Density Dot Density Variation

Index

3D scene 35–39, 76, 83, 148, 151, 156, 176

absolute value 1, 7, 10, 51
adjacency 170–173, 175, 177, 179–183, 185, 187, 189–191, 201–203
algebra 127, 137, 146, 148, 151, 154
angle 33, 69, 77–78, 81–82, 85, 140, 143, 146, 153–154, 168, 184–185, 191
approximation 6–7, 72, 150, 158–161
ArcGIS Online 26, 35–37, 45, 63, 75–78, 83, 88, 90, 94–96, 101, 121–122, 134, 166–167, 173–175, 186–187, 191, 195–196
area under a curve 150, 158
aspect 187–190
attributes 16, 35, 45–46, 52, 103–105, 118, 121–122, 200
axiom 127–129, 152, 183, 195
axiom of choice 127, 128
axiom of infinity 127, 128
azimuth 78–79

basemap 48, 83, 88–89, 94, 96, 99, 166, 174, 179, 197
bearing 77–79
binary 19, 30–32, 128–129
binning 110–111
bivariate 104, 121–122, 124, 126, 182
Bouncy Maps 201–203

cancellation 19, 22–23
cardinal 5, 28, 70, 77–79, 140, 142
Cartesian 33–35, 56, 70–72, 74, 81, 131–133, 139, 142, 152, 171
cartogram 193, 195, 201–206
central tendency 41, 51, 55, 118, 120
choropleth 104, 120–121, 162, 204
circular measurement 20, 32–33
classification method 3, 52, 109, 123, 141
clustering 9, 51, 62, 103, 110, 114, 196, 200
coastline 113, 150, 166–167, 172
coefficient of determination 42, 61, 63
coloring 119–120, 170, 179–182, 191–192
compact 17, 24, 31, 55, 129, 171, 175–176, 185, 192
contour 156, 162–163
coordinate system 35, 56, 71–73, 79–81, 88, 131–132, 152, 184, 191, 204
correlation 42, 57–58, 61
cosecant 82–83
cosine 82–83, 90–91
cotangent 82–83

dashboard 114–116
data scale 103–104
data types 103–104, 117–118, 120, 128
decimal degrees 33, 40, 74–75, 88, 90
decimal minutes 36, 80
defined interval 103, 108
degrees minutes seconds 36, 80, 87–88
denominator 20–22, 24, 49
digitizing 151, 168
dimension 150, 156, 164–166, 186
direction 77
distributive law 1–3, 5, 7, 9, 11, 13–15, 17, 22–23, 32, 59, 71, 137, 145, 147–148
divisibility 19, 23
dot density 162, 193, 195–201

ellipse 54–55, 98, 146–148
empty set 127, 129, 131
English 20, 32, 35, 48, 50, 52
EPA EnviroAtlas 29
equal interval 103, 109
equation 50, 138–143, 145, 147
equator 28, 33–36, 73–75, 79, 85–87, 89, 99, 112, 148, 151, 159
Eratosthenes 71, 85–87, 129, 168
error 9–10, 43, 62, 69, 86–87, 151, 168–169, 180–181, 196, 200
estimation 41–44, 68
Euclid 72, 85, 129, 150–152
Euclidean 37, 71, 81, 150, 152, 165–168, 185–186, 195
excluded middle 1, 9, 127–128
existence 127–128, 137–138, 183
exponent 24–25
extruded 66–67, 103, 106, 154, 186

factor 15, 23, 48, 62, 110, 154, 190
Fibonacci 42, 67–68
filtering 103, 106–107, 163–164
fractal 165–167
fraction 20, 22–25, 32–33, 40, 42, 44, 68, 87, 92, 94–95, 142
fractional dimension 150, 165

Gapminder 7–9
generalization 103, 105–106, 114, 158
geocoding 70, 75
Geographic Information System 13, 135
geometrical interval 103, 109
GIS 18, 75, 79, 88, 104–106, 108–109, 114–115, 135, 137, 156, 168, 172, 187–190, 203
global scale 104, 126, 191
globe 35, 37–38, 72, 75, 79, 85, 92, 94, 97, 103, 106–107, 148, 151, 190
golden ratio 42, 67–68
goodness of fit 42, 61
GPS 86, 87, 90
graph theory 170–172, 179–180
graphs 3, 9, 127, 138–140, 143–144
great circle 72–73
grid cells 103, 106, 190
grouping 2–3, 9, 47, 65, 103, 109–110, 129, 179, 194

heat map 104, 124
hexadecimal 20, 30–32, 184
histogram 42, 56, 59–61, 122–123
hot spot 65–66

inequalities 127–128, 138, 141–144, 148
infinity 1, 6, 8, 34, 36, 127–128, 176
integers 1, 5–7, 23, 129, 137, 184
interactive mapping activity 1–2, 41–42, 44, 49, 59
International Date Line 33, 75, 112
interval attribute 103, 105
interval data type 104, 120
interval scale survey 104, 118
irrational numbers 1, 5—7
isolines 150, 162–163

Jenks 103, 109
Jordan curve theorem 151, 168–169

Königsberg 170, 172–173, 180

labeling 117, 164
latitude 14, 26, 35–40
legend 1, 13, 16, 65, 66, 92, 112, 116, 124–126, 135, 163, 196–198
linear equations 127, 138–139, 141–142
linear regression 42, 58, 61
local scale 71, 94, 104
logarithm 25
longitude 14, 26, 35–40
longitude at sea 33

manual classification 103, 108, 123
map overlay 135
map projection 71, 97, 170, 190, 191
map scale 71, 91–92, 94–97
MapMaker 111–112
mass 32, 99, 113, 150, 157–158, 166
mean center 41, 54–55
median 8, 41, 51–52, 64–65, 118, 205
meridian 33–35, 50, 72–75, 79–80, 85, 87, 90, 101, 112
metric 20, 32, 95
mode 8, 41, 43, 52, 117–118, 168
model 7, 61, 70–72, 74, 81, 83–85, 87, 171, 183–185

natural breaks 4, 59, 103, 109, 123, 141
natural numbers 1, 4–6, 67, 128
nested hierarchy 1, 5–6, 193, 200–201
nominal attribute 103, 105
nominal data type 104, 120
nominal scale 92, 104, 117–118
non-Euclidean 81, 150, 167, 168
non-nested 5, 193, 200–201
number line 1, 6–7, 10
numerator 20, 22

octal 19, 31–32
order of operations 13, 15, 25, 32, 132
ordinal attribute 103, 105
ordinal data type 104
ordinal scale 104, 117–118

parentheses 2, 15–16
percent 30, 41–44, 50, 87, 122–125
percentage 43–44, 46–48, 50, 67–68, 83, 99, 101, 113, 115–116, 122–123, 154–155, 163–164, 167, 179, 206
perimeter 99, 154–156, 166, 176–178
pixels 186, 189
plane geometry 71, 155
polar coordinates 70, 80–81
popups 163–164
Prime Meridian 33–35, 72–75, 101, 112
prime numbers 1, 7, 19, 23
proof 137–138, 154, 180–181
proportion 41–42, 46, 48, 62, 109
proximate 171, 173, 196
proximity 170–173, 175–177, 179, 181, 183, 185, 187, 189, 191, 206–207
p-value 42, 62
Pythagorean theorem 84–85, 154

quadratic equations 127, 143–145, 147
quantile 103, 109, 123

rapid mental calculation 19, 23–24, 147
raster 106–107
rate 30, 49–50, 174, 205
ratio attribute 103, 105
ratio data 104–106, 117, 121
ratio scale 104, 118
rational numbers 1, 5–7, 166
real numbers 1, 5–7, 132, 145
reference information 104, 119
regression 42, 58, 61
relationship map 104, 124–125
resolution 94, 106–107, 128, 179
resources 50–51, 58, 69, 115
rounding 19, 26–27, 29, 41–44, 47, 68, 161, 165

satellite 43, 48, 75, 83, 95, 99, 114, 120, 168, 175, 179
scatter 42, 56–58, 61, 63–64, 200–201
secant 82–84
selection 72, 103, 106–107, 129, 182
set fundamentals 127, 129
set operations 127, 130–132
set theory 127, 129, 131–133, 134–135, 137
simplification 103, 106, 108–109
Simpson's rule 150, 158–161
sine 82–83
slope 139–140, 146–148, 187–190
solid geometry 150, 155
spatial thinking 1, 17–19, 41, 70–71, 103, 127, 150, 170, 193
standard deviation 41, 51–55, 61, 103, 109, 123
standard deviational ellipse 54–55
standardization 61, 104, 119
state plane 79
statistical significance 42, 61–62
subsets 127, 129, 182
survey 18, 45–46, 63, 79, 91, 104, 114–118, 122, 161
Syene 85
symbolization 104, 106, 114, 164
synthetic geometry 150–152, 153, 171

tangent 72, 82–83, 145, 192
terms 2, 14–15, 68, 77–78, 101, 140, 143, 145, 152, 168, 183
terrestrial 170, 187
tessellation 65–67
thematic 113, 115, 117–121, 123, 125, 182, 195
theorem 51, 84–85, 151, 154, 168–170, 172, 180–181
Tissot Indicatrix 71, 97
topographic 71, 92–94, 153, 162, 169

topology 170–172, 175, 182
trace 167–168, 174–175
transparency 46, 195–196
trigonometry 70, 81, 85, 148, 161, 169

uncertainty 104, 119–120, 167
unique factorization 19, 23
universal set 127, 129
universal transverse Mercator 70, 79
urban litter 41, 44–45
USGS 71, 92–94, 101, 174, 175
UTM 79–80, 86–88, 140

variation 41–42, 48, 55–56, 61–62, 97, 118, 171, 182
vector 106
Venn 127, 130, 132, 134, 185
visibility 13–14
volume 25, 100, 105, 150, 156–158, 169, 186

weighted 55, 160, 187–190
whole numbers 1, 4–7, 20–21, 25–26
word problems 48–49

z-score 42, 61–62, 66

For Product Safety Concerns and Information please contact our EU representative GPSR@taylorandfrancis.com Taylor & Francis Verlag GmbH, Kaufingerstraße 24, 80331 München, Germany